DIGITAL COMPUTATION
AND NUMERICAL
METHODS

McGraw-Hill Series in Information Processing and Computers

J. P. Nash, Consulting Editor
Richard W. Hamming, Consulting Editor
Norman Scott, Consulting Editor

DIGITAL COMPUTATION AND NUMERICAL METHODS

RAYMOND W. SOUTHWORTH

Associate Professor,
Department of Engineering and Applied Science,
Yale University

SAMUEL L. DELEEUW

Chairman and Professor,
Department of Civil Engineering,
University of Mississippi

McGraw-Hill Book Company

NEW YORK   ST. LOUIS   SAN FRANCISCO   TORONTO   LONDON   SYDNEY

Digital Computation and Numerical Methods

PREFACE

This book was developed from notes used in an undergraduate engineering course offered at Yale University. The aims of this course are (1) to introduce the student to numerical methods, as applied to the analysis and solution of engineering problems and (2) to develop enough facility in the programming of computers to allow him to solve problems on a digital computer. Most engineering schools include this type of material in their curricula, and there appears to be a trend toward offering a course in this area at an early level. The authors felt that a book for such a course should contain the following:

1. Programming, which includes flow charting and preferably the FORTRAN language
2. Numerical methods at a level understandable and meaningful to undergraduate engineering students
3. An ample supply of problems
4. Many engineering applications distributed throughout the text and problems

These four items then determined the format of this book which is divided into two main sections: programming (Chapters 2 through 4) and numerical methods (Chapters 5 through 12). However, the principles of programming are further amplified in the second main section through flow charts of certain numerical methods and problems to be solved on a digital computer. The main topic of the programming section is the FORTRAN language (Chapter 3), which is probably the most popular programming language now being used. The version of the language presented here is FORTRAN IV. The authors have attempted to present this material in a style favorable to learning, as opposed to a "manual style." Thus, input has been separated from output, the presentation of subscripted variables has been deferred, and an abundant supply of examples has been presented.

Just as the *free-body diagram* is an essential step in solving certain problems in engineering mechanics, flow charting is a vital part of programming. Accordingly, one complete chapter (Chapter 2) is devoted to flow charting. Further, flow charts accompany the examples in Chapter 3 and are given for many of the numerical methods presented in the other chapters. The student should realize that the programming of a problem consists of two basic steps, the determination of an algorithm and the writing of the actual statements of the program. Constructing a flow chart aids in checking the algorithm and in writing the program. Separating the two steps also relieves the programmer of the problem of simultaneously thinking about the logic of the problem and the details of the individual program statements. Finally, it is much easier to determine what a program does by looking at a flow chart than by looking at the actual program statements.

Chapter 4 presents a synthetic machine language. The purpose here is to convey to the student some idea of how a computer works and what happens to a FORTRAN program when it is compiled. The authors felt that something should be included to keep the student from having only a black-box concept of the computer.

The second portion of the book includes the basic topics of numerical methods: errors in computation, solution of equations, interpolation, numerical differentiation and integration, and the fitting of experimental data. Even though interpolation perhaps does not have as many engineering applications as the other areas, it is a classical part of numerical analysis and can be used as a base for many of the more applicable numerical methods. Material on matrices is included in the chapter on simultaneous equations, since this topic has been recently finding many useful applications in engineering. Flow charts are presented for many of the numerical methods covered; there seemed to be no reason to include FORTRAN programs as well.

All topics are presented at an introductory level, following which some are developed to a considerable depth, e.g., the use of matrices in solving systems of simultaneous equations in Chapter 7. Each chapter considers one general area and is fairly independent of the others. Of course, interpolation, numerical differentiation and integration, Taylor's series, and the numerical solution of differential equations are all closely related to one another. However, an attempt has been made to write these chapters in such a way that any one of the topics can be studied with a minimum of reference to the other topics. On the other hand, if all four chapters are studied, the student will easily be able to see the interconnections of the various topics.

At the end of each chapter several engineering applications are presented. Further applications are given in examples and problems. A large number of problems, some of which can be done by hand and others on a computer, follow each chapter.

The course at Yale is a three-hour, half-year course. Obviously, it would take more time to cover all the given material. Instead, a choice of topics can be made from the second main section. The students at Yale take a concurrent course in calculus. Some of the topics require calculus, while others require little or none at all. Many of the numerical methods are derived geometrically, but, naturally, many derivations need calculus.

As a guide for a course outline, the following time requirements are suggested on the basis of the authors' experience. About twelve class hours are needed to cover the introductory material and FORTRAN programming of the first three chapters. Chapters 4 and 5 on machine language and errors may either be omitted altogether, in a short course, or else included at whatever point is found convenient; they may be covered in six to ten class hours. Chapters 6 and 7, on roots of equations and simultaneous linear equations, may serve as the main material on numerical methods in a one-term course. From six to twelve periods may be devoted to each, depending on how far the instructor goes into the more advanced sections.

Chapters 8 through 10 are concerned with interpolation, numerical integration and differentiation, and Taylor's series. Again, the amount of time to be spent on each topic can vary widely, perhaps from three to six periods. Considerably more time may be devoted to ordinary differential equations, the topic of Chapter 11, with up to twelve periods being used.

Chapter 12 covers the elementary aspects of the least-squares method of empirical curve fitting and also the use of orthogonal polynomials in connection with least squares. This material does not depend on the material of Chapters 8 through 11 and can be presented earlier if desired. A minimum of nine periods is needed.

The authors gratefully acknowledge the helpful suggestions of their colleagues and students. Those who have also contributed much by teaching from these notes include Professors Arturs Kalnins, Trenchard More, Jr., and Gerald Smith, and Messrs. John W. McCredie, Jr., and Joseph Schofer. Acknowledgement is also made to Dr. Richard W. Hamming for his review of the manuscript and constructive criticism.

Raymond W. Southworth

Samuel L. DeLeeuw

CONTENTS

DIGITAL COMPUTATION
AND NUMERICAL
METHODS

1   INTRODUCTION

1.1   INTRODUCTION

Most engineering problems require numerical calculations in their solution, perhaps in the substitution of parameters into design equations, the estimation of the cost of a change in plant operation, or the statistical analysis of quality-control data. Until a few years ago these calculations were done by the engineer on a slide rule or occasionally on a desk calculator. Many of the problems that are encountered today, however, often require that millions or hundreds of millions of calculations be done accurately within a reasonable length of time and at a reasonable cost. The engineer can no longer use a slide rule; he must use a high-speed automatic computer.

It is important to note that the computer is only a tool of the engineer—it does not replace him. He still must formulate the correct solution to his problem, gather whatever data are needed, and look at the results and judge their value. In addition he must know how to use a computer, the tool that has taken over the tedious job of his arithmetic calculations.

The steps in solving a problem on a computer are shown in Fig. 1.1. After a suitable mathematical model has been formulated, such as a set of differential equations, a technique must be developed for arriving at numerical answers. This procedure is usually put in the form of a diagram, called a *flow chart*. A set of instructions for the computer, called a *program*, is then prepared and executed on the computer, giving the desired results.

Several of the procedures that arise in this process are due to language barriers. While a "natural" language, such as English, is suitable for everyday communication among persons, it lacks the precision and conciseness for the easy formulation of mathematical equations. Hence we have the languages of algebra and logic. These, in turn, are still too abstract and general for use in directing a computer calculation. In fact, the language suitable for computer instructions, usually called *machine*

1

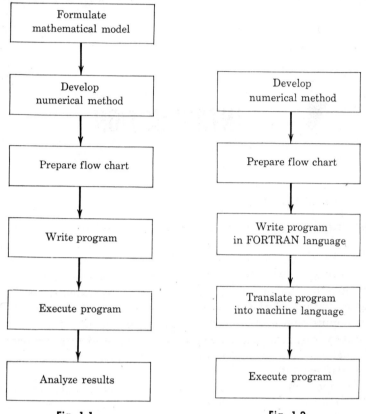

Fig. 1.1   **Fig. 1.2**

language, is a set of elementary arithmetic and logical functions and simple
choices, expressed in a coded numerical form.   It is possible for an engineer or mathematician to write the machine-language program corresponding to the mathematical formulation of a problem, but the process is
tedious and prone to error because of the dissimilarity of the languages.
An "in-between" language is a convenience, if not an absolute necessity.
Several such languages have been invented, some suited more to one kind
of problem than to another.   For engineering and scientific calculations
the language in widest usage is called FORTRAN, an acronym for
*Fo*rmula *Tran*slation.[1]   As we shall see later, the language of FORTRAN

[1] The original FORTRAN language and its associated translator were written in 1956
for the IBM 704 as a joint development of IBM, R. A. Hughes of the University of
California, and R. Nutt of the United Aircraft Corporation.   The language has since
been expanded and modified, and translators have been written by most computer
manufacturers for their particular machines.   The versions of the language now most
widely in use are FORTRAN-II and FORTRAN-IV.   The latter is somewhat more

is rather similar to that of algebra and as a result is relatively easy for an engineer to learn.

The steps in writing and executing a FORTRAN language program are shown in Fig. 1.2. It is not immediately apparent that we have gained anything by inserting another language into the process. Obviously, if it were necessary for us to perform another translation, from FORTRAN to machine language, we would have gained only extra work. Fortunately, the computer can do this routine work, and translator programs are available for most computers. They are called *translators*, or *compilers*. What we have gained, then, is a language that is suitable for expressing engineering calculations in the form of a computer program but that is not a computer language itself. A program written in the FORTRAN language is largely independent of any particular computer and, with the proper FORTRAN translator program, it can be executed on any computer.

1.2   COMPUTER ORGANIZATION

The modern computer may be thought of as consisting of five sections: an input unit, an output unit, an arithmetic and logical unit, one or more storage (memory) units, and a control unit. A typical set of interconnections between the sections is shown in Fig. 1.3. The solid lines indicate

---

flexible than the former. It is the language used in this text. Important differences between the two versions are pointed out in footnotes.

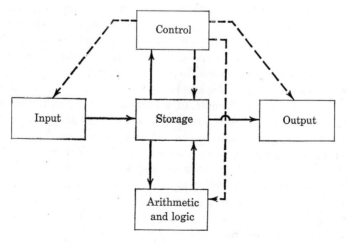

Fig. 1.3

flow of data and instructions, whereas the dashed lines simply are meant to show control functions.

The *input* and *output units* serve to convert information on input from a convenient external form, such as punched cards, paper tape, or magnetic tape, into machine form and on output to reverse the process.   The rate at which information is transferred covers a very wide range. Punched cards, for example, are read at typical rates of 250 to 1000 digits per second, while magnetic tape may be read or written at rates of 15,000 to 300,000 digits per second.   Direct printing of output may also be obtained, at up to 1000 lines of 130 characters per minute.

The *arithmetic* and *logical unit* contains the circuitry for performing the basic arithmetic operations, such as adding, subtracting, multiplying, and dividing, and for certain logical operations, such as comparing two numbers to find whether they are equal or unequal.   There are usually two or more registers for holding the numbers being operated upon and the result of the operation, much as in a desk calculator.   There may also be several counters, or index registers, for controlling the execution of instruction loops.

The *storage unit* serves to hold initial information to be processed as it comes from the input devices, intermediate results from calculations, and final results being retained for output.   Last, and by no means least, it holds the instructions for processing the data.

The capability of the machine to hold the set of instructions needed to solve an entire problem is very important.   It differentiates the computer from a desk calculator, in which each instruction must be entered by hand and cannot be "remembered" by the machine, even if the same instruction is repeated over and over.   In the computer, on the other hand, any of the instructions are equally accessible, since all are in storage simultaneously. In many cases an instruction, which in coded form looks like a number, can be treated as a piece of data and modified algebraically, giving rise to another instruction.   This gives the possibility of changing the course of action of a program of instructions when desired.

It may be convenient to think of the storage unit as an array of pigeonholes or post-office boxes, in each of which there is stored a number, say of 10 digits in length, often referred to as a *word*.   The boxes, or storage cells, are identified by a number, called an *address*.   In a machine containing 1000 storage cells, the addresses would be 000 to 999.   Typical storage capacities range from 1000 to over 250,000 cells.

The number stored in a cell may be a piece of data or it may be an instruction.   Usually the two functions are kept in separate parts of storage.   Instructions might be stored in cells 000 to 250, for example, and data in cells 600 to 900.

The final unit, the *control unit*, is the "master mind" of the entire system. It contains an *instruction counter*, which keeps a running tally of the address of the next instruction to be executed, and equipment for making decisions based on the results of previous operations. Ordinarily, the counter proceeds in sequence through the instructions stored in memory. If there is a break in the sequence, called a *branch*, the instruction counter is reset to the new address and the program continues in operation from that point.

The rate at which instructions are executed again varies widely among computers, from about only 5000 to over 3,000,000 sec^{-1}.

1.3 HISTORICAL DEVELOPMENT OF COMPUTERS

No attempt will be made here to give a complete or thorough history of the digital computer. Instead only the high points will be mentioned.

The first successful digital computer was built by a French mathematician, Blaise Pascal, in 1642. It was, of course, mechanical in nature; it employed wheels on which were placed the decimal digits 0 through 9 and which were rotated on an axle by means of gears and ratchets. It was the forerunner, through much development, of the adding machine.

The antecedent of the modern stored-program computer was begun in 1812 by a British mathematician, Charles Babbage. It was called the *difference engine*, because it was designed to calculate values of a polynomial by starting from the constant high-order difference and accumulating intermediate differences back to the value of the function itself. Many tables used in navigation and astronomy involve polynomial functions, and in Babbage's time the tables were derived by the hand calculations of large teams of mathematicians.

Babbage completed a small model of his difference engine that could calculate with 6-digit accuracy, and then he proposed and designed a much more elaborate machine that would handle up to sixth-order differences with an accuracy of 20 digits. Work was begun with the financial backing of the British government, but it was discontinued for various reasons after the expenditure of about $75,000.

Babbage also envisioned a far more complex machine, called an *analytical engine*. It was capable of obeying many instructions, both arithmetic and logical, and it was to include a storage of 1000 words. It was to have an arithmetical unit, called a *mill*, and units for input and output. Punched cards were to be the means of transferring information. Unfortunately, Babbage was too far ahead of his time, and the mechanical techniques were not available for building his analytical engine. It was not until 1944 that such a machine came into existence.

This machine, which was the first large-scale general-purpose digital computer, was designed by Dr. Howard Aiken at Harvard University and built by the International Business Machines Corporation (IBM). It was called the *automatic sequence controlled calculator*, or more commonly the Mark I. It was electromechanical in nature, using relays and mechanical counters. The storage capacity was 132 words, and it took 3 sec to add two numbers or 60 sec to divide them. Instructions were not stored internally but were fed to the machine by punched paper tape.

A great advance in speed was made possible by the use of vacuum tubes to replace the relays of Mark I. The ENIAC (*E*lectronic *N*umerical *I*ntegrator *a*nd *C*alculator) was completed at the University of Pennsylvania in 1946, as designed by J. P. Eckert and Dr. J. M. Mauchly.

The important idea of placing the program instructions in storage, along with the data being processed, was suggested by Dr. John von Neumann in a classic proposal for computer design in 1946.[1] It included nearly all the features of present-day computers, together with the design of circuits for the use of the binary-number system for storage and arithmetic. A machine built on these principles, called the EDVAC (*E*lectronic *D*iscrete *V*ariable *A*utomatic *C*omputer), began operation in 1952. A direct descendant of ENIAC and EDVAC was the UNIVAC I, which was placed on the market in 1951.

Since 1951 there have been continuing advances in the design and components of computers, with such developments as the use of transistors in place of vacuum tubes, larger storage capacity, greatly increased speeds of calculation, and a further diversification of instructions. Recently, research has been directed toward the development of facilities for remote computing and time sharing on computers. With this hardware it is possible to transmit programs and data from remote locations by telephone lines to a central, high-speed processing unit, which performs the desired calculations and then transmits back the results. The master control programs that operate the main computer are comparable in their complexity with the hardware itself.

1.4   NUMERICAL ANALYSIS

Engineers are often concerned with obtaining numerical answers to physical problems. The first step in such a problem is to transform the physical problem into a mathematical one. Often, by using various mathematical methods, the solution can be expressed in terms of one or several

---

[1] A. W. Burke, H. H. Goldstine, and J. von Neumann, Preliminary Discussion of the Logical Design of an Electronic Computing Instrument, a report prepared for the Ordnance Dept., U.S. Army, at the Institute for Advanced Study, Princeton, N.J., June, 1946. Reprinted in *Datamation*, vol. 8, nos. 9 and 10, 1962.

algebraic formulas. Such a solution is said to be in a *closed form*. Many mathematical problems, however, do not possess a closed-form solution. Numerical solutions to such problems can often be obtained by a trial-and-error scheme. If the trial-and-error procedure follows some sort of pattern whereby the results of one or more previous calculations are used to calculate a new trial, the procedure is an *iterative procedure*.

As an example, consider the equation

$$ax^2 + bx + c = 0 \tag{1.1}$$

The closed-form solution to this equation is

$$x = \frac{-b \pm \sqrt{b^2 - 4ac}}{2a} \tag{1.2}$$

and the *numerical method* associated with obtaining numerical answers to this problem consists in simply substituting values of a, b, and c into Eq. (1.2). On the other hand, the equation

$$a_1x^7 + a_2x^6 + a_3x^5 + a_4x^4 + a_5x^3 + a_6x^2 + a_7x + a_8 = 0 \tag{1.3}$$

has no closed-form solution, but a numerical solution to the equation

$$x^7 - 4x^6 + 6x^5 - 11x^4 + 7x^3 - 4x^2 + 4x - 3 = 0 \tag{1.4}$$

can be obtained by an iteration scheme such as the one shown in Table 1.1. Here, trial values of x were substituted into the polynomial until $f(x)$ changed sign, at which point the new trial was obtained by averaging the previous two trials.

Table 1.1

x	$f(x) = x^7 - 4x^6 + 6x^5 - 11x^4 + 7x^3 - 4x^2 + 4x - 3$
0	-3
2	-67
4	3725
3	0

Devising an iterative procedure for obtaining a numerical solution to Eq. (1.3), given numerical values for the coefficients a_1, a_2, . . . , a_8, is an example of a problem in the field of *numerical analysis*. The iterative procedure itself is an example of a *numerical method*. Obtaining methods of solving mathematical problems which do not possess closed-form solutions is the principal goal of numerical analysis.

As another illustration of the methods of numerical analysis, consider the following problem from the field of mechanics: Figure 1.4 shows a pendulum, of which it is desired to determine the angle θ as a function of the

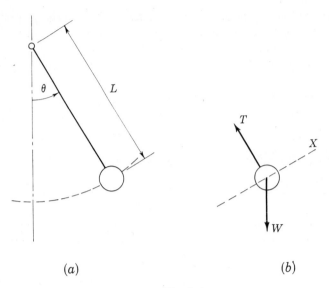

(a)   (b)

Fig. 1.4

time t.   If θ is small enough so that $\sin \theta$ can be replaced by θ in radians with sufficient accuracy, the problem has the closed-form solution

$$\theta(t) = \theta_0 \cos \sqrt{\frac{g}{L}}\, t \qquad (1.5)$$

where

θ_0 = angle at which pendulum is released to initiate the motion at time
         $t = 0$
g = acceleration due to gravity
L = length of pendulum

In a given problem where θ_0, g, and L are known quantities, the angle θ can be determined at any time t by direct substitution into Eq. (1.5). For example, when $t = (\pi/3) \sqrt{L/g}$,

$$\theta = \frac{\theta_0}{2}$$

If the angle θ is not always small, it cannot be expressed as a function of time by some simple algebraic formula.[1]   Instead, an approximate

---

[1] The problem actually has a closed-form solution in terms of elliptic functions.   However, these functions are rather complex, and the solving of this problem by using numerical methods is justifiable.

relation involving three discrete values of θ, as shown in Fig. 1.5, is

$$\theta_3 = 2\theta_2 - \theta_1 - \frac{g(\Delta t)^2}{L} \sin \theta_1 \tag{1.6}$$

where

Δt = an arbitrarily chosen small interval of time

$\theta_1, \theta_2, \theta_3$ = angles at times t, $t + \Delta t$, and $t + 2\Delta t$, respectively

If two values θ_1 and θ_2 are known, a third value θ_3 can be determined from Eq. (1.6). A fourth value θ_4 can then be obtained by replacing θ_1, θ_2, and θ_3 in the equation by θ_2, θ_3, and θ_4, respectively. More values of θ are obtained in a similar manner. Table 1.2 shows the results of applying this method when using the following data:

$g = 32.2$ ft/sec^2

$L = 2$ ft

$\Delta t = 0.10$ sec

$\theta_1 = \theta_2 = 0.8$ rad, starting values needed to initiate the procedure

Table 1.2

t	0	0.10	0.20	0.30	0.40	0.50
θ	0.8	0.8	0.68	0.44	0.10	−0.31

This solution differs from the closed-form solution given by Eq. (1.5) in two ways. First of all, if one desired the value of θ at some given time,

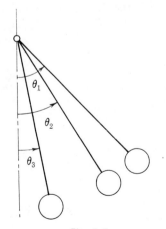

Fig. 1.5

say 0.5 sec, simple substitution of $t = 0.5$ into an algebraic equation is not possible. Instead, an iterative procedure must be employed whereby Eq. (1.6) is successively solved for values of θ at $t = 0.2$, 0.3, and 0.4 before $\theta(0.5)$ is determined.

The second way in which this solution differs from a closed-form solution is that Eq. (1.6) is an approximate relation. A solution which is more accurate than that given in Table 1.2 can be obtained by using a smaller value of Δt. Table 1.3 gives a numerical solution obtained by using $\Delta t = 0.05$. More results can be obtained by the further halving of Δt, and a *good* result is obtained when two sets of values agree to within an allowable difference. The results given in Tables 1.2 and 1.3 agree to within 0.08 rad. This method must be used with caution, however, since Eq. (1.6) is very susceptible to round-off errors caused by the subtraction of two quantities of approximately the same size.

Table 1.3

t	0	0.05	0.10	0.15	0.20	0.25	0.30	0.35	0.40	0.45	0.50
θ	0.8	0.8	0.77	0.71	0.62	0.50	0.36	0.20	0.03	−0.15	−0.33

The deriving of approximate relations, such as Eq. (1.6), for engineering problems which have no closed-form solutions and the use of these relations to obtain numerical solutions with given data are two of the principal aims of this text. The above example illustrates the fact that the numerical solution of a problem which does not have a closed-form solution may involve a long, tedious, and repetitive set of calculations. This is the type of problem for which the digital computer is very useful.

2   FLOW CHARTING

2.1   ALGORITHMS AND FLOW CHARTS

In Chap. 1 we noted that the third step in solving a problem is to determine a procedure to be followed and to prepare logical flow charts. To be usable on a computer, the procedure must possess two very important attributes. It must be unambiguous, with each possible point of decision considered and the steps defined for all possible occurrences. It must lead to an answer in a *finite* number of calculations. Such a procedure, guaranteed to give an answer (or to indicate that an answer cannot be obtained), is called an *algorithm*. A block diagram of the algorithm is called a *logical* flow chart.

2.1.1   An Algorithm

As an example of the construction of an algorithm, consider the following problem: A given set of values of a, b, and c are to be read into a computer, and the quadratic equation

$$ax^2 + bx + c = 0$$

is to be solved for the values of x. The problem is trivial and the algorithm obvious—at least so they seem. By the well-known quadratic formula

$$x = \frac{-b \pm \sqrt{b^2 - 4ac}}{2a}$$

The program should simply instruct the computer to substitute into the formula and then print out the result.

On second thought, however, we see that the formula is not an algorithm, because it is *not* unambiguous with respect to the sign in front of the radical. We can easily remedy this defect by calculating two values of x, according

11

to the formulas

$$x_1 = \frac{-b + \sqrt{b^2 - 4ac}}{2a}$$

$$x_2 = \frac{-b - \sqrt{b^2 - 4ac}}{2a}$$

There is still an unresolved problem: What happens if the discriminant $(b^2 - 4ac)$ is negative? The answer to this question depends on the method by which square roots are calculated in the computer, but the most likely result is that the computer will indicate that there is an error and the remainder of the program will not be executed. In order to avoid this possibility, our algorithm must include steps for checking the sign of the discriminant. In the case of a minus sign it may then either note an error or else change the sign and handle the roots as a complex conjugate pair.

Now do we have an algorithm? Unfortunately, the answer is still no, because we have not considered the possibility of division by a equal to zero. The above formulas become indeterminate, of the form $0/0$, and the resolution of the indeterminacy is a problem for calculus rather than the arithmetic of a computer. Instead, we recognize that we should use the formula

$$x = \frac{-c}{b}$$

with the provision that if b is also equal to zero we must bypass the division and note that there is no solution to the original equation. Now we do have a procedure that is defined for all possible occurrences, and we may say that we have an algorithm. There are several other algorithms for this same problem that are equally valid or better.

2.1.2   A Flow Chart

A flow chart presents an overall view of a procedure that is frequently much easier to grasp than is a written description. If it is sufficiently detailed, it serves as a convenient and rapid means of communication between programmers, in much the same way as a blueprint is used by engineers. Since the logical steps are usually not dependent on the particular computer for which the program is finally written, the flow chart of a procedure can be understood by another programmer, or an engineer, even though he may not understand the computer language itself. The language of flow charts is as universally understood by persons who work with computers as is the language of mathematics.

In the programming of long and complex procedures, a large number of programmers may be working together. Flow charts are then an absolute necessity. On small problems they may not seem necessary, but it is strongly recommended that they be used for all except the most trivial programs. This will help to eliminate errors in the program and will also prepare the student to handle more complex problems.

In constructing a flow chart for the above algorithm, we shall connect together several boxes, each of which contains one or more steps of the algorithm. The flow of the program is shown by the connecting arrows. The shape of the box may be used to indicate the kind of operation being performed. Several shapes are relatively standard; of these we shall use the following:

1. The beginning of the program and points at which the computer is stopped are indicated by an oval.

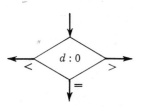

2. Declarative statements, i.e., those in which the computer is instructed to calculate a certain quantity or to set it equal to another quantity, are indicated by a rectangle.

Calculate:
$d = b^2 - 4ac$

3. A point at which a decision is made and the program must choose between two or three branches to other parts of the program is indicated by a diamond.

$$d : 0$$

The contents of the diamond may be put in the form of a question (Is $d < 0$?) or perhaps more simply and conveniently as a comparison ($d:0$). The appropriate branch is followed depending on the answer to the question.

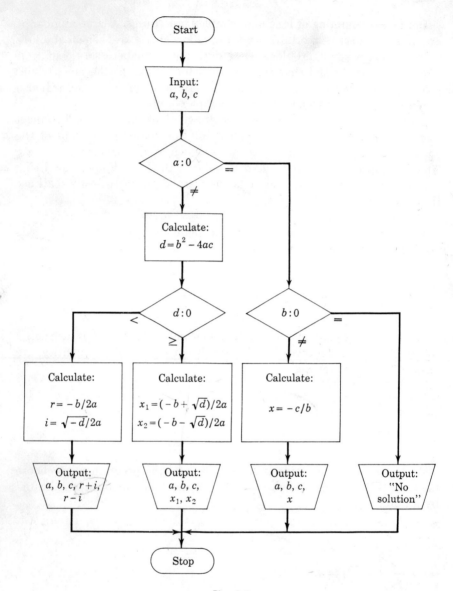

Fig. 2.1

4. Points at which data values are either read into the computer or printed out are indicated by a trapezoid.

Input:
a, b, c

5. If a flow chart is too large to fit on a single page, or if a branch line is difficult to draw between two points, a circle with a letter serves as a link between the sections.

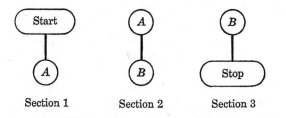

Section 1   Section 2   Section 3

A flow chart of the algorithm for the example problem is shown in Fig. 2.1. After looking at this diagram, the student might well conclude that it would be faster and much easier to solve this particular problem by hand than to figure out an algorithm for doing it on a computer. He would be right, unless this same equation had to be solved many times with different sets of data.

2.2   LOOPS AND INDEXES

All problems that are sufficiently complex or time-consuming to warrant the use of a computer in their solution involve the repetition of one or more operations on different data. A section of a program that is used repetitively is called a *loop*. As a very simple example, consider a program for getting the sum of n values of a variable x and the mean or average value. It will be convenient to refer to the data in terms of subscript notation. Thus, the first value is x_1, the second value is x_2, the ith value is x_i, etc. Upon letting S be the sum and μ the mean, the operations to be performed are

$$S = \sum_{i=1}^{n} x_i$$
$$\mu = \frac{S}{n}$$

where the sigma notation means addition of the values of x_i for the range of i that is indicated. Here, i is being used not only to specify which piece of data is to be added into the sum but also to *count* the number of times the operation is executed. It is called an *index*.

A flow chart for this example is given in Fig. 2.2, in which the steps for the summation form a loop. An important point to note is that the loop includes those operations which depend on the index i. The value of i must be defined for each pass through the loop, or the operations within the loop will be ambiguous. The first time through, for example, the value of i is 1, and the operations are performed on x_1. Then, i is incremented by 1, the result is compared with the limit n, and if it is less than or equal to n, the loop is executed again, now by use of x_2.

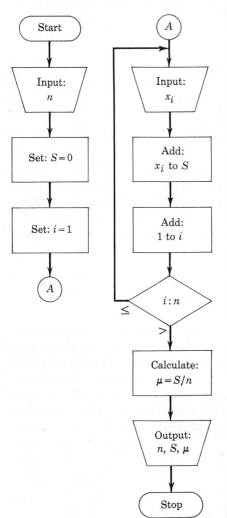

A second point is that the loop should *not* contain operations that do *not* depend on the index. The result of including such operations varies from a waste of computer time to getting incorrect answers or even no answers. If, in this flow chart, the loop returned one box higher, so as to include the step Set $i = 1$, the result would be a *permanent loop* for all values of $n > 1$. If the steps Set $S = 0$ and Set $i = 1$ were interchanged and the loop included Set $S = 0$, the loop would be executed the correct number of times but the value of S would be only x_n. On the other hand, if the calculation of μ were inside the loop, all answers would be correct but unnecessary work would be done.

To save writing, we may omit such words as Calculate and Set from a flow chart. Also, instead of writing such statements as ADD x_i TO S, we may use a

Fig. 2.2

shorter form, $S = S + x_i$.   Unfortunately the statement now looks like an equation, which it is not.   To avoid this confusion, many programmers use an arrow in place of the equals sign, for example, $S \leftarrow S + x_i$. In either case the statement should be read as "The current value of S is replaced by the sum $S + x_i$."   The student should have no difficulty if he remembers that the computer is doing arithmetic, not algebra, and that the steps of a flow chart tell what arithmetic operations are to be performed.

2.3   PLAYING COMPUTER

This example also affords an opportunity for the student to learn to "play computer," an ability that is invaluable in finding errors in programs and in initially writing relatively error-free programs.   The technique is simple: just follow the flow chart (or program), doing *exactly* as directed. It is best to make a column headed by the name of each variable and to record each change as it is calculated.   *A lways use the current values of all variables.*

Suppose that the value of n is 3 and that $x_1 = 5$, $x_2 = 22$, and $x_3 = 18$. After the first three steps of the flow chart have been executed, the record would be

n	S	i
3	0	1

After three more steps we would have

n	S	i	x_i
3	$\emptyset$	$\not{1}$	5
	5	2	

and at the end of the calculations

n	S	i	x_i	μ
3	$\emptyset$	$\not{1}$	$\not{5}$	15
	$\not{5}$	$\not{2}$	$\not{22}$	
	27	$\not{3}$	18	
	45	4		

2.4   LOOPS WITHIN LOOPS

Loops often occur within other loops, sometimes to a depth of several levels.   An example is presented in Fig. 2.3, which is a flow chart for

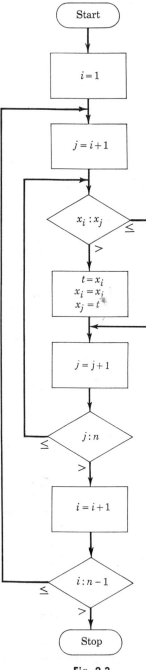

Fig. 2.3

sorting a group of n numbers, i.e., arranging them in an ascending sequence. It is assumed that the numbers x_i and the value of n have already been read into the computer.

The algorithm used here for sorting has the sole virtue of being simple; much faster but more complex routines are used in practice. Briefly, the method consists in finding the lowest number and placing it at the head of the list, reducing the length of the list by one, and repeating until the list is down to only one number. In more detail, the first number is compared with the second, and they are exchanged if the second is lower than the first, the *new* first number is compared with the third, and they are exchanged if necessary, and so on, until the first number is compared with the last number in the list. This set of comparisons comprises the inner loop. Then the process is repeated, this time a start being made with a comparison of the second and third numbers, second and fourth, etc. The final comparison made, just before dropping out of both loops, is between the $(n - 1)$st and the nth numbers.

2.5   ITERATION

A loop is also used in an *iteration*, or *iterative procedure*, in which a quantity is being calculated by a trial-and-error method that uses the results of one or more previous calculations to arrive at an improved value. The iteration is stopped when two successive values are within a prescribed tolerance or after it has been allowed to run for a preset maximum number of times.

An example is the use of Newton's method for calculating the approximate

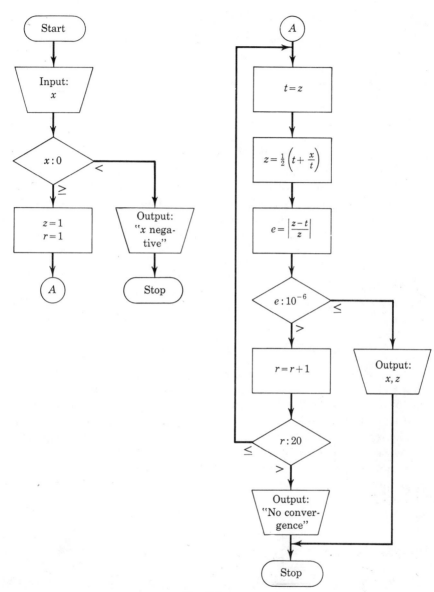

Fig. 2.4

square root z of a number x. The iteration is defined by

$$z^{(r+1)} = \frac{1}{2}\left(z^{(r)} + \frac{x}{z^{(r)}}\right)$$

where the superscripts are iteration indexes. Thus, the $(r + 1)$st value of z is a function of x and the rth value of z.   A suitable starting value is $z^{(0)} = 1$.   The iteration converges very rapidly as shown for $x = 4$.

$$z^{(0)} = 1$$

$$z^{(1)} = \frac{1}{2}\left(1 + \frac{4}{1}\right) = 2.5$$

$$z^{(2)} = \frac{1}{2}\left(2.5 + \frac{4}{2.5}\right) = 2.05$$

$$z^{(3)} = \frac{1}{2}\left(2.05 + \frac{4}{2.05}\right) = 2.0006097$$

and one more iteration gives the correct root to eight digits.

In designing a flow chart for Newton's algorithm (Fig. 2.4) a variable t is used as a temporary quantity, representing $z^{(r)}$, while z is the improved value.   The tolerance and maximum value of r are arbitrarily set at 10^{-6} and 20, respectively.   The relative error between z and t is defined as

$$e = \left|\frac{z - t}{z}\right|$$

2.6   PROGRAMMING IN TERMS OF PARAMETERS

If a program is to be of the greatest utility, it should not be restricted to the use of particular numerical values, in so far as practicable.   Consider the flow chart of Fig. 2.2, for example.   It will work equally well for the addition of 1000 numbers as for 100 numbers, because the value of n has been kept as a parameter in the program.   Similarly, in the flow chart for sorting (Fig. 2.3) the number of items to be sorted is not specified.   In this manner programs may be written that are very flexible in their application and that need not be rewritten to handle different sets of data.

PROBLEMS

2.1   In Sec. 2.1.1, an algorithm is presented for the solution of a quadratic equation by the use of the quadratic formula.   Derive an alternative formula in which the radical appears in the denominator.   Under what conditions does it have an advantage over the standard form?   Draw a flow chart in which it is used.

♦**2.2**   Construct a flow chart that can be used to solve the simultaneous

♦ Answers to problems marked with a diamond symbol are listed at the end of the book.

equations

$$ax + by = c$$
$$dx + ey = f$$

for x and y. If $ae - bd = 0$, the equations do not have a unique solution. The flow chart should include a check for that condition.

2.3 Given the coordinates (x_1,y_1) and (x_2,y_2), construct a flow chart that will determine the linearly interpolated value of y_3, where $x_1 \leq x_3 \leq x_2$.

2.4 Given the coordinates of the vertices of a triangle in the xy plane, construct a flow chart that will determine whether or not it is a right triangle. Since there may be a small error in the coordinates, allow for a tolerance within which the triangle is considered to be right-angled.

2.5 Two circles lying in the same plane are specified by giving the coordinates of the center (x_i,y_i) and the radius r_i. Construct a flow chart that will determine whether or not (*a*) the circles intersect; (*b*) all of one lies within the other.

2.6 Construct a flow chart that will evaluate an nth-order polynomial

$$y = a_1 x^n + a_2 x^{n-1} + \cdots + a_n x + a_{n+1}$$

The flow chart should include statements for reading into the computer the value of n, the coefficients a_i, and the value of x. Why is it preferable to rearrange the polynomial to the form

$$y = \{[(a_1 x + a_2)x + a_3]x + \cdots + a_n\}x + a_{n+1}$$

2.7 Construct a flow chart that will calculate the n terms of the subscripted variable z, according to the relation

$$z_i = x_i + y_i$$

2.8 Construct a flow chart that will calculate the *inner product* of two subscripted variables, where the inner product is defined by the relation

$$z = \sum_{i=1}^{n} x_i y_i$$

◆**2.9** Construct a flow chart that will exchange the components of two subscripted variables.

◆**2.10** The sine of an angle x, where x is expressed in radians, may be calculated from the Taylor's series expansion

$$\sin x = x - \frac{x^3}{3!} + \frac{x^5}{5!} - \frac{x^7}{7!} + \cdots + \frac{x^{2n-1}}{(2n-1)!}$$

Construct a flow chart for calculating $\sin z$, where z is given in degrees. Continue the series until the last term used is less than 10^{-6} times the sum of the previous terms.

2.11 Play computer, using the flow chart constructed in Prob. 2.8. Use the following data: $n = 5$, $x_1 = 1$, $x_2 = 3$, $x_3 = 5$, $x_4 = -2$, $x_5 = -1$, $y_1 = 2$, $y_2 = -1$, $y_3 = 1$, $y_4 = 2$, $y_5 = -3$.

2.12   Construct a flow chart that will:

(*a*) Average a set of *n* grades.
(*b*) Count the number of grades of 90 or above.
(*c*) Count the number of grades of 80 or above and less than 90.
(*d*) Count the number of grades of 70 or above and less than 80.
(*e*) Count the number of grades of 60 or above and less than 70.
(*f*) Count the number of grades below 60.

2.13   Play computer, using the flow chart of Fig. 2.3, to sort the numbers 8, 6, 3, 2, 1, 5, 7, 4.   What is the total number of comparisons made during the sorting?   Derive an expression for the total number of comparisons made in sorting *n* numbers.

2.14   Revise the flow chart of Fig. 2.3 so as to sort the numbers in the order of their absolute values.

♦**2.15**   A method of sorting a group of *n* numbers that is different from that described in Sec. 2.4 is as follows: In the first pass through the list, the first number is compared with the second and exchanged with it if it is greater, the third is compared with the fourth, the fifth with the sixth, and so on.   If there are an odd number of numbers in the list, the last number is neglected in this pass.   In the second pass, the comparisons are between the second and third, fourth and fifth, sixth and seventh, and so on.   The passes are repeated alternately until there are no exchanges in two successive passes.   The list is then in sequence.

Construct a flow chart for this algorithm, and play computer with a list of nine numbers not originally in sequence.

Under what conditions is this algorithm superior to that described in Sec. 2.4?

2.16   Construct a flow chart for *merging* two lists of sorted numbers to form a single sorted list.   Play computer with the lists 1, 3, 5, 6, 8, 9 and 2, 4, 7.

2.17   At a given instant in time, there are *n* airplanes flying, the positions of which may be specified by coordinates (x_i, y_i, z_i).   Each airplane carries an identifying number P_i.   Construct a flow chart that will check the distance between every pair of planes and will print a message if any distance is less than a specified safe minimum.   The message should include the identifying numbers of the airplanes.

2.18   Referring to Prob. 2.17, assume that the data for each airplane are given as two sets of coordinates, representing the ends of a line segment that will be occupied by the airplane during the next 10 sec.   Construct a flow chart that will determine the *minimum* distance between every pair of line segments and will print a message if any distance is less than a specified safe minimum, as in Prob. 2.17.

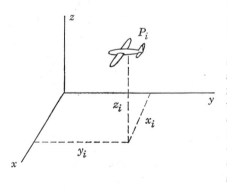

Fig. P2.17 and P2.18

2.19 Construct a flow chart that will find the median value of a set of n grades. The median is such that there are as many grades below the median as above it. If n is an even number, take the average of the two central grades as the median. Do not assume that the grades are initially sorted. Is there any way to find the median without sorting the grades?

2.20 One of the methods often used to check a number x to find whether or not it is prime is the use of the Eratosthenes sieve. The number is divided by all the primes up to the largest prime below or equal to the square root of x; if no prime divides exactly into x, then x itself is prime. Construct a flow chart that will generate the first n prime numbers, starting with 2.

♦**2.21** The accompanying flow chart was drawn to calculate values of the doubly subscripted variable z_{ij}, according to the relation

$$z_{ij} = 0.25a_1 + 3.5x_i + 6.5y_j + 0.79a_2x_i^2y_j$$

for $i = 1$ to n and $j = 1$ to m. Construct a flow chart that will accomplish the same result and use less computer time.

2.22 Play computer, using the flow chart of Fig. 2.4, to obtain the square root of (a) 4, (b) 25, and (c) 0.25. Retain eight significant digits at each step, and continue the calculation until the iteration converges.

♦**2.23** The approximate cube root z of a number x may be calculated by Newton's method as

$$z^{(r+1)} = \frac{1}{3}\left(2z + \frac{x}{z^2}\right)^{(r)}$$

where r is an iteration index. Construct a flow chart for this calculation.

2.24 The quadratic equation

$$ax^2 + bx + c = 0$$

can be rearranged into the form

$$x = -\frac{b}{a} - \frac{c}{ax} = -\frac{1}{a}\left(b + \frac{c}{x}\right)$$

Construct a flow chart for determining a root by iteration on this equation. Then play computer, using the data $a = 1$, $b = -5$, $c = 6$ and a first trial $x = 4$. What happens if the first trial is taken as $x = 1$?

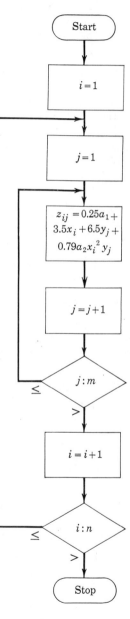

Fig. P2.21

2.25 It is desired to construct a computer program which will help prevent accidents between railroad trains.   Let

n = number of trains on a particular track
m_i = identification number of a train
P_i = position of front of engine of ith train, measured from end of track
L_i = length of ith train
D = minimum allowable free distance between trains

Construct a flow chart which checks the spacing of the trains and prints a message if two trains are too close to each other.   The subscript i varies from 1 to n, but the trains are not necessarily arranged in that order along the length of the track.

3   THE FORTRAN LANGUAGE

3.1   A SAMPLE PROGRAM

Before the details of the FORTRAN language are discussed, an example will be presented illustrating the basic parts of the language. Suppose that it is desired to calculate the value of Q in the equation

$$Q = \frac{1.49 A^{\frac{5}{3}} S^{\frac{1}{2}}}{E P^{\frac{2}{3}}} \qquad (3.1)$$

where

$$P = 2C + D \qquad (3.2)$$

for given values of A, S, E, C, and D. Actually, this problem is not worthy of a computer, since it could be solved quite readily by hand. However, a simple example such as this serves to illustrate the FORTRAN language. The problem could become worthy of the computer if one wanted more accuracy than could be easily obtained by a hand calculation or if the equation were to be solved many times for various values of A, S, E, C, and D.

It is now desired to write a program which solves Eq. (3.1). The program must provide for the following three steps to be performed by the computer:

1. Storage of data in the computer
2. Execution of the required calculations, using the stored data
3. Printing of the desired results

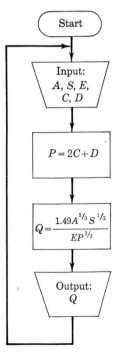

Fig. 3.1

A flow chart for programming this problem is given in Fig. 3.1.   A complete FORTRAN program follows which will cause the computer to execute the above three steps:

```
1   READ 2, A, S, E, C, D
2   FORMAT (5F10.4)
    P = 2.0*C + D
    Q = 1.49*A**(5.0/3.0)*S**0.5/(E*P**(2.0/3.0))
    PRINT 3, Q
3   FORMAT (F10.1)
    GO TO 1
    END
```

| 120.0000 | 0.0001 | 0.0300 | 8.0000 | 20.0000 |
| 115.0000 | 0.0001 | 0.0300 | 7.6000 | 21.5000 |

Before studying in detail the individual parts of the program, consider first some of its overall features.   The block of lines

```
1   READ 2, A, S, E, C, D
        .
        .
        .
    END
```

is called the *program*, while the last two lines of numbers represent *data*. Each line in the program is called a *statement*.   The program and data are generally transmitted to the computer from punched cards, paper tape, or magnetic tape.   On the assumption that the input is to be from cards, each line in the program and data is punched on a separate card, as illustrated in Fig. 3.2.

Only one character may be punched in each column of the card.   Therefore, fractions cannot be written in the form

$$\frac{A}{B}$$

but must be written as

A/B

Consider now the statements one at a time.   The statement

```
1   READ 2, A, S, E, C, D
```

when executed causes the data to be stored in the computer.   The statement

```
2   FORMAT (5F10.4)
```

specifies the manner in which the data are to be punched on the data cards. The expression 5F10.4 signifies that there are five numbers on each data card, each of which has 10 columns reserved for it on the card, and that there are four digits to the right of the decimal point. The numbers appearing to the left of the READ and FORMAT statements are called *statement numbers*. Note that the READ statement refers to the FORMAT statement by using the number 2. Execution of the READ statement causes data to be transmitted to the computer from the first

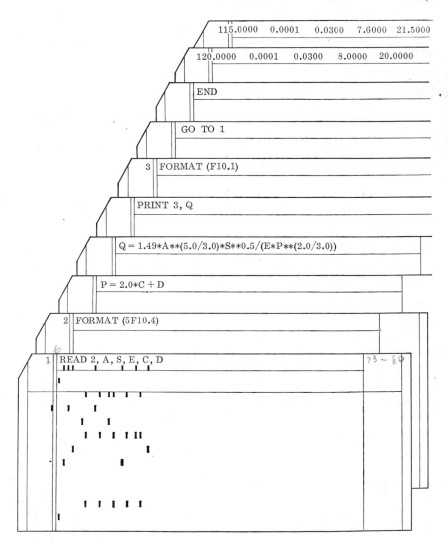

115.0000   0.0001   0.0300   7.6000   21.5000

120.0000   0.0001   0.0300   8.0000   20.0000

END

GO TO 1

3 | FORMAT (F10.1)

PRINT 3, Q

Q = 1.49*A**(5.0/3.0)*S**0.5/(E*P**(2.0/3.0))

P = 2.0*C + D

2 | FORMAT (5F10.4)

1 | READ 2, A, S, E, C, D

Fig. 3.2

data card.   Thus, the numbers listed in Table 3.1 would be stored in the computer.

Table 3.1

A	S	E	C	D
120.0000	0.0001	0.0300	8.0000	20.0000

The statements which cause the desired calculations to be performed are

P = 2.0*C + D
Q = 1.49*A**(5.0/3.0)*S**0.5/(E*P**(2.0/3.0))

where the symbols for the arithmetic operations used in these statements are defined as

+   addition
−   subtraction
*   multiplication
/   division
**   exponentiation

Parentheses are used in the FORTRAN language in much the same manner as in algebra.   These two statements are the FORTRAN equivalent of the two algebraic equations (3.1) and (3.2).
The statement

PRINT 3, Q

will cause the value of Q to be printed on the output paper.   The statement

3   FORMAT (F10.1)

specifies the manner in which Q is to be printed in much the same way as the FORMAT statement numbered 2 specified how the data were to appear on the data cards.   The statement

GO TO 1

causes the control in the computer to branch to statement 1, the READ statement.   Data from the second data card will then be transmitted to the computer, calculations with the new data performed, the results printed, and the control again returned to the READ statement.   Since there are no more data cards to be read, execution of the program is com-

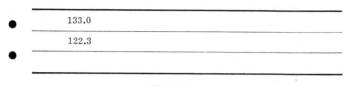

Fig. 3.3

plete. The results of the calculations will be the two values of Q, printed as shown in Fig. 3.3.

The statement

END

indicates the end of the program cards.

The order of the cards is important, since the computer will perform the operations indicated by the statements sequentially, unless a statement indicates otherwise. The GO TO statement exemplifies a method of causing the computer not to perform the operations sequentially. It should also be noted that not every statement must be numbered. The purpose of assigning a number is to provide a means of referring to a statement in another statement. In the sample program the READ and PRINT statements refer to the FORMAT statements in this way, and the statement

GO TO 1

refers to statement 1, the READ statement.

The FORTRAN language contains two different types of statements, *executable* and *nonexecutable*. Executable statements correspond to the actual operations performed by the computer while the program is running. All the steps in a flow chart can be specified with this type of statement. Nonexecutable statements are generally concerned with information needed by the FORTRAN translator to assign storage locations for the data or with the format being used for input and output. In the sample program all the statements are executable except FORMAT and END. The last two lines of data are not program statements.

Although this program does not illustrate all the features of the FORTRAN language, it does indicate the general manner in which the computer can be made to perform arithmetic and logical operations.

3.2 CONSTANTS

Constants are numbers which are used in the statements of a program. An *integer* constant is formed simply by writing a decimal whole number of not over 11 digits and including no decimal point.   Integer constants are also called *fixed-point* constants.   *Real* constants are distinguished by the inclusion of a decimal point and may not exceed 9 digits.   Examples of real constants appear in the following statement from the program given in Sec. 3.1:

Q = 1.49*A**(5.0/3.0)*S**0.5/(E*P**(2.0/3.0))

Real constants are also called *floating-point* constants.   For clarity, the words *integer* and *floating point* will be used throughout the text.

Discussion of three other types of constants, *double-precision, complex,* and *logical,* is deferred to later sections of the chapter.

3.2.1   E Form for Floating-point Constants

Floating-point constants can also be written in an alternative form which is generally quite convenient for comparatively large or small numbers. For example, the numbers

12,330,000
0.00015

can be written in the form

12.33E6
1.5E−4

The letter E when used in a floating-point constant signifies *multiplied by* 10 *to the power of.*.   The number to the left of the letter E contains up to nine significant digits and can be any decimal number.   The number to the right of the letter E is called the *exponent* and can be any integral number.

3.2.2   Use of Constants

In FORTRAN programming, integer constants are used as subscripts, index values, and integer exponents.   Floating-point constants are used for the actual numerical calculations to be performed.   As explained in Sec. 3.5.1 on the rules for writing arithmetic statements, it is sometimes necessary to express an integer in floating-point form.   This is illustrated by the statement from the sample program

P = 2.0*C + D

In such an instance it is not necessary to include the zero after the decimal point, and thus the above statement may be shortened to

P = 2.*C + D

A plus or minus sign preceding a constant indicates whether the constant is positive or negative. A constant with no sign preceding it is understood to be positive. The magnitudes of integer and floating-point constants must be less than 2^{35} and 10^{38}, respectively. Unless it is zero, the magnitude of a floating-point constant must be greater than 10^{-38}. Eight significant digits are retained in the computer.

3.3   VARIABLES

A *variable* is a quantity which is referred to by a name or symbol. In FORTRAN the name of a variable may be one to six characters in length. The first must be alphabetic, and the remaining may be any combination of alphabetic or numerical characters. Often variables can be given names that indicate the physical quantities which they represent. For example, a computer program concerned with the trajectory of a rocket may have variables named THRUST, SPEED, FUEL, ANGLE, TIME, etc. Examples of variables appearing in the sample program are C, D, A, S, E, P, and Q. The limits placed upon the magnitudes and number of significant digits for constants apply also to values of variables.

3.3.1   Type Statements[1]

As with constants, there are five types of variables. *Type* statements are used to indicate to the FORTRAN translator the type of each variable. Integer and floating-point variables are specified in type statements of the form

INTEGER   *list*
REAL   *list*

respectively, where *list* represents a list of variables separated by commas. The type statement is not executable and must appear in the program before any of the variables in the list appear. The sample program could contain the statement

REAL A, S, E, C, D

Since no integer variables appear in the program, no INTEGER type statement is required.

If a variable does not appear in a type statement, it must be named according to the following rules: The first character in the name of an integer variable must be I, J, K, L, M, or N. The name of a floating-

[1] Type statements are not available in FORTRAN-II.

point variable must begin with an alphabetic character other than these. Since the variables in the program given in Sec. 3.1 conform to these rules, the above type statement is not needed. However, the use of type statements is a good means of avoiding errors.

3.4   INPUT

Generally, the first step of a computer program is the reading in of data. This is accomplished through the use of two statements: the READ statement, which causes the data to be read into and stored in the computer, and the FORMAT statement, which specifies how the data appear on the data cards.

3.4.1   READ Statement

The general form of the READ statement is the word READ followed by the number of an input unit and the number of a FORMAT statement, separated by commas and enclosed within parentheses, and a list of the variable names for which values are to be transmitted. Each variable name in the list must be separated from the others by a comma. An example of a READ statement is

READ (5,2) A, S, E, C, D

which means "Read the data from the input unit 5 according to the FORMAT statement numbered 2." Input unit 5 is the standard unit for reading data in a FORTRAN program.

If data are to be read directly from cards, an input-unit number is not needed and this number and the parentheses are omitted from the READ statement. A comma must follow the FORMAT statement number. An example of such a READ statement appears in the program given in Sec. 3.1.

1   READ 2, A, S, E, C, D

The READ statement is executable and causes data to be read from the next sequential data card.[1]

3.4.2   FORMAT Statement

The general form of the FORMAT statement is the word FORMAT followed by the *specifications* enclosed in parentheses. The specifications give the precise manner in which the data must appear on the data

[1] In FORTRAN-II, the second form of the READ statement given above is used.

cards. Table 3.2 presents the most common specifications. The FORMAT statement is not executable.

Table 3.2

Specification	Form
Integer............................	Iw
Floating point......................	Fw.d
Floating point (exponent form)........	Ew.d
Alphanumeric......................	Aw
Skip..............................	wX

3.4.3   Specifications for Numerical Data

Numerical data to be read into the computer can be transmitted in either an integer or a floating-point form. Integer variables are transmitted by the specification Iw, where w signifies the *width of the field*. By width of field is meant the number of columns reserved on the data card for a particular piece of data. Floating-point variables are transmitted by the specification Fw.d, where w again specifies the width of the field and d represents the number of positions in the field which are to the right of the decimal point. The value of d must be between 0 and 9 inclusive. The specification for floating-point data in exponent form is Ew.d, where in this case the d represents the number of positions in the field which are to the right of the decimal point before the letter E. As an example, suppose that the following data are to be transmitted:

125
32.17
29.8×10^6
-11.5×10^{-6}

The READ and FORMAT statements could be written as

      READ (5,1) M, GRAV, YOUNGS, STRAIN
1   FORMAT (I5, F10.5, E10.4, E10.4)

The data should then appear on a data card as shown in Fig. 3.4. All blanks on a data card are taken to be zero in using specifications for

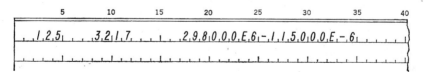

Fig. 3.4

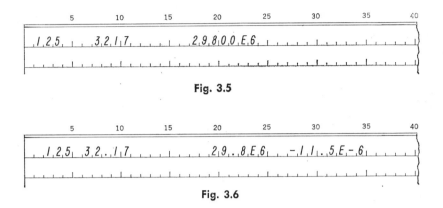

Fig. 3.5

Fig. 3.6

numerical data. Thus, care must be taken in the placement of the numbers on the data card. For example, if the first three numbers were erroneously punched as shown in Fig. 3.5, the numbers would be transmitted as 1250, 321.7, and 2.98×10^{60}. To avoid this type of error with floating-point numbers, it is allowable to punch the decimal point on the data card. With the F specification the exact location of the number on the card is then not important, provided that it is within its field. Note that integer and E-specified floating-point data must be right-adjusted in their fields. The data card for the previous example could appear as shown in Fig. 3.6.

If several variables of the same kind are to be read in successively, only one specification need be used, provided that the specification is preceded by the number of variables. In this case, the FORMAT statement in the previous example could read

FORMAT (I5, F10.5, 2E10.4)

In the sample program there are five successive floating-point variables transmitted. Thus, the corresponding specification is 5F10.5.

There are 80 columns available on each data card. Consequently, it may take more than one card to transmit an entire list in a READ statement. Suppose that the list in a READ statement contained seven integer variables and seven floating-point variables. The FORMAT statement could have the form

FORMAT (7I10/7F10.5)

A slash (/) in a FORMAT statement indicates that the next data to be read are located on the next sequential data card. Here, the seven integer numbers would be punched on the first card and the seven floating-point numbers on the second card. Each piece of data has a field width of 10 columns. Not all 80 columns on a card must be used, as illustrated by this example.

If the end of the specifications in a FORMAT statement is reached and there are still data to be transmitted according to the list in the READ statement, the FORMAT is reused, beginning from the right-most left parenthesis, and data are read from the next sequential data card. In the previous example, if all 14 of the variables were floating-point, the FORMAT statement could have the form

FORMAT (7F10.5)

Another application of the above is illustrated by the following example: Suppose that the list in a READ statement contained 7 integer variables and 14 floating-point variables. The FORMAT statement for this case could have the form

FORMAT (7I10/(7F10.5))

The 7 integer numbers would be punched on the first data card and the 14 floating-point numbers on the second and third cards, 7 per card. A maximum of three levels of parentheses is permitted.

Transmission of the data ceases when the end of the list on the READ statement is reached, regardless of the position in the FORMAT statement. Thus, the previous FORMAT statement could be used for the transmission of 7 integer variables followed by any number of floating-point variables. For example, if there were 10 floating-point variables, 7 would be punched on the second data card and 3 on the third data card.

Occasionally, it is desired to repeat a group of specifications. This can be done by enclosing the specifications in parentheses and preceding the parentheses with the number of repetitions. For example, the FORMAT statement

FORMAT (I10/3(F10.5, E12.6), I10)

is completely equivalent to the statement

FORMAT (I10/(F10.5, E12.6, F10.5, E12.6, F10.5, E12.6, I10))

3.4.4   Alphanumeric Specification

The Aw specification is used to transmit alphanumeric characters, which include the entire set of characters in the FORTRAN language (see Table 4.9 for list of all characters utilized in FORTRAN). As in the other specifications, w indicates the width of the field. Suppose that a program is being written to analyze data taken from several different

Fig. 3.7

alloys and that it is necessary to read in the name of the alloy. Statements in the program could appear in the form

READ (5,122) ALLOY, WT
122   FORMAT (A5, F10.5)

and the corresponding data card might appear in the form shown in
Fig. 3.7. In this example ALLOY is a variable name in which are
stored the characters TIN-2. The maximum field width (often six)
is limited by the number of characters that can be stored in a single
location in the computer. This varies with different computers. However, larger field widths can be obtained by using more than one storage
location for a given field. Subscripted variables are convenient for this,
and use of them for this purpose will be considered in detail in Sec. 3.9.4.

3.4.5   Skip Specification

A number of columns on a data card can be skipped by using the specification wX, where w indicates the number of columns to be skipped
(field width). Thus, a statement of the form

FORMAT (I10, 10X, F10.7)

signifies that an integer number appears in columns 1 to 10 on the data
card and a floating-point number in columns 21 to 30. Anything
punched in columns 11 to 20 will not be transmitted. It is allowable,
but not necessary, to separate a skip specification and the succeeding
specification with a comma.

3.4.6   Execution of the READ Statement

At the beginning of this section it was implied that the first step of a
program is to transmit the data to the computer. This is not necessary,
however. Data may be read into the computer anywhere in the program
and can even be read in at several different points. Each time a READ
statement is encountered, the transmission of data will begin with the

next sequential data card.   For example, in the program

        READ (5,1) A, B, C
    1   FORMAT (3F12.4)
        .
        .
        .

        READ (5,12) I, J, K
    12 FORMAT (I5, 2I7)
        .
        .
        .

the values of A, B, and C would be punched on the first data card and the values of I, J, and K on the second data card.   Each READ statement must refer to a FORMAT statement, although several READ statements could refer to the same FORMAT statement.   As another example, consider the statements

        READ (5,10) VOLT, AMP, OHM
        .
        .
        .

        READ (5,10) WATT, EFFIC, DIAM
        .
        .
        .

    10 FORMAT (3F8.3)
        .
        .
        .

Since a FORMAT statement is not executable, it can be placed anywhere in the program, as illustrated.

3.5   ARITHMETIC STATEMENTS

The general form of an arithmetic statement is

variable = expression

where the *variable* must be unsigned and the *expression* consists of operations with constants and variables.   The operations consist of addition, subtraction, multiplication, division, and exponentiation and are symbolized by $+$, $-$, $*$, $/$, and $**$, respectively.   The meaning of the statement is as follows: The present value of the variable on the left-hand side of the equals sign is replaced by the value calculated from the expression on the right-hand side of the equals sign.   A better way of writing the statement, to show more clearly its meaning, would be

$variable \leftarrow expression$

Unfortunately, the familiar algebraic symbol $=$ is used in place of $\leftarrow$. An arithmetic statement should not be thought of as an equation, however.   It is not permissible, for example, to write

$P + Q = R + S$

since the left-hand side is not a single variable.   On the other hand, it is permissible to write

$N = N + 1$

meaning that the value of N is to be incremented by 1.   Written as

$N \leftarrow N + 1$

there would be no confusion.   There are two arithmetic statements in the program given in Sec. 3.1.

$P = 2.0*C + D$
$Q = 1.49*A**(5.0/3.0)*S**0.5/(E*P**(2.0/3.0))$

An arithmetic statement is executable.

3.5.1   Constructing Expressions

There are several rules which must be followed in constructing expressions.   An expression has been defined as consisting of operations on constants and variables, although it does not necessarily contain any explicit operators.   In the statement

$RATE = SPEED$

SPEED is an expression.   Further, no two operators can be placed in succession, nor can an operator be placed in succession with the sign of a constant.   Thus, the statement

$A = B** - 3.0$

is not allowable but could be expressed correctly in the form

$A = B**(-3.0)$

The sign of a constant can, however, be included at the beginning of an expression. For example,

$$F = -2. + G$$

is permissible.

In constructing expressions, it is important to know the order in which the computer will perform the operations. The hierarchy of operations in descending order is

**	exponentiation
−	unary minus[1]
*/	multiplication and division
+−	addition and subtraction

Operations on the same hierarchal level are performed from left to right in an expression. For example, in the expression

$$1.49*A**3*S**0.5$$

the computer will perform the operations in the order

$$A^3$$
$$S^{\frac{1}{2}}$$
$$1.49 \times A^3$$
$$1.49 \times A^3 \times S^{\frac{1}{2}}$$

Parentheses are used in FORTRAN in the same manner as they are in mathematical equations. The mathematical expression

$$ab(c - d)$$

as a FORTRAN expression becomes

$$A*B*(C - D)$$

If parentheses were not available, this expression would have to be written as

$$A*B*C - A*B*D$$

In the hierarchy, operations within parentheses take precedence over the other operations. Hence, the expression

$$A*B*(C - D)$$

---

[1] In an expression such as $-B*C$ the computer will execute $(-B)*(C)$ rather than $-(B*C)$. On the other hand, the expression $A - B*C$ executes as $(A) - (B*C)$. The expression $-B**2$ executes as $-(B**2)$.

causes calculations to be performed in the order

C − D
A × B
A × B × (C − D)

For parentheses within parentheses, operation begins within the inner-most parenthesis pair and continues outward. Consider, for example, the expression

A*(B + (C*D) − E)

The order of the operations is

C × D
B + (C × D)
B + (C × D) − E
A × (B + (C × D) − E)

If A = 2., B = 3., C = 4., D = 5., and E = 6., the arithmetic calculations are performed in the following order:

4. × 5. = 20.
3. + 20. = 23.
23. − 6. = 17.
2. × 17. = 34.

An expression of the form

B(C − D)

is not valid and must be constructed

B*(C − D)

Care must also be taken in dividing. The expression

A*B/C*D

is equivalent to the algebraic expression

$$\frac{A \times B}{C} \times D$$

A correct FORTRAN expression for the algebraic expression

$$\frac{A \times B}{C \times D}$$

is

A*B/(C*D)

Extra parentheses may be put into an expression for clarity or to specify a desired order of operations.   For example, in the expression

(A*B)/(C*D)

the order of operations is

A × B
C × D
(A × B)/(C × D)

while in the expression

A*(B/(C*D))

the order of operations is

C × D
B/(C × D)
A × (B/(C × D))

   In dividing integer numbers, any remainder that may occur is lost in the operation.   Thus, the expression

5/2

produces a result of 2.   The expression

(I/2)*2

gives a result of I only if I is an even number.

   An important concept to remember in constructing expressions is that the value of every variable appearing in the expression must be previously *defined* in the program.   The value of a variable is said to be defined if it either appears on the left-hand side of an arithmetic statement or is read in as data.   As an example, consider a program which starts with the statements

1   N = M + 4
     READ (5,2) M
2   FORMAT (I3)

and contains the data card shown in Fig. 3.8.   The value of N will not

Fig. 3.8

be calculated as 20. The computer will execute statement 1 before the READ statement and thus will not have access to the desired value of M.

If the value of a variable is defined at two different points in a program, only the value determined by the statement which is executed last is retained. After the execution of the two statements

N = 5

.

.   .

.

N = 6

only the value 6 is stored in the computer, and 5 is erased.

With one exception, all constants and variables appearing in an expression must be of the same type. Thus, an expression such as

2*C + D

is not allowed. The one exception to this rule occurs in exponentiation, whereby it is permissible for the exponent of a floating-point quantity to be of the integer type. Therefore, an expression such as

AREA**5

is allowed even if AREA is a floating-point variable.

The variable on the left-hand side of an arithmetic statement need not be of the same type as the expression on the right-hand side. If they are different, an automatic conversion will be performed after the expression has been evaluated. Thus,

A = 2 − I
J = C/D
X = 8/3
K = 3.5*2.75

are all legitimate arithmetic statements. In the third example the floating-point number 2. is stored in X, and in the fourth example the integer number 9 is stored in K. This process of dropping all the digits to the right of the decimal point, regardless of the values of these digits, is called *truncation*.

3.6   OUTPUT

After reading data into a computer and performing calculations by using arithmetic statements, the next step in a program is to print out the results. To accomplish this, the two statements WRITE and FORMAT are used in the same manner as were the READ and FORMAT statements. The form of the WRITE statement is the same as that of the READ statement. The standard output unit is numbered 6. If a direct printing of the output is desired, a PRINT statement, which has the same form as the card READ statement, should be used. If it is desired to produce the output on punched cards, a PUNCH statement, which has the same form as the card READ statement, should be used.

At most computer centers, input and output operations are handled by auxiliary computers, disks, magnetic tapes, or some combination of these. For example, if a program submitted on cards is first transferred to magnetic tape by means of an auxiliary computer, the main computer reads this tape, compiles and executes the program, and writes the output on another magnetic tape. The auxiliary computer then causes the output to be printed on paper. Thus, even though a programmer submits a deck of cards and desires a printed output on paper, he must use the statements which refer to input-output units.

READ (i,n) *list*
WRITE (i,n) *list*

The specifications defined in the section on Input are also utilized with the FORMAT statement for output. However, there are several special considerations for printing out results. One of these is that output paper generally has more columns (usually either 120 or 134) in width, as compared with the 80 columns available on data cards. The discussion of other special considerations will follow. An example illustrating output statements occurs in the program given in Sec. 3.1.

PRINT 3, Q
3   FORMAT (F10.1)

As in the case of the READ statement, the WRITE, PRINT, and PUNCH statements are executable.

3.6.1   Numerical Output

Numerical output is printed by using the Iw, F$w.d$, and E$w.d$ specifications. Care must be taken in specifying field widths, since often the order of magnitude of the output numbers is not known in advance. Another important consideration is that a space must be allowed for a decimal point. To illustrate numerical output, consider the statements

WRITE (6,52) SODIUM, CHLOR, J
52   FORMAT (2F10.5, I4)

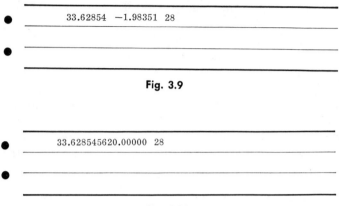

33.62854 −1.98351 28

Fig. 3.9

33.628545620.00000 28

Fig. 3.10

If the values of SODIUM, CHLOR, and J are 33.628539, −1.9835146, and 28, respectively, they will appear on the output paper as illustrated in Fig. 3.9. However, if the value of CHLOR were −15620.000, the output would appear as shown in Fig. 3.10. Notice that the minus sign and the most significant digit were not printed for CHLOR. A wider field should have been used.

If the E specification is used, four columns must be reserved for the exponent, including the letter E. The decimal point is always placed just to the left of the most significant figure. The statements

        WRITE (6,53) SODIUM, CHLOR
53   FORMAT (2E15.7)

where the value of SODIUM is 33.628539 and CHLOR is −0.015639841, produce output in the form shown in Fig. 3.11. The d in the specification represents the number of significant digits printed out. The total width of the field should be at least seven more columns than the value of d (four for the exponent, one for the decimal point, one for the leading zero, and one for the sign).

0.3362854E 02 − 0.1563984E − 01

Fig. 3.11

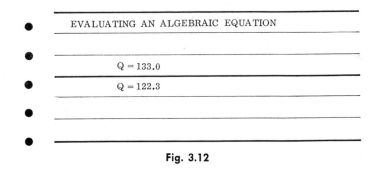

Fig. 3.12

3.6.2   Skip Specification

The skip specification wX is used for output in the same manner as for input.

3.6.3   Alphanumeric Specification

Often it is desirable to print out titles, column headings, etc., as a part of the output. This can be done with either the Aw specification or the specification wH, where w represents the number of characters (including blank spaces) to be printed. In the wH specification, the actual characters to be printed follow the H. Suppose that the program given in Sec. 3.1 were altered as shown below.

```
      WRITE (6,10)
10    FORMAT (5X,
      $32HEVALUATING AN ALGEBRAIC EQUATION//)¹
1     READ (5,2) A, S, E, C, D
2     FORMAT (5F10.4)
      P = 2.*C + D
      Q = 1.49*A**(5.0/3.0)*S**0.5/(E*P**(2.0/3.0))
      WRITE (6,3) Q
3     FORMAT (10X, 2HQ=, F5.1)
      GO TO 1
      END
```

The output would then appear as shown in Fig. 3.12. Notice that the two slashes in the first FORMAT statement cause a line to be skipped; n slashes will cause $n - 1$ lines to be skipped.

[1] Statement 10 is too long to fit on one line. Its continuation on the following line is denoted by the $ character preceding the continuation. A general discussion of the card format for FORTRAN statements is given in Sec. 3.13.1.

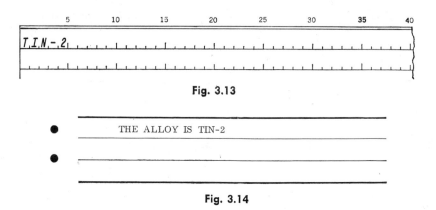

Fig. 3.13

Fig. 3.14

To illustrate how the Aw specification can be utilized for the output of characters, consider the following statements:

        READ (5,122) ALLOY
122 FORMAT (A5)

        .
        .
        .

        WRITE (6,356) ALLOY
356 FORMAT (5X, 12HTHE ALLOY IS, A6)

If the data card appears in the form shown in Fig. 3.13, the output will be printed out as illustrated in Fig. 3.14.

3.6.4   Carriage Control

Control over the vertical spacing of the output is accomplished with carriage control, which is obtained by placing any of the characters listed in Table 3.3 into *the first column of each line of output* with the wH specification.

Table 3.3

Character	*Control*
blank (b)†	Print on next line
0 (zero)	Skip a line before printing
1 (one)	Start printing on new page

† In order to indicate a blank character in FORMAT carriage control, the letter b is used throughout this text. The corresponding card column should be left blank.

The statements

WRITE (6,30) NUMB, POUND, GRAM
30   FORMAT (1Hb, I5/1Hb, 2F20.9)

cause the value of NUMB to be printed on the first line and the values of POUND and GRAM to be printed on the second line.   The skip specification can also be used to print on the next line, since this likewise places a blank in the first output column.   Thus, the following FORMAT statement is equivalent to the one used above:

30   FORMAT (1X, I5/1X, 2F20.9)

Skipping a line is obtained by the specification 1H0.   Similarly, if it is desired to start printing on a new page, the specification 1H1 can be used. For example, the FORMAT statement

30   FORMAT (1H1, I5/1H0, 2F20.9)

will cause printing on the first and third lines of a new page.

The carriage-control characters are not printed on the output paper. Further, if any other character is inadvertently placed in column 1, it will not be printed but may cause the printer to skip large amounts of paper.   For example, if a program contained the statements

WRITE (6,12) POWER
12   FORMAT (F6.3)

and the value of POWER were 26.394, the digit 2 would not be printed. Thus, it is of extreme importance that no character other than a legitimate carriage-control character be placed in the first output column.

3.7   CONTROL STATEMENTS

In any program it is necessary to control the sequence in which the statements of the program are executed, i.e., the *flow* of the program. Statements which provide this facility are called *control statements*.

3.7.1   GO TO Statements

There are two statements of this type, one of which appears in the sample program.   The first of these is the GO TO statement, which has the form

GO TO n

where n is a statement number.   Execution of this statement causes the program to branch to the statement numbered n.

The second statement of this type is the computed GO TO, which has the general form

GO TO $(n_1, n_2, \ldots, n_j, \ldots, n_m)$, i

where the n_j's are statement numbers and i is an integer variable. A branch is made to statement n_j, where j is the value of i. An example of the computed GO TO is

GO TO (5,13,129,6,42), K

Upon execution of this statement, the next statement to be executed will be 5, 13, 129, 6, or 42, depending upon whether the value of K is 1, 2, 3, 4, or 5, respectively. As many statement numbers as desired can appear within the parentheses, but there is a maximum number for each type of computer. The value of the integer variable i must have been previously defined before execution of this statement.

The flow-chart symbol for a computed GO TO is shown in Fig. 3.15, illustrating a seven-way branch. Both types of GO TO statements are executable.

3.7.2   Arithmetic IF Statement

The computer has the power to make very simple decisions, such as branching to one instruction or another depending on the sign or magnitude of a specified quantity. In English one might say, "If the value of z is positive or zero, then take the square root of z; otherwise stop the calculation." In FORTRAN the arithmetic IF statement is used. The general form is

IF (*expression*) n_1, n_2, n_3

where n_1, n_2, and n_3 are the numbers of executable statements to which the program will branch if the value of the *expression* is respectively

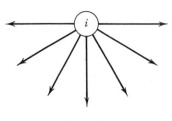

Fig. 3.15

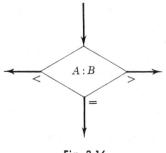

Fig. 3.16

negative, zero, or positive.   An example of an IF statement is

IF (A − B) 10, 20, 30

If the value of (A − B) is negative, the statement numbered 10 is executed next.   If the value of (A − B) is zero, the statement numbered 20 is executed next.   If the value of (A − B) is positive, the statement numbered 30 is executed next.

The flow-chart symbol for the IF statement is shown in Fig. 3.16.

As another example, consider the following variation of the program given in Sec. 3.1.   Suppose that the actual value of Q is not needed but that it is desired to have a message printed out if Q is less than 125.   The flow chart is given in Fig. 3.17, and the FORTRAN program follows. STANUM is an identification number indicating the time and place of the data reading.   Type statements are incorporated for illustration.

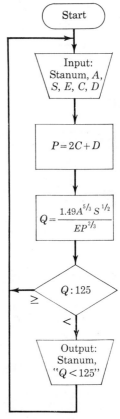

```
  INTEGER STANUM
  REAL A, S, E, C, D, P, Q
1 READ (5,2) A, S, E, C, D, STANUM
2 FORMAT (5F10.4, I5)
  P = 2.*C + D
  Q = 1.49*A**(5./3.)*S**0.5/(E*P**(2./3.))
  IF (Q − 125.) 3, 1, 1
3 WRITE (6,4) STANUM
4 FORMAT (5X,
  $21 HIDENTIFICATION NUMBER,
  $I5/10X, 18HQ IS LESS THAN 125)
  GO TO 1
  END
```

3.7.3   DO Statement

One of the greatest advantages of a computer is its ability to perform a set of operations over and over again.   The DO statement causes a series of steps to be repeated a specific number of times.   It has the general form

DO n i = m_1, m_2, m_3

where n is a statement number, i is an integer variable, and the m_j's are integer variables or positive integer constants.   The meaning of the DO statement is as follows:

Fig. 3.17

Starting with an initial value of i at m_1, all statements after the DO, up to and including the one numbered n, are executed. Then, the value of i is incremented by m_3, and if it is not greater than m_2, the same statements are again executed. The process continues until the value of i exceeds the limit m_2, at which time control passes to the next statement after statement n. The group of statements following the DO, up to and including statement n, is called the *range* of the DO. The integer variable i is called the *index,* m_1 the *initial value,* m_2 the *upper limit,* and m_3 the *increment*. If an integer variable is used for m_1, m_2, or m_3, its value must be defined before execution of the DO statement and must be positive. If the increment m_3 is not explicitly stated, it is taken to be 1. The index, initial value, upper limit, and increment can be used in statements within the range of the DO, provided that their values are not redefined. At the completion of a DO, the value of the index is no longer defined. The statement numbered n cannot be an IF or GO TO statement. An example of a DO statement is

DO 23 I = 5, 36, N

If the value of N had been previously defined as 2, the statements in the range of the above DO would be executed sixteen times. If it were desired to have an increment of 1, the statement could be written simply

DO 23 I = 5, 36

The flow charting of a DO is illustrated in Fig. 3.18.

As an example illustrating the use of a DO, consider the following problem concerned with determining the orbit of a satellite: Figure 3.19 shows an xy coordinate system fixed to the center of the earth, with the path of the satellite indicated by the solid curve. The points labeled 1, 2, and 3 indicate the positions of the satellite at equal time intervals. That is, the time it takes for the satellite to move from position 1 to position 2 is the same as the time from position 2 to position 3. If the coordinates of positions 1 and 2 are known, the coordinates of position 3 can be approximately determined from the equations

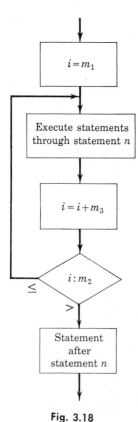

Fig. 3.18

$$x_3 = 2x_2 + x_1 \left[\frac{c}{(x_1{}^2 + y_1{}^2)^{\frac{3}{2}}} - 1 \right]$$

$$y_3 = 2y_2 + y_1 \left[\frac{c}{(x_1{}^2 + y_1{}^2)^{\frac{3}{2}}} - 1 \right]$$

(3.3)

These are obtained by using finite differences with equations derived from Newton's laws of motion and gravitation. The constant c depends upon the time interval mentioned above and the gravitational attraction of the earth. After point 3 has been determined, a fourth point can be determined by using points 2 and 3, a fifth point can be determined by using points 3 and 4, etc. A flow chart for a program which determines the coordinates of n additional points and prints out these values is given in Fig. 3.20. A FORTRAN program is then easily written from the flow chart.

```
        INTEGER N
        REAL X1, X2, X3, Y1, Y2, Y3, C
100     READ (5,1) X1, X2, Y1, Y2, C, N
  1     FORMAT (4F10.5, E10.5, I5)
        WRITE (6,2) X1, Y1, X2, Y2
  2     FORMAT (9X, 12HX-COORDINATE, 8X,
        $12HY-COORDINATE/(5X, 2F20.9))
        DO 3 J = 1, N
        X3 = 2.*X2 + X1*(C/((X1*X1 + Y1*Y1)**1.5) − 1.)
        Y3 = 2.*Y2 + Y1*(C/((X1*X1 + Y1*Y1)**1.5) − 1.)
        WRITE (6,4) X3, Y3
  4     FORMAT (5X, 2F20.9)
        X1 = X2
        X2 = X3
        Y1 = Y2
  3     Y2 = Y3
        GO TO 100
        END
```

3.7.4   CONTINUE Statement

Occasionally one may want to bypass the last statement in the range of a DO but to continue the indexing. A dummy statement CONTINUE is provided for this and other purposes. As the last statement in the range of a DO, it serves as a convenient point for branching from within the range. To illustrate, suppose that it is desired to print out only the coordinates of every rth point in the previous program, where r divides into n evenly. The flow chart for accomplishing this is given

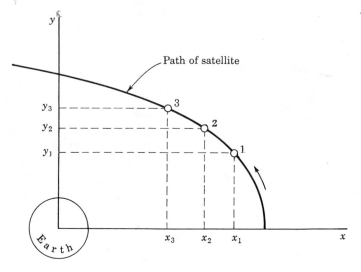

Fig. 3.19

in Fig. 3.21.   The FORTRAN program is as follows:

```
        INTEGER N, R, MARK
        REAL X1, X2, X3, Y1, Y2, Y3, C
100     READ (5,1) X1, X2, Y1, Y2, C, N, R
  1     FORMAT (4F10.5, E10.5, 2I5)
        WRITE (6,2) X1, Y1, X2, Y2
  2     FORMAT (9X, 12HX-COORDINATE, 8X,
        $12HY-COORDINATE/(5X, 2F20.9))
        MARK = 0
        DO 3 J = 1, N
        X3 = 2.*X2 + X1*(C/((X1*X1 + Y1*Y1)**1.5) − 1.)
        Y3 = 2.*Y2 + Y1*(C/((X1*X1 + Y1*Y1)**1.5) − 1.)
        X1 = X2
        X2 = X3
        Y1 = Y2
        Y2 = Y3
        MARK = MARK + 1
        IF (MARK − R) 3, 5, 5
  5     WRITE (6,4) X3, Y3
  4     FORMAT (5X, 2F20.9)
        MARK = 0
  3     CONTINUE
        GO TO 100
        END
```

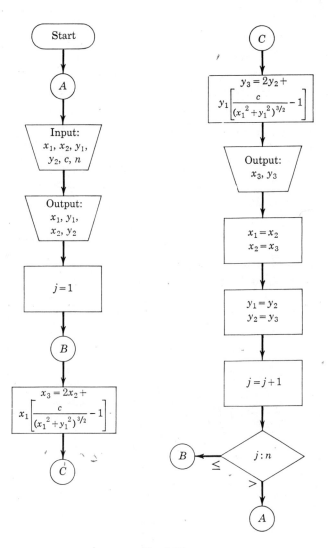

Fig. 3.20

Note that CONTINUE means to continue with the DO until the indexing is complete. At that time, control passes to the next sequential statement.

Although the CONTINUE statement finds its greatest application with a DO, as illustrated above, it may be used at any point in a program if needed. The CONTINUE statement is executable.

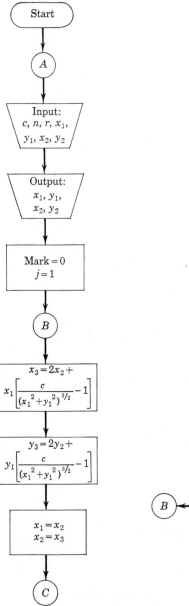
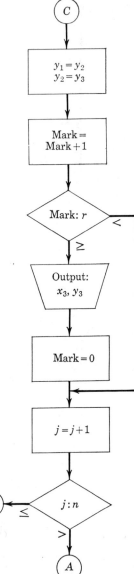

Fig. 3.21

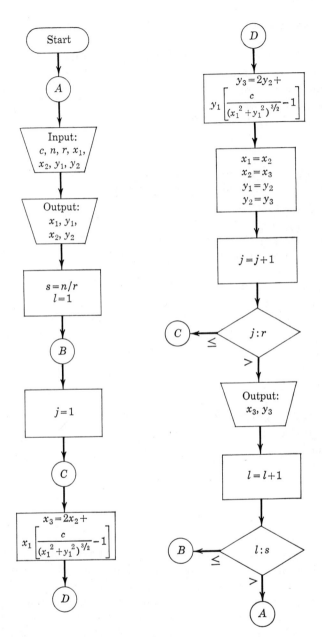

Fig. 3.22

3.7.5   Nested DO's

It is possible to have a second DO within the range of another DO.   A
third DO can be within the range of the second DO, a fourth within the
third, etc.   Such a set of DO's is called a *nest* of DO's.   As an illustration, the previous example is programmed in a different manner.   The
flow chart is shown in Fig. 3.22.

```
      INTEGER N, R, S
      REAL X1, X2, X3, Y1, Y2, Y3, C
100   READ (5,1) X1, X2, Y1, Y2, C, N, R
  1   FORMAT (4F10.5, E10.5, 2I5)
      WRITE (6,2) X1, Y1, X2, Y2
  2   FORMAT (9X, 12HX-COORDINATE, 8X,
      $12HY-COORDINATE/(5X, 2F20.9))
      S = N/R
      DO 5 L = 1, S
      DO 3 J = 1, R
      X3 = 2.*X2 + X1*(C/((X1*X1 + Y1*Y1)**1.5) − 1.)
      Y3 = 2.*Y2 + Y1*(C/((X1*X1 + Y1*Y1)**1.5) − 1.)
      X1 = X2
      X2 = X3
      Y1 = Y2
  3   Y2 = Y3
  5   WRITE (6,4) X3, Y3
  4   FORMAT (5X, 2F20.9)
      GO TO 100
      END
```

3.7.6   Ending a Program

In the examples presented thus far, programs have been ended by branching to a beginning READ statement.   This is desirable because it allows
results to be obtained for several groups of data with only one pass
through the computer.   When the data are exhausted, one of the following two possibilities will occur: If the program is one of a sequence of
programs to be run (monitor control), an error message will be printed
and the next sequential program started.   If the program is not under
monitor control, the computer will stop.   The proper way of ending a
program will now be discussed.

3.7.7 CALL EXIT and STOP Statements

If the program is run on a monitor-control system, the last executable statement should be either

CALL EXIT   or   STOP

These statements cause the control of the computer to pass to the next sequential program.

A seemingly strange, but usually very convenient, place for one of these statements is shortly after an early READ statement. This is illustrated by the following modification of the program given in Sec. 3.1. It is assumed that the variable A will never be negative.

```
1   READ (5,2) A, S, E, C, D
2   FORMAT (5F10.5)
    IF (A) 6, 7, 7
6   CALL EXIT
7   P = 2.*C + D
    Q = 1.49*A**(5.0/3.0)*S**0.5/(E*P**(2.0/3.0))
    WRITE (6,3) Q
3   FORMAT (F10.1)
    GO TO 1
    END
```

An extra card with a negative value of A must then be placed at the end of the data cards.

3.7.8 PAUSE Statement

If it is desired to halt the execution of a program temporarily, the executable statement

PAUSE

can be used. This may be convenient, for example, if the programmer is operating the computer and desires to check some intermediate results. Execution of the program can be resumed by pushing the START button on the computer.

3.7.9 Restrictions on Use of Control Statements

Care must be taken to observe certain rules in using the DO with other control statements. The first of these is that branching out of the range of a DO is permissible, as illustrated by the flow chart given in Fig. 3.23.

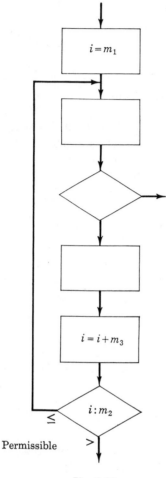

Fig. 3.23

Permissible

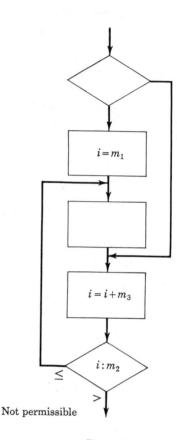

Fig. 3.24

Not permissible

The branch statement can be anywhere within the range of the DO except as the last statement. This restriction can be circumvented by using the CONTINUE statement. In branching out of the range of a DO, the value of the index remains defined, and the index can be used as an integer variable.

The second general rule is that branching into the range of a DO is not permissible, as illustrated by the flow chart given in Fig. 3.24. There is one exception to this rule: branching out of and back into the range of a DO is permissible, as illustrated in Fig. 3.25. In the statements labeled by box A, the values of i, m_1, m_2, and m_3 cannot be redefined.

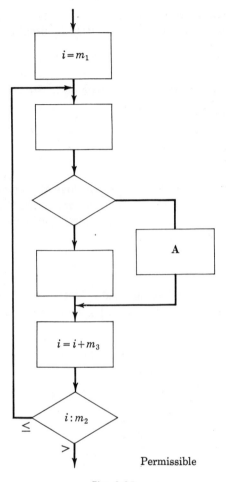

Permissible

Fig. 3.25

3.8   LOGICAL STATEMENTS[1]

A logical decision is based upon whether a given item or quantity is true or false or whether a quantity is less than, equal to, or greater than another quantity.   Methods of making such decisions are needed to control the flow of a computer program.   The arithmetic IF statement can be used to make such decisions.   However, FORTRAN provides a complete set of logical constants, variables, operators, and statements which give great flexibility in logical decision making.   In addition to

[1] Logical statements are not available in FORTRAN-II.

controlling the flow of programs, the use of logical statements can be applied to problems in the design of switching circuits, in theorem proving, and in logic.

3.8.1   Constants, Variables, and Arithmetic Statements

There are two *logical constants,*

.TRUE.
.FALSE.

The two periods are used to distinguish the logical constants from variable names and must be included.   A *logical variable* is named in the same manner as other types of variables and must appear in the list of a type statement of the form

LOGICAL *list*

The value of a logical variable is either .TRUE. or .FALSE.
  Two kinds of operators are used in logical expressions.   The first kind includes the *logical operators,*

.AND.
.OR.
.NOT.

The second kind includes the *relational operators,*

.GT.   greater than
.GE.   greater than or equal to
.LT.   less than
.LE.   less than or equal to
.EQ.   equal to
.NE.   not equal to

Again the two periods are necessary.
  Logical expressions consist of certain combinations of logical operations with logical constants, logical variables, and arithmetic expressions. Clearly, the quantities on both sides of the .AND. and .OR. operators and the right-hand side of the .NOT. operator must have logical values.   The quantities on both sides of a relational operator must have numerical values and, further, must be of the same type, either integer or real.   The value of a logical expression is either .TRUE. or .FALSE.   The following examples illustrate the use of all the logical operations.

Expression	*Value*
A.AND.B	.TRUE. if *both* A *and* B are .TRUE., otherwise .FALSE.
A.OR.B	.TRUE. if *either* A *or* B is .TRUE., otherwise .FALSE.
.NOT.A	.TRUE. if A is .FALSE., .FALSE. if A is .TRUE.
A.GT.B	.TRUE. if A > B, otherwise .FALSE.
A.GE.B	.TRUE. if A ≥ B, otherwise .FALSE.
A.LT.B	.TRUE. if A < B, otherwise .FALSE.
A.LE.B	.TRUE. if A ≤ B, otherwise .FALSE.
A.EQ.B	.TRUE. if A = B, otherwise .FALSE.
A.NE.B	.TRUE. if A ≠ B, otherwise .FALSE.

More complicated expressions can easily be constructed. For example,

A+B.LT.3.*C.OR.X   .TRUE. if either $(A + B) < 3.*C$ or if X is .TRUE., otherwise .FALSE.

3..LT.B.OR.X.EQ.Y.AND..NOT.C   .TRUE. if $(3. < B)$ or $(X = Y$ and C is .FALSE.), otherwise .FALSE.

The hierarchy of logical operations in descending order is

.LT., .LE., .EQ., .NE., .GT., .GE.
.NOT.
.AND.
.OR.

Parentheses may be used in logical expressions, as illustrated by

$(A + B)/C.LT.(A - B)/C$

Quantities within parentheses and arithmetic operations precede logical operations in the hierarchy. The one exception to this occurs in the use of the .NOT. operator, whereby the logical expression following the .NOT. must be enclosed within parentheses if it contains two or more quantities.

The expression in an arithmetic statement may be a logical expression. This implies that the variable on the left-hand side of the equals sign must be a logical variable. For example, the value of QUEST in the arithmetic expression

QUEST = A.NE.B

is .TRUE. if A ≠ B and .FALSE. if A = B. In the statement

ANSW = QUEST

ANSW has a value of .TRUE. or .FALSE. depending upon whether QUEST has a value .TRUE. or .FALSE.

3.8.2   Logical IF

The general form of the logical IF statement is

IF (*logical expression*) *s*

where *s* is any executable FORTRAN statement except another logical IF or DO. If the value of the logical expression is .TRUE., *s* is executed. Then, the next sequential statement is executed, unless *s* is a branching statement. If the value of the logical expression is .FALSE., *s* is not executed and control passes to the next sequential statement. An example of a logical IF statement is

```
   IF (A.AND.B) WRITE (6,5)
5  FORMAT (1X, 16HA AND B ARE TRUE)
   G = 3.*A − X
```

The message given in the FORMAT statement will be printed only if the values of both A and B are true. In any case the statement G = 3.*A − X will be executed.

3.8.3   Use of Logical IF

The logical IF can be used to great advantage in the control of a program. As an illustration, consider the following variation of the program given in Sec. 3.7.2:

```
   INTEGER STANUM
   REAL A, S, E, C, D, P, Q
1  READ (5,2) A, S, E, C, D, STANUM
2  FORMAT (5F10.4, I5)
   IF (A.LT.0.) CALL EXIT
   P = 2.*C + D
   Q = 1.49*A**(5./3.)*S**0.5/(E*P**(2./3.))
   IF (Q.LT.125.) WRITE (6,4) STANUM
4  FORMAT (5X, 21HIDENTIFICATION NUMBER, I5/10X,
   $18HQ IS LESS THAN 125)
   GO TO 1
   END
```

The logical IF statement is certainly more convenient than the arithmetic IF statement and avoids the possibility of making an error in statement numbers.

The logical IF statement can be the last statement in the range of a DO. For example,

DO 5 I = 1, 5

.

.

.

5   IF (I.EQ.3) WRITE (6,29)

causes the information specified in the FORMAT statement numbered 29 to be printed the third time through the loop.   The DO is executed five times.

If *s* is a branching statement and the logical expression is .TRUE., the transfer is executed even if this takes the control of the program out of the loop.   If the last statement in the above example is changed to

5   IF (I.EQ.3) GO TO 12

a branch to statement 12 occurs at the end of the third time through the loop.   If statement 12 is outside of the range of the DO, the DO is executed only three times.

3.9   SUBSCRIPTED VARIABLES

Subscripted variables are often encountered in mathematical and engineering problems.   Such a variable is also known as an *array*, or a *matrix*.   A singly subscripted variable can be referred to as a *one-dimensional* array and is sometimes called a *vector*.   Doubly subscripted variables are referred to as *two-dimensional* arrays.   Mathematically, it is possible to have three-, four-, or five-dimensional arrays or arrays of any dimension. FORTRAN provides for one-, two-, and three-dimensional arrays.

In FORTRAN a subscripted variable is formed by naming in the usual way and following the name with the subscripts in parentheses.   If there is more than one subscript, they are separated by commas.   A subscript must be of the integer type and can take any of the following forms:

1. Constant	Examples: POWER(5); REV(3,13); TORQUE(8,1,4)
2. Variable	Examples: ION(M); BETA(LL,JK); GAMMA(NN,5,J)
3. Variable $\pm$ constant	Examples: STRESS(I + 3); STRAIN(KL − 1, JJ); FORCE(KK + 2, 5, KK − 2)
4. Constant*variable	Examples: AMP(2*L); OHM(5*KL, 3); VOLT(J, 2*MAT, 4*MAT)
5. Constant*variable $\pm$ constant	Examples: SOD(5*L + 2); MAGN(M, 2*KL − 4); HCL(2*KL + 4, 8, N − 3)

Each constant and the value of each variable appearing in a subscript must be positive, and the evaluated subscript must be greater than zero. Subscripts themselves cannot be subscripted.

Subscripted variables are used in arithmetic statements in the same manner as nonsubscripted variables.   An example is

DIST(I,J) = 0.5*GRAV*TIME(J)**2

3.9.1   Singly Subscripted Variables

The coefficients of a polynomial provide an excellent illustration of subscripted variables.   A general expression for an nth-order polynomial is

$$a_1 x^n + a_2 x^{n-1} + \cdots + a_n x + a_{n+1}$$

where the coefficients of the powers of x are represented by the subscripted variable a_i.   The subscript i represents any integer between 1 and $n + 1$. The individual a_i's are called *elements* of the subscripted variable.   In the numerical example

$$5.0x^4 - 3.0x^3 + 12.9x^2 + 189.2$$

the elements of the subscripted variable are

$a_1 = 5.0$
$a_2 = -3.0$
$a_3 = 12.9$
$a_4 = 0.0$
$a_5 = 189.2$

As an illustration of the use of singly subscripted variables, the following statements evaluate any nth-order polynomial:

```
      VALUE = A(N + 1)
      DO 104 I = 1, N
104   VALUE = VALUE + A(I)*X**(N - I + 1)
```

Note that, as I increases from 1 to N, the quantity (N − I + 1) decreases from N to 1.

3.9.2   Doubly Subscripted Variables

Variables may be multisubscripted. For example, the coefficients of three simultaneous linear algebraic equations could be represented by a doubly subscripted variable b_{ij}.

$$b_{11}x + b_{12}y + b_{13}z = b_{14}$$
$$b_{21}x + b_{22}y + b_{23}z = b_{24}$$
$$b_{31}x + b_{32}y + b_{33}z = b_{34}$$

If the simultaneous equations are considered without the unknowns and algebraic symbols, the subscripted variable b_{ij}, where i represents the row and j the column in which the element appears, is given below.

$$b_{11} \quad b_{12} \quad b_{13} \quad b_{14}$$
$$b_{21} \quad b_{22} \quad b_{23} \quad b_{24}$$
$$b_{31} \quad b_{32} \quad b_{33} \quad b_{34}$$

In the numerical example

$$x + 2y - 3z = 8$$
$$4x - 12y + 9z = 13$$
$$3x + y + z = 0$$

the elements of the subscripted variable are

$$b_{11} = 1 \qquad b_{12} = 2 \qquad b_{13} = -3 \qquad b_{14} = 8$$
$$b_{21} = 4 \qquad b_{22} = -12 \qquad b_{23} = 9 \qquad b_{24} = 13$$
$$b_{31} = 3 \qquad b_{32} = 1 \qquad b_{33} = 1 \qquad b_{34} = 0$$

N simultaneous equations may be expressed in the following form if the unknowns are expressed as a singly subscripted variable x_i:

$$b_{1,1}x_1 + b_{1,2}x_2 + \cdots + b_{1,N}x_N = b_{1,N+1}$$
$$b_{2,1}x_1 + b_{2,2}x_2 + \cdots + b_{2,N}x_N = b_{2,N+1}$$
$$\cdots \cdots \cdots \cdots \cdots \cdots \cdots \cdots \cdots \cdots$$
$$b_{N,1}x_1 + b_{N,2}x_2 + \cdots + b_{N,N}x_N = b_{N,N+1}$$

As an example, the following part of a program multiplies every element in each row of an M × N two-dimensional array by its row number.

```
      DO 54 K = 1, M
      AK = K
      DO 54 L = 1, N
   54 B(K,L) = AK*B(K,L)
```

Note that the conversion of the integer variable K to its floating-point equivalent AK was necessary to avoid two types of variables in the last statement.

3.9.3   DIMENSION Statement

To ensure that the FORTRAN translator reserves enough storage locations for all the elements of a subscripted variable, the name of each subscripted variable must appear in a DIMENSION statement. It has the general form

DIMENSION v_1( ), v_2( ), v_3( ), . . . , v_n ( )

where each v_i is the name of a subscripted variable and within the parentheses are one, two, or three integer positive constants which indicate the maximum size of the array. If a program uses three subscripted variables, COPPER, IRON, and STEEL, the DIMENSION statement might be

DIMENSION COPPER(12), IRON(10,15), STEEL(6,8,5)

In this example COPPER is a singly subscripted variable which may have up to 12 elements. IRON is a doubly subscripted variable which can have a maximum of 10 rows and 15 columns. The FORTRAN translator thus reserves 150 storage locations for the variable IRON. STEEL is a triply subscripted variable, with the maximum values of the first, second, and third subscripts 6, 8, and 5, respectively. The DIMENSION statement is not executable but must be placed in the program before the appearance of the subscripted variables.

3.9.4   Input-Output of Subscripted Variables

Subscripted variables can be read into the computer in several ways, depending upon the FORMAT statement and the list in the READ statement.

If the name of the array occurs in the list in unsubscripted form, the entire array, as defined by the corresponding DIMENSION statement, will be read into the computer. The order in which the elements are transmitted is such that the first subscript varies through its entire range for each change of the second subscript and the second varies through its entire range for each change of the third. The elements are stored in increasing storage locations in the same order. The elements must be punched on the data cards in that order. As an example, consider a doubly subscripted variable named HEAT, and assume that the following statements appear in a program:

```
    DIMENSION HEAT(10,15)
    READ (5,1) HEAT
1   FORMAT (6E12.5)
```

Data are punched as six elements per card in the order

```
HEAT(1,1)
HEAT(2,1)
HEAT(3,1)
    .
    .
    .
HEAT(10,1)
HEAT(1,2)
HEAT(2,2)
    .
    .
    .
```

All 150 elements must be transmitted.

It is also possible to read in only selected elements of an array or to deviate from the normal order by using the DO or subscripting within the READ list. The following example illustrates a second method of transmitting the values of the subscripted variable HEAT:

```
    READ (5,1) N, M
1   FORMAT (2I5)
    DO 2 I = 1, N
    DO 2 J = 1, M
2   READ (5,3) HEAT (J,I)
3   FORMAT (E12.4)
```

Data are transmitted in the same order as in the previous example. However, only one piece of data is punched per card. The integer variables M and N represent the number of rows and columns, respectively, to be transmitted. The value of M can be any integer not greater than 10 and N not greater than 15. M and N must have positive values, of course.

A third method, which transmits the array in the same order, is illustrated below:

```
    READ (5,1) N, M, ((HEAT(J,I), J = 1, M), I = 1, N)
1   FORMAT (2I5/(6E12.4))
```

The values of N and M are punched on the first data card, and then, starting with the second card, elements are punched six per card as in the first method.

A fourth method combines the features of the second and third methods.

```
    READ (5,1) N, M
1   FORMAT (2I5)
    DO 2 I = 1, N
2   READ (5,3) (HEAT(J,I), J = 1, M)
3   FORMAT (6E12.4)
```

The second, third, and fourth methods have the advantage of not making it necessary to punch all 150 elements of HEAT. In the fourth method if N = 4 and M = 6, data must be punched in the following order:

```
     4        6
A(1,1)  A(2,1) · · · A(6,1)
A(1,2)  A(2,2) · · · A(6,2)
A(1,3)  A(2,3) · · · A(6,3)
A(1,4)  A(2,4) · · · A(6,4)
```

Previously it was stated that subscripted variables are convenient in transmitting character information. Assume that it is desired to transmit information up to 24 characters in length but that each storage location can hold only 6 characters. The following statements will read in 24 characters by using the subscripted variable PROC.

```
    DIMENSION PROC(4)
    READ (5,1) PROC
1   FORMAT (4A6)
```

If the characters to be read in are CHEMICAL PROCESS *1, they should appear on a data card in the form shown in Fig. 3.26. The characters will be stored in the elements of the variable PROC as follows:

```
PROC(1) = CHEMIC
PROC(2) = AL PRO
PROC(3) = CESS *
PROC(4) = 1
```

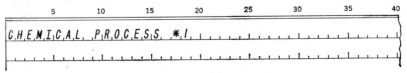

Fig. 3.26

The fortran language   69

Subscripted variables can be printed out with methods similar to those used for input.

3.10   FUNCTIONS[1]

Many types of functions, such as polynomial functions, the logarithm function, trigonometric functions, etc., are used in the solution of numerous engineering problems. Some of the most common are "built into the FORTRAN language," and these are referred to as "built-in" functions. They are available for use in any FORTRAN program. If a programmer desires to use a function which is not built in, he must write his own program for evaluating the function.

The advantages of using functions in programming are numerous. Built-in functions save the programmer much work. Other types are advantageous if a specific set of calculations involving certain arguments is to be performed at several different points in a program. Also, it is frequently helpful to break a large program into several smaller programs, called *subprograms*, which have the advantage of being compiled independently. Making several small, error-free programs is usually easier than making one very long, error-free program. After all the errors in the subprograms have been removed, the subprograms can then be combined into a complete program.

In FORTRAN there are three types of functions the programmer can form himself. They are:

1. Arithmetic-statement function
2. FORTRAN FUNCTION
3. SUBROUTINE

The evaluation of a function can be thought of as an operation. When a function appears in a statement of a program, it is said to have been *called* by the program. The details of forming and calling functions will now be discussed.

3.10.1   Built-in Functions

The built-in functions represent those which are automatically available to the programmer. A list of the functions which are generally available

[1] In FORTRAN-II, all functions have names ending with the letter F. Also, some of the built-in functions have different names from those given in Table 3.4.

is given in Table 3.4. The symbol x represents the argument of the function.

Table 3.4

Function	Description	Type	
		Argument	*Function*
ABS(x)	$\|x\|$	Floating	Floating
IABS(x)		Integer	Integer
AINT(x)	Sign of x multiplied by the largest integer which is $\leq \|x\|$	Floating	Floating
INT(x)		Floating	Integer
AMOD(x_1,x_2)	$x_1 - (x_1/x_2)x_2$, where x_1/x_2 equals the integral part of x_1/x_2	Floating	Floating
MOD(x_1,x_2)		Integer	Integer
AMAX0(x_1,x_2, . . . ,x_n)	Maximum of x_1, x_2, . . . , x_n	Integer	Floating
AMAX1(x_1,x_2, . . . ,x_n)		Floating	Floating
MAX0(x_1,x_2, . . . ,x_n)		Integer	Integer
MAX1(x_1,x_2, . . . ,x_n)		Floating	Integer
AMIN0(x_1,x_2, . . . ,x_n)	Minimum of x_1, x_2, . . . , x_n	Integer	Floating
AMIN1(x_1,x_2, . . . ,x_n)		Floating	Floating
MIN0(x_1,x_2, . . . ,x_n)		Integer	Integer
MIN1(x_1,x_2, . . . ,x_n)		Floating	Integer
FLOAT(x)	Floating-point equivalent of x	Integer	Floating
IFIX(x)	Integer equivalent of x	Floating	Integer
SIGN(x_1,x_2)	Sign of x_2 multiplied by $\|x_1\|$	Floating	Floating
ISIGN(x_1,x_2)		Integer	Integer
DIM(x_1,x_2)	x_1 − minimum of x_1 or x_2	Floating	Floating
IDIM(x_1,x_2)		Integer	Integer
ALOG(x)	Natural logarithm of x	Floating	Floating
ALOG10(x)	Common logarithm of x	Floating	Floating
SIN(x)	Sine of x, where x is in radians	Floating	Floating
COS(x)	Cosine of x, where x is in radians	Floating	Floating
EXP(x)	e^x, exponential function	Floating	Floating
SQRT(x)	Square root of x	Floating	Floating
ATAN(x)	Arctangent of x in radians	Floating	Floating
TANH(x)	Hyperbolic tangent of x	Floating	Floating

Functions are used in arithmetic expressions in the same manner as variables. This is known as *calling* the function. Any legal FORTRAN expression can be used as the argument. Examples of how functions appear are given below:

```
ABS(3.*X + Y)
SIN(A - B/(C + D))
SQRT(B**2 - 4.*A*C)
EXP(3./B - ATAN(FOOT - 3.))
```

As an example illustrating the calling of a function, consider the following statement taken from the program in Sec. 3.1:

Q = 1.49*A**(5.0/3.0)*S**0.5/(E*P**(2.0/3.0))

This statement could be revised to

Q = 1.49*A**(5.0/3.0)*SQRT(S)/(E*P**(2.0/3.0))

3.10.2   Arithmetic-statement Functions

The arithmetic-statement function is used when the desired function can be defined by a single arithmetic statement. It has the general form

name $(x_1, x_2, \ldots, x_n)$ = *expression*

Name is the symbolic name of the function. The rules for naming arithmetic-statement functions are the same as those for naming variables. Either the name appears in a type statement, or its type is determined by the first letter. The x_i's are names of variables which represent the arguments of the function. They are separated by commas and enclosed within parentheses. Each argument must be a nonsubscripted variable and can be of the integer, floating-point, or logical type. There must be at least one argument. The expression can be any FORTRAN expression with the one exception that it cannot contain subscripted variables. All the arguments must appear in the expression. Other variables, which are usually called *parameters*, may also appear in the expression. The function statement must appear in the program before the first executable statement. Type statements indicating the types of the function and its arguments must appear in the program before the function statement.

As an example, suppose that it is desired to form a function of the quadratic

Thrust = $ax^2 + bx + c$

The FORTRAN statement could appear as

THRUST(X) = (A*X + B)*X + C

Arithmetic-statement functions are called in a program in the same manner as built-in functions. An example which illustrates calling the above function is

FORCE = THRUST(AI*T) − WEIGHT − DRAG

The function is evaluated by replacing the argument X by the expression AI*T. Of course, the values of the variables AI, T, A, B, and C must have been defined previously to the execution of the above statement.

In the above examples A, B, and C were parameters. They could be made arguments by putting the function in the form

THRUST(X,A,B,C) = (A*X + B)*X + C

When called in an arithmetic statement, the function could then appear in the form

FORCE = THRUST(AI*T, D, E, F) − WEIGHT − DRAG

The variables D, E, and F replace the arguments A, B, and C, respectively, when the function is being evaluated.

The arguments which are used in the statement that defines a function are only *dummy* variables. They are replaced by the expressions actually used when the function appears in an arithmetic statement. Correspondence between dummy and actual variables is given simply by position in the list; i.e., the first dummy variable is replaced by the first actual expression, the second dummy variable by the second actual expression, etc.

For example, the arithmetic statement

FORCE = THRUST(AI*T, C, B, A) − WEIGHT − DRAG

with the previous definition of THRUST function will cause FORCE to be evaluated by

FORCE = (C*(AI*T) + B)*(AI*T) + A − WEIGHT − DRAG

It is allowable for the expression of one arithmetic-statement function to contain other arithmetic-statement functions. These latter functions, however, must be defined before the function in which they appear. The following statements illustrate this idea:

THRUST(X) = (A*X + B)*X + C

.
.
.

FORCE(X) = THRUST(X) − WEIGHT − DRAG

.
.
.

In this example FORCE(X) could not be placed before THRUST(X).

3.10.3   FORTRAN FUNCTION

The FORTRAN FUNCTION can be used when more than one arithmetic statement is required to define a function. This function is formed as a separate program by itself, and for this reason is called a subprogram. It has the general form

Type FUNCTION *name* $(x_1, x_2, \ldots, x_n)$

.

.

.

name $= \cdot \cdot \cdot$

.

.

.

RETURN

.

.

.

END

Name is the symbolic name of the function. The rules for naming FORTRAN FUNCTIONS are the same as those for naming variables. *Type* is any of the words INTEGER, REAL, or LOGICAL, depending upon whether the value of the function is to be of the integer, floating-point, or logical type. If none of these three words appears in the statement, the type is determined by the first letter of *name*, according to the rules for naming variables. *Name* must appear on the left-hand side of at least one arithmetic statement in the subprogram. This statement defines the value of the function. The x_i's are names of variables which represent the arguments of the function. These variables may be of the integer, floating-point, or logical type. If an array name is used as an argument, it must be written in unsubscripted form. The arguments, of which there must be at least one, are separated by commas and enclosed within parentheses. Other variables may also appear in these statements. As with arithmetic-statement functions, the arguments should be thought of as dummy variables.

The RETURN statement is the last executable statement in the subprogram. Its purpose is to return the control in the computer to the program which called the subprogram. There must be at least one RETURN statement in a FORTRAN FUNCTION. There could be more if there were several branches in the subprogram.

The physically last statement of a FORTRAN FUNCTION must be

END

As an illustration, the FORTRAN FUNCTION which follows evaluates an *n*th-order polynomial. The program statements are taken from the example in Sec. 3.9.1.

```
        REAL FUNCTION VALUE(A,X,N)
        DIMENSION A(100)
        INTEGER N
        REAL A, X
        VALUE = A(N + 1)
        DO 104 I = 1, N
104     VALUE = VALUE + A(I)*X**(N − I + 1)
        RETURN
        END
```

FORTRAN FUNCTIONS are called in exactly the same manner as built-in and arithmetic-statement functions. An example of an arithmetic statement which calls the above function is

FORCE = THRUST(AI*T) − WEIGHT − VALUE(C, V + DV, N)

In the program in which the above statement appears, the values of the variables C, V, DV, and N must be defined previously to the execution of the statement. Further, the variable C must be a subscripted variable. When the function is evaluated, the expressions C, V + DV, and N replace the dummy variables A, X, and N, respectively.

3.10.4   SUBROUTINE

All the functions considered thus far have the restriction that they return only one value to the calling program. Sometimes, however, it is necessary to form a function which evaluates several quantities. Examples of this are functions which calculate the n roots of an nth-order polynomial and those which sort a group of numbers. In FORTRAN they can be formed with a SUBROUTINE, the general form of which is

SUBROUTINE *name* $(x_1, x_2, \ldots, x_n)$
.
.
.

RETURN
.
.
.

END

The rules governing the name and the arguments of a SUBROUTINE are the same as those of the FORTRAN FUNCTION with the important exception that the name refers only to the SUBROUTINE and does not represent a functional value. The variable names that represent the functional values are included in the list of arguments. In this way the SUBROUTINE can return any number of functional values to the calling program. Since *name* does not have a value, it has no type associated with it. Further, *name* cannot appear on the left-hand side of an arithmetic statement within the SUBROUTINE. Instead, the values of all the arguments which represent functional values must be defined within the SUBROUTINE.

A SUBROUTINE is presented below which determines two real roots, x_1 and x_2, of the quadratic equation

$$ax^2 + bx + c = 0$$

on the assumption that $a \neq 0$ and that imaginary roots are not to be considered.

```
    SUBROUTINE ROOTS(A,B,C,X1,X2)
    REAL A, B, C, D, X1, X2
    D = B**2 - 4.*A*C
    IF (D.GE.0.) GO TO 2
    WRITE (6,1)
1   FORMAT (5X, 10HD NEGATIVE)
    RETURN
2   X1 = (-B + SQRT(D))/(2.*A)
    X2 = (-B - SQRT(D))/(2.*A)
    RETURN
    END
```

Since a SUBROUTINE returns more than one value, it must be called in a manner different from the other types of functions, i.e., by a statement of the form

CALL *name* $(x_1, x_2, \ldots , x_n)$

where *name* is the name of the SUBROUTINE. The x_i's are the arguments, which include the names of functional values. They must be separated by commas and enclosed within parentheses and can be any FORTRAN expression. The following statement calls the SUBROUTINE given previously.

CALL ROOTS(5.0, B, 3.*F - G, R1, R2)

In the program in which the above statement appears, the values of the variables B, F, and G must be defined previously to the execution of the statement. After execution, the values of the variables R1 and R2 are defined and equal the functional values obtained from the SUBROUTINE.

3.10.5   Use of Functions

The following general rules apply to functions:

1. Functions which are called by arithmetic statements are evaluated before any of the other arithmetic operations. Thus, the evaluation of functions is on the highest level in the hierarchy of the operations.

2. Names of all variables and functions must be unique. Even if a built-in function is not used in a program, its name must not be used as a variable name or the name of another function. For example, there should never be a variable named SQRT.

3. All functions except SUBROUTINE need at least one argument. In subprograms, arguments can alternatively be transmitted between programs by use of the COMMON statement, to be discussed in Sec. 3.12.2.

4. The arguments given in the calling statement of a function must be in the same order and of the same type as the corresponding dummy variables which appear in the function statements. Furthermore, if an argument in the calling statement of a function is an array name, the corresponding dummy variable must also be an array name of the same dimensions.

5. An argument of a subprogram may be a dummy name of another subprogram. The actual subprogram transmitted is determined by the calling statement. In such a case, the statement

EXTERNAL *list*[1]

must appear in the calling program. *List* is a list of the subprograms, separated by commas, which appear as arguments in the calling statements of the calling program. The following built-in functions may also be transmitted in this manner: ALOG, ALOG10, SIN, COS, EXP, SQRT, ATAN, and TANH.

6. In a subprogram, the dimensions of a subscripted variable may be left undetermined by using integer variables in the DIMENSION statement, e.g., DIMENSION IRON(M,N). The dimensions M and N as well as the variable IRON must be transmitted to the subprogram as arguments.

A complete program must include a main program and can, in addition, include several subprograms. The subprograms may be called either by the main program or by other subprograms, but a subprogram cannot be called by itself. The main program cannot be called by any program.

[1] In FORTRAN-II, a different method is used to list the subprograms that may be called by the main program.

An example is presented here illustrating the use of the different kinds of functions. Suppose that it is desired to calculate the velocity, at various times, of a rocket fired vertically from the surface of the earth. Newton's law of motion is expressed by the equation

$$F = ma$$

where F is the resultant force acting on the rocket, m the combined mass of the rocket and fuel, and a the acceleration of the rocket. Acceleration is defined as rate of change of velocity with respect to time and is approximated by

$$\frac{\Delta V}{\Delta t}$$

where ΔV represents the change in velocity of the rocket as it moves from one point (1) to another (2) during the time interval Δt, as shown in Fig. 3.27. Letting V_1 and V_2 be the velocities of the rocket at positions 1 and 2, respectively,

$$\Delta V = V_2 - V_1$$

and the law of motion becomes

$$V_2 \approx V_1 + \frac{F}{m} \Delta t \tag{2}$$

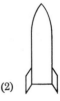

If V_1, F, m, and Δt are known, V_2 can be determined. The velocity at a third point V_3 can then be determined from V_2, the velocity at a fourth point V_4 from V_3, etc. The resultant force acting on the rocket is equal to the thrust T minus the weight W minus the drag D, as indicated in Fig. 3.27. It is assumed that the thrust varies with time according to the relation

$$T = at^2 + bt + c$$

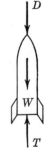

where a, b, and c are constants which have been determined from test firings of the rocket. Burnout time (time when thrust becomes zero) is determined by the smaller root of the quadratic if both roots are positive or the positive root if they are of opposite sign. The combined weight of the rocket and fuel is given by the formula

$$W = W_0 - qt$$

where W_0 is the initial combined weight of the rocket and fuel and q is the rate at which the fuel is being consumed. It is assumed that the drag force D has been experimentally determined to vary with the velocity V as an

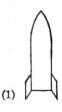

Fig. 3.27

nth-order polynomial,

$$D = d_1 V^n + d_2 V^{n-1} + \cdots + d_n V + d_{n+1}$$

The combined mass of the rocket and fuel is obtained by dividing the weight W by g, the acceleration of a freely falling body. Lift-off time is designated as $t = 0$, at which time the velocity of the rocket is instantaneously zero. The velocity is to be evaluated at equal time intervals until the time after lift-off reaches a final value t_f.

A flow chart of the main program is given in Fig. 3.28. A FORTRAN program constructed from the flow chart is given below. The examples from Secs. 3.10.2 to 3.10.4 are utilized.

```
      INTEGER N
      REAL DT, A, B, C, W0, Q, G, TF, D, R1, R2, TB, TIME,
     $VLCTY, T, WEIGHT, FORCE, THRUST, VALUE
      THRUST(X) = (A*X + B)*X + C
      DIMENSION D(100)
14    READ (5,1) DT, A, B, C, W0, Q, G, TF, N, (D(I), I = 1, N)
1     FORMAT (4F10.5/4F10.5, I10/(7F10.5))
      IF (DT.LE.0.) CALL EXIT
      CALL ROOTS(A,B,C,R1,R2)
      TB = R1
      IF (R1.LE.0..OR.R2.GT.0..AND.R1.GT.R2) TB = R2
      TIME = 0.
      VLCTY = 0.
      WRITE (6,6) TIME, VLCTY
6     FORMAT (1H1, 10X,
     $37HVELOCITY OF A ROCKET FIRED VERTICALLY/
     $13X, 4HTIME, 14X, 8HVELOCITY/5X, 2F20.9)
11    T = 0.
      IF (TIME.LT.TB) T = THRUST(TIME)
      WEIGHT = W0 - Q*TIME
      FORCE = T - WEIGHT - VALUE(D, VLCTY, N - 1)
      VLCTY = VLCTY + FORCE*DT*G/WEIGHT
      TIME = TIME + DT
      WRITE (6,10) TIME, VLCTY
10    FORMAT (5X, 2F20.9)
      IF (TIME.LT.TF) GO TO 11
      WRITE (6,13)
13    FORMAT (15X, 19HEND OF CALCULATIONS)
      GO TO 14
      END
```

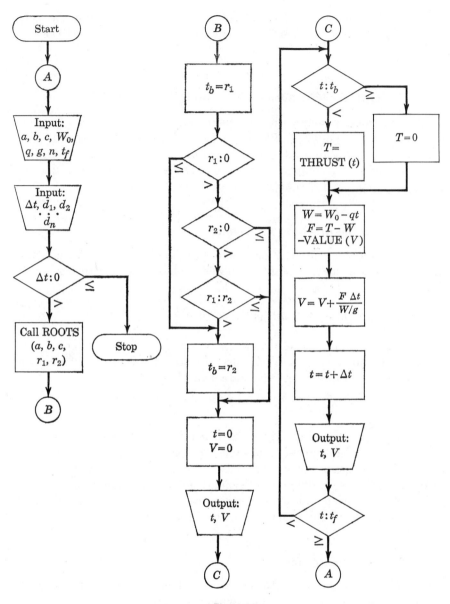

Fig. 3.28

The complete program consists of this main program and the subprograms REAL FUNCTION VALUE and SUBROUTINE ROOTS.

3.11   DOUBLE-PRECISION AND COMPLEX ARITHMETIC[1]

A floating-point number occupies one storage location in the computer and possesses 8 significant digits. A *double-precision* floating-point number occupies two adjacent storage locations and thus possesses 16 significant digits. A complex number has two parts, a real part and an imaginary part, and also occupies two adjacent storage locations.

3.11.1   Double-precision Arithmetic

A double-precision constant may have up to and including 17 digits and must contain a decimal point. The letter D is used for the exponent form. Thus,

$$6.83972144369D6 = 6.83972144369 \times 10^6$$

The magnitude of a constant must be between 10^{-29} and 10^{38} or be zero. Names of double-precision variables are formed in the usual manner, but they must appear in a type statement of the form

DOUBLE PRECISION *list*

where *list* is a list of variables. Arithmetic expressions are formed in the usual manner, and it is allowable for them to contain both floating-point and double-precision quantities. In all cases, the operations are carried out by use of double-precision arithmetic. The input and output of double-precision variables are accomplished with the specification D$w.d$, which is used in exactly the same manner as E$w.d$.

3.11.2   Complex Arithmetic

A complex constant is formed by enclosing two floating-point constants, which are separated by a comma, within parentheses. For example,

$$(63.9, \ -5.21E\text{-}3) = 63.9 - 0.00521i$$

[1] In FORTRAN-II, double-precision and complex arithmetic facilities are available but are denoted in a different manner.

Names of complex variables are formed in the usual manner, but they must appear in a type statement of the form

COMPLEX *list*

where *list* is a list of variables. Arithmetic expressions are formed in the usual manner, and it is allowable for them to contain both floating-point and complex quantities. In all cases, the operations are carried out by means of complex arithmetic. One special rule, here, is that the exponent of a complex quantity must be of the integer type. The input and output of complex variables are accomplished by using two successive floating-point specifications in the FORMAT statement.

A simple program is given below, illustrating the use of double-precision and complex arithmetic:

```
   REAL A
   DOUBLE PRECISION B, D
   COMPLEX C, E
   READ (5,1) A, B, C
 1 FORMAT (F10.5, D25.16, 2F10.5)
   D = 3.*B − A
   E = (2.,3.)*C − A
   WRITE (6,2) D, E
 2 FORMAT (5X, 2HD=, D25.16/5X, 2HE=, E15.8, 2X, E15.8, 1X,
   $1HI)
   CALL EXIT
   END
2.0    2.0000000001D0    2.0    −4.0
```

The output from this program would appear as shown below:

D = 0.4000000000300000D 01
E = 0.14000000E 02   −0.20000000E 01 I

3.11.3 Double-precision and Complex Functions

Table 3.5 presents a list of built-in functions which use double-precision and complex arithmetic. An arithmetic-statement function can be a double-precision or a complex function. In such a case the function name must appear in a corresponding type statement. A FORTRAN FUNCTION can be made double-precision or complex by including the proper type word in the definition statement. For example, such a statement might be

COMPLEX FUNCTION FLOW(VORTEX, PSTN)

Table 3.5

Function	Description	Argument	Function
SNGL(x)	Most significant part of x	Double	Floating
REAL(x)	Real part of x	Complex	Floating
AIMAG(x)	Imaginary part of x	Complex	Floating
DABS(x)	$|x|$	Double	Double
IDINT(x)	Sign of x multiplied by the largest integer which is $\leq |x|$	Double	Double
DMAX1($x_1,x_2,$ . . . ,x_n)	Maximum of x_1, x_2, . . . , x_n	Double	Double
DMIN1($x_1,x_2,$ . . . ,x_n)	Minimum of x_1, x_2, . . . , x_n	Double	Double
DSIGN(x_1,x_2)	Sign of x_2 multiplied by $|x_1|$	Double	Double
DBLE(x)	x in double-precision form	Floating	Double
CMPLX(x_1,x_2)	$x_1 + x_2i$	Floating	Complex
CONJG(x)	Conjugate of x	Complex	Complex
DMOD(x_1,x_2)	$x_1 - (x_1/x_2)$, where x_1/x_2 equals the integral part of x_1/x_2	Double	Double
DLOG(x)	Natural logarithm of x	Double	Double
DLOG10(x)	Common logarithm of x	Double	Double
DSIN(x)	Sine of x, where x is in radians	Double	Double
DCOS(x)	Cosine of x, where x is in radians	Double	Double
DEXP(x)	e^x, exponential function	Double	Double
DSQRT(x)	Square root of x	Double	Double
DATAN(x)	Arctangent of x in radians	Double	Double
CABS(x)	Absolute value of x	Complex	Complex
CLOG(x)	Natural logarithm of x	Complex	Complex
CSIN(x)	Sine of x	Complex	Complex
CCOS(x)	Cosine of x	Complex	Complex
CEXP(x)	e^x	Complex	Complex
CSQRT(x)	Square root of x	Complex	Complex

3.12   EQUIVALENCE, COMMON, DATA, AND BLOCK DATA[1]

There are several other items in the FORTRAN language which can be used by a programmer at his convenience. These include the statements EQUIVALENCE, COMMON, DATA, and the subroutine BLOCK DATA.

3.12.1   EQUIVALENCE Statement

During compilation each variable is assigned to a storage location. Two or more variables can be assigned to the same location by using an

---

[1] Common *blocks*, DATA statements, and BLOCK DATA subprograms are not available in FORTRAN-II.

EQUIVALENCE statement of the form

EQUIVALENCE $(\cdot \ \cdot), (\cdot \ \cdot), \ldots, (\cdot \ \cdot)$

Within each parenthesis pair is a list of the variables which are equivalent to each other. The order of the variables is unimportant. The statement

EQUIVALENCE (C,G,Q), (A,B)

means that C, G, and Q have a single storage location and A and B have a single storage location. Obviously, all variables which are equivalent to each other have the same value.

The EQUIVALENCE statement is useful in large programs which may exceed the capacity of the computer. Otherwise, it is not needed unless the programmer has a particular reason for making the values of several variables equal. For example, the sample program given in Sec. 3.1 could have the statement

EQUIVALENCE (P,C), (D,Q)

since, once the value of P is determined, the values of C and D are no longer needed. This would save two storage locations. In such a short program, however, this really serves no useful purpose.

The EQUIVALENCE statement is not executable and can be placed anywhere in a program.

Two or more subscripted variables may be made equivalent by simply placing their names in unsubscripted form in an EQUIVALENCE statement. If the arrays are dimensioned exactly the same, all the corresponding elements of the arrays are equivalent.

Arrays can be made equivalent in a staggered form by including subscripts with the array name in the EQUIVALENCE statement. For example, the statement

EQUIVALENCE (A(3), B(2,2))

means that storage locations are allotted as follows:

```
   .        .
   .        .
   .        .
A(1)     B(3,1)
A(2)     B(1,2)
A(3)     B(2,2)
A(4)     B(3,2)
   .        .
   .        .
   .        .
```

Three special rules concerning the EQUIVALENCE statement are:

1. Variable names not listed in an EQUIVALENCE statement are given unique storage locations.

2. Dummy arguments of a function or subroutine cannot appear in an EQUIVALENCE statement.

3. The equivalence of a double-element variable (double-precision or complex) with a single-element variable (any other type) means that the single-element variable is equivalent to either the more significant part or the real part of the double-element variable, whichever applies.

3.12.2   COMMON Statement

FORTRAN allows a programmer to place variables in specific storage locations within the computer, called *common storage*. Let it be assumed that common storage starts at location 32,000 and that data are stored in the order of increasing location numbers. The data are stored in common storage by using a COMMON statement which has the general form

COMMON $x_1, x_2, \ldots, x_n$

where the x_i's are names of variables. For example, the statement

COMMON PRESS, VOLUM, DENS

causes the placement of

PRESS in storage location 32,000
VOLUM in storage location 32,001
DENS in storage location 32,002

If an array name is included in a COMMON statement, it may be written in unsubscripted form. Information concerning the number of locations needed for the array is obtained from the DIMENSION statement. To illustrate, assume that the following statement appears in the same program with the above COMMON statement:

DIMENSION VOLUM(5)

The common statement then causes the placement of

PRESS in storage location 32,000
VOLUM(1) in storage location 32,001
VOLUM(2) in storage location 32,002
VOLUM(3) in storage location 32,003
VOLUM(4) in storage location 32,004
VOLUM(5) in storage location 32,005
DENS in storage location 32,006

The dimensions of an array can also be stated in a COMMON statement.   Thus,

COMMON PRESS, VOLUM(5), DENS

accomplishes the same storage allocation as the previous example.   In such a case, the name of the subscripted variable, i.e., VOLUM, must not appear in a DIMENSION statement.

The arrangement of multiply subscripted variables in common storage follows the same form as that discussed in Sec. 3.9.4.

The use of a COMMON statement is an alternative way of providing a communication of variables between the main program and subprograms. COMMON statements placed in the main program and subprograms place the respective variables in the same locations.   Suppose that the statements

COMMON E, R1, R2
CALL ROOTS(5.0, 3.*F − G)

appear in a main program which calls SUBROUTINE ROOTS, given in Sec. 3.10.4.   SUBROUTINE ROOTS is altered as shown below:

SUBROUTINE ROOTS(A,C)
COMMON B, X1, X2

.

.

.

END

The two COMMON statements place

E and B in storage location 32,000
R1 and X1 in storage location 32,001
R2 and X2 in storage location 32,002

The COMMON statement is not executable and can be placed anywhere within a program.

3.12.3   COMMON Blocks

In order to avoid the necessity of making a complete list in the COMMON statements of every subprogram, when particular subprograms do not use all the variables in the list, COMMON blocks can be formed.   This is accomplished by naming the blocks with one to six characters, the first of which is alphabetic, and placing these names between slashes in the COMMON statement.   The variables in the block are listed after the slash at the end of the block name.   The COMMON statement in a subprogram or the main program need contain only the COMMON blocks required in the program.   For example, suppose that the statement

COMMON G/ABCR/A, B, C, R1, R2

appeared in a main program and a subroutine called ROOTS needed access only to the storage locations defined by A, B, C, R1, and R2. The COMMON statement in the subroutine could be

COMMON /ABCR/A, B, C, X1, X2

ABCR is the name of the COMMON block.

Six special rules concerning the use of the COMMON statement are:

1. There can be any number of separate blocks.

2. Several COMMON statements appearing in the same subprogram (or main program) act as a single, continuous COMMON statement.

3. A COMMON block appearing in several subprograms must have the same length in each block.

4. Variables in COMMON statements may not be made equivalent to each other by means of the EQUIVALENCE statement.

5. Variables in COMMON statements may be made equivalent to variables not in any COMMON statements.   In such a case, care must be taken that a COMMON block is not extended at its beginning.   It may be extended at its end in this manner.

6. A double-element variable must be placed in a COMMON statement in such a way that it is an even number of locations away from the first element in the list.

3.12.4   DATA Statement

This statement provides an alternative way of storing initial data in the computer.   It has the general form

DATA $list/n_1, n_2, \ldots /, list/n_1, n_2, \ldots /, \ldots$

where *list* is as defined before and $n_1, n_2, \ldots$ are the values corresponding one by one to the variables in the list.   The DATA statement is not executable.

This statement is particularly useful for initializing variables that represent constants that appear many times in a program.   As an example, suppose that the constants 3.14159, 32.17, 29.8E6, and 6600. are all needed at several points in a program.   Obviously, it would be convenient to assign variable names to these constants.   One means of accomplishing this is by using arithmetic statements,

PI = 3.14159
G = 32.17
E = 29.8E6
HP = 6600.

An alternative method consists in using the one statement

DATA PI, G, E, HP/3.14159, 32.17, 29.8E6, 6600./

or, in another form,

DATA PI/3.14159/, G/32.17/, E/29.8E6/, HP/6600./

For convenience in assigning the same value to several variables, n_i can be prefixed by

(integer constant)*

Thus, the statement

DATA A, B, C, D/3.14159, 3*32.17/

signifies that A has a value of 3.14159 and B, C, and D all have a value of 32.17.

Subscripted variables can also be stored in the computer by using the DATA statement. The method of accomplishing this is illustrated below:

DATA ((HEAT(J,I), J = 1, 10), I = 1, 15)/150*2.0/

This causes the value 2.0 to be stored in all 150 elements of the subscripted variable HEAT. It should be pointed out that the indexing limits must be given as integer constants. A statement such as

DATA ((HEAT(J,I), J = 1, M), I = 1, N)/150*2.0/

is not permissible.

It should be emphasized that, if any variable which appears in a DATA statement is redefined in the program, it will no longer have the value assigned to it by the DATA statement.

3.12.5   BLOCK DATA Subprogram

The purpose of this subprogram is to make it possible to enter data into common storage by means of the DATA statement. The exact form of this subprogram is as follows:

BLOCK DATA
COMMON (statements which refer only to COMMON blocks)
Type statements
DATA statements
END

No other kinds of statements may appear in the subprogram. It is not executable and cannot be called. An illustration of a BLOCK DATA subprogram is

BLOCK DATA
COMMON /ABCR/A, B, C, R1, R2
REAL A, B, C
DATA A/1.0/, B/−2.0/, C/1.0/
END

A statement such as

COMMON G/ABCR/A, B, C, R1, R2

is not allowed in a BLOCK DATA subprogram since G is not in a COMMON block.

3.13   PREPARING A PROGRAM DECK

One of the commonest methods of having a program run at a computer center is by submitting a deck of punched cards called the *program deck*. The computer-center personnel then usually transfer the information on the program deck to a magnetic tape, which becomes the input to the computer. The programmer receives output in a printed form from the computer-center personnel. This section is concerned with the preparation of a program deck and includes a discussion of the form of the output received by the programmer.

3.13.1   FORTRAN Cards

The cards used in a program deck contain 80 columns. FORTRAN statements must be punched within certain fields of the cards as listed in Table 3.6. If a statement requires the use of more than one card, it can be continued on a second card by punching a character other than zero in column 6 of that card. If more cards are needed, a nonzero character is punched in column 6 of those cards also. The maximum number of cards which can be used for a single FORTRAN statement is 19.

Table 3.6

Columns	*Information*
1–5	Statement numbers
6	Continuation code
7–72	FORTRAN statement
73–80	Ignored by translator. May be used by programmer for identification information

Blanks on a FORTRAN statement card are ignored by the translator. The two forms of the statement illustrated in Fig. 3.29 are equivalent. The one exception to this is within the field width of a wH specification in a FORMAT statement.

3.13.2   Comments

It is often useful to place comments at certain points in a program to identify its various parts.   This is done by placing a C in column 1 of a card and then punching the comment.   Such a card serves only the convenience of the programmer and is ignored by the translator.   A comment which might appear in the sample program of Sec. 3.1 is illustrated in Fig. 3.30.   Comments are nonexecutable and may be placed anywhere in the program.

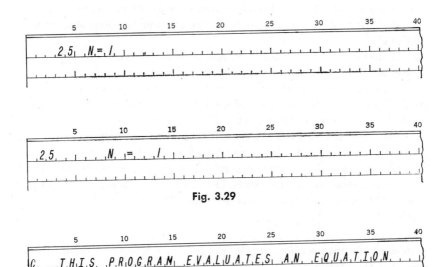

Fig. 3.29

Fig. 3.30

3.13.3   END Statement

The last card in a FORTRAN program must be the statement

END

The purpose of this statement is to signal the translator that there are no more statements to be translated. It must not be confused with the statements which end the execution of a program, CALL EXIT and STOP. The END statement is not executable. Since END is a FORTRAN statement, it must be punched between columns 7 and 72.

3.13.4   Monitor-control Cards

In addition to the FORTRAN program and the data cards, several others, called *monitor-control cards*, may have to be included in the program deck. These vary according to the monitor system used and will not be discussed here.

3.13.5   Output

After a program has been run on a computer, the programmer will receive output in a printed form. It will include:

1. A list of all the FORTRAN statements in the program.
2. A listing of the corresponding coded machine-language program (optional).
3. A listing called a *map* which, among other things, gives the storage locations of all the variables appearing in the program.
4a. A diagnostic report if there are FORTRAN errors in the program. In this case no listing of a machine-language program or map will be printed. (The diagnostic report will be discussed in the next section.)
4b. If there are no FORTRAN errors, a print-out of the results called for by the program. If there are no logical errors in the program, these results represent the computer solution to the given problem. Otherwise, no results, incomplete results, or incorrect results will be printed out.
5. A listing of the numbers stored in each location used by the program (optional).

Besides the printed output, the programmer receives a deck of cards, provided that the FORTRAN program is translated. This deck, called a *binary deck*, is the machine-language program corresponding to the FORTRAN program. If it is desired to make additional runs with new data, the binary deck can be used in place of the FORTRAN deck, saving some time, since the translation of a program will not be necessary.

3.14 "DEBUGGING"

Generally speaking, there are two types of errors that can be made in programming. The first is an error in the use of the FORTRAN language itself, such as misspelling, an unequal number of left and right parentheses, an incorrect count of characters in using the *w*H specification in a FORMAT statement, a misplaced comma, a duplication of statement numbers, a branch to a nonexecutable statement, an arithmetic statement which contains quantities of different types, etc. The second is an error in the logic of the program, causing incorrect results to be produced when the program is executed.

3.14.1 Diagnostic Report

The first type of error is easily corrected, since it will be detected by the translator. Before translating a FORTRAN program into machine language, the translator checks the FORTRAN statements for errors. This is done in two phases. During the first phase all the statements are checked for FORTRAN-language errors. However, if one error is found in a statement, that particular statement is not checked further. If any errors are found, a description of each error is printed. This printed output is called the *diagnostic report*. As an example, consider the statement

DO 2, I = 1, NDIV

The diagnostic report for this statement would be

DO 2, I = 1, NDIV
UNEXPECTED PUNCTUATION AFTER STATEMENT NUMBER

As another example, if the program contained the statement

10   FORMAT (2I5/(3F10.3)

the diagnostic report would be

10   FORMAT (2I5/(3F10.3)
      PARENTHESES DO NOT BALANCE

If no errors are found in the first phase, the translator goes to the second phase of the diagnostic procedure. In this phase errors in the flow of the program, such as incorrect nesting of DO's, illegal branching, etc., are detected. As an example, consider the following statements:

```
   IF (A − B) 6, 7, 6
   S = 3
6  S = 4
   GO TO 12
7  S = 5
```

One can easily see that the second statement would never be executed. The diagnostic report would be printed as follows:

ERROR IN PROGRAM AT OR NEAR STATEMENT NUMBER 6
A PART OF THE PROGRAM CANNOT BE REACHED

When a programmer has a diagnostic report printed with his output, he simply corrects the errors and resubmits the program for another run on the computer. Although the diagnostic report is very convenient, time is saved if the programmer takes some care to avoid these types of error. Before submitting a program to be run on a computer, a good programmer checks each statement for possible FORTRAN errors and the flow of the program with the flow chart. A check list of common errors is given in the next section.

The types of error detected and the diagnostic report vary widely with different FORTRAN translators. The translator used by the student may print out different messages for the errors given in the above examples. In fact, it may not even detect some of these errors! Generally, a programmer quickly becomes aware of the diagnostic reports of the particular computer he is using.

3.14.2   FORTRAN Check List

1. General
   a. Any statement containing parentheses must contain the same number of left and right parentheses. Further, a right parenthesis cannot appear before its corresponding left parenthesis.
   b. Statement numbers must be unique.
   c. Do not confuse the numbers 1 (one) and 0 (zero) with the letters I and O. For example, F10.5 might be mistakenly punched FI0.5 or a variable CI might mistakenly be punched C1.
   d. Check punctuation. Statements in which errors are likely to occur are READ, WRITE, FORMAT, IF, DO, and GO TO ($\cdot \cdot \cdot$), I.
   e. All statement numbers referred to in READ, WRITE, IF, GO TO, and DO statements must be present in the program.

f. It must be logically possible for every section of the program to be executed. In particular, the statement after a GO TO or an arithmetic IF must be numbered and referred to in another statement.

g. The arguments of a function must be enclosed within parentheses and separated by commas both in the definition statement and in the calling statement.

h. The actual arguments in a function calling statement must be in one-to-one correspondence with the dummy arguments of the function with respect to number, order, type, and array size.

i. Check that all variables which are to be transmitted to a subprogram from the main program or another subprogram appear as arguments or in a COMMON statement.

j. Check consistency and order of variables appearing in COMMON statements within the main program and subprograms.

2. Constants and variables

a. Integer constants must be written without a decimal point. Floating-point constants must include a decimal point.

b. Variable names cannot be more than six characters in length. The first character must be alphabetic.

c. A variable may not have the same name as a function.

d. The names of all subscripted variables must appear in either a DIMENSION or a COMMON statement. If a COMMON statement is used for this purpose, the dimensions must be stated there and the name cannot appear in a DIMENSION statement.

e. A negative or zero subscript is not permitted.

f. The variable part of a subscript must never be negative; e.g., in the subscript $I + 4$, I must never be negative.

g. Subscripts of subscripts are not allowed.

h. Subscripted variables must be used only with the same number of subscripts indicated in the DIMENSION or COMMON statement.

3. Arithmetic statements

a. The left side of an arithmetic statement must be a variable. It may not be a constant or an expression.

b. Use the * for all multiplications. $A(B - C)$ is incorrect; $A*(B - C)$ must be used.

c. No two operation symbols can appear together unless the second of them is .NOT..

d. Beware of truncation in integer arithmetic; for example, $A**(7/3)$ is computed as $A**2.0$.

e. The quantities in an expression must be of the same type. In particular, a constant in a floating-point expression must contain a decimal point. The exceptions to this rule are:

(1) Integer exponents of floating-point, complex, or double-precision quantities are permitted.

(2) Real and complex quantities may appear in the same expression.

(3) Real and double-precision quantities may appear in the same expression.

4. Input-output statements

  a. The types of the variables in the list of a READ or WRITE statement must correspond with their specifications in the FORMAT statement.

  b. A constant may appear in the list of a READ or WRITE statement only as a subscript or an indexing parameter.

  c. Check that there is the correct number of characters in every wH specification.

  d. In F, E, and D FORMAT specifications, for example, F$w.d$, d should not exceed 9.

  e. Check that each piece of data is in its correct field on the data card.

  f. Data transmitted by Iw, E$w.d$, and D$w.d$ specifications must be right-adjusted in their fields.

  g. The first entry in each line of an output FORMAT statement must be a valid carriage-control character.

  h. The total number of characters to be printed in a given line as specified in a FORMAT statement must not exceed the capacity of the printing machine.

5. Control statements

  a. A branching statement must branch to an executable statement. (The nonexecutable statements are FORMAT, DIMENSION, COMMON, EQUIVALENCE, END, DATA, type statements, function statements, and comment statements.)

  b. The indexing parameters in a DO statement must be integer constants or integer variables and must have positive values. They may not be expressions; for example, DO 5 I = 1, N + 4 is not allowed. The upper limit must be greater than the lower limit; for example, DO 5 I = 10, 1 is not allowed.

  c. The indexing parameters in a DO statement cannot be redefined within the loop.

  d. A transfer into a DO is not permitted.

  e. The ranges of two or more DO's may not overlap.

  f. The last statement in the range of a DO may be any executable statement except a GO TO, an arithmetic IF, or another DO. It cannot be a nonexecutable statement.

  g. The executable statement within a logical IF statement must not be either another logical IF or a DO.

h. The physically last statement of a program or subprogram must be END.

3.14.3 Correcting Errors after Program Has Been Executed

If a program has been translated and execution begun, there are two possible occurrences.   The first is that execution of the program is halted with none or only a part of the answers printed out.   The second is that the desired number of answers is printed out, some of which may be incorrect.

If the execution of a program is halted before a complete set of answers is obtained, there are several possible causes.   One is that the program may be *looping*, as illustrated in the flow chart in Fig. 3.31.   If A never becomes greater than B, the program will be in an infinite loop. The given flow chart may represent some trial-and-error scheme to solve an equation, and the scheme may not converge to the correct solution for every case.   On the assumption that 100 is a sufficient number of trials, the flow chart can be altered to avoid the infinite loop, as shown in Fig. 3.32.   Other possible causes include overflow or underflow due to generation of numbers of a size

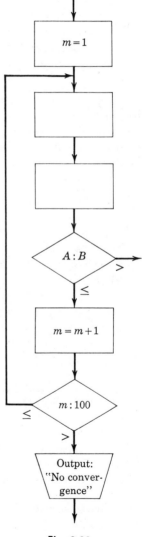

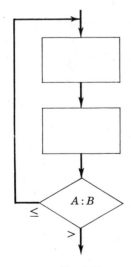

Fig. 3.31

Fig. 3.32

beyond the capacity of the machine or incorrect indexing in an input statement.

If all the answers are printed out, they may be either correct or incorrect. If they are correct, the program has been run successfully. A good question at this point is: How does one know whether answers are correct or incorrect? This brings up an important principle in programming. *A program should be run first with a test case for which the solution is known.* This solution can be obtained by any of the following methods:

1. It appears in a book, article, etc., or is available from another checked program.

2. It is obtained by hand calculations.

3. It is obtained from another computer program which solves the problem in a different way.

There are several possible causes for obtaining incorrect results. An arithmetic statement may be incorrectly formed. A branching statement may branch to the wrong point. The FORMAT for a WRITE statement may not allow the most significant digits to be printed. The statements in the range of a DO may be executed one too many or one too few times. These types of error are usually the hardest to find. Three general techniques of finding such errors are given below:

1. Extra WRITE statements can be inserted into the program at various strategic points to print out intermediate results. Once the program has been tested successfully, these WRITE statements can be removed.

2. On all computers it is possible for information stored in every location used by the program to be printed. Such output is called a *dump*. With this information, the programmer can easily determine the current value of any or all of the variables used in the program. Two statements cause a dump to be printed.

CALL DUMP$(A_1,B_1,F_1, \ . \ . \ . \ ,A_n,B_n,F_n)$
CALL PDUMP$(A_1,B_1,F_1, \ . \ . \ . \ ,A_n,B_n,F_n)$

On the assumption that the variable names are compiled in alphabetic order, the meaning of A_i, B_i, F_i is

$A_i \leq$ (all variables to be dumped) $\leq B_i$

according to the format F_i, where A_i and B_i are variable names. The code for F_i is

0   dump in octal
1   dump in floating point
2   dump in integer

Execution of the program is terminated with a CALL DUMP statement, while execution is continued after a CALL PDUMP statement.

3. The programmer can play computer, i.e., execute each step in the program by performing the operation with a hand calculation.

With these techniques, the programmer can check the result of each calculation performed by the computer until the error is discovered. It should be emphasized that a flow chart is invaluable at this time for checking the logic of the program.

To illustrate the process of debugging, suppose that the first trial of the program given in Sec. 3.7.5 was

```
      INTEGER N, R, S
      REAL X1, X2, X3, Y1, Y2, Y3, C
100   READ (5,1) X1, X2, Y1, Y2, C, N, R
  1   FORMAT (4F10.5, E10.5, 2I5)
      WRITE (6,2) X1, Y1, X2, Y2
  2   FORMAT  (9X,  12HX-COORDINATE,  8X,
      $12HY-COORDINATE/(5X, 2F20.9))
      S = N//R
      DO 5 L = 1, S
      DO 3 J = 1, R
      X3 = 2.+X2+X1*(C/((X1*X1+Y1*Y1)**1.5) - 1.
      Y3 = 2.+Y2+Y1*(C/((X1*X1+Y1*Y1)**1.5) - 1.
      WRITE (6,101) X3, Y3
101   FORMAT (50X, 2E15.8)
      X1 = X2
      X2 = X3
      Y1 + Y2
  3   Y2 = Y3
  5   WRITE (6,4) X3, Y3
  4   FORMAT (5X, 2F20.9)
      CALL DUMP(C,C,1,N,S,2)
      GO TO 100
      END
4060.  4060.  0.  500.  −9.55E8  4  2
```

Notice the extra WRITE statement.

While the program is being processed at the computer center, hand

calculations for this test case can be performed with a slide rule.   The
results of such calculations are given in Table 3.7.   The output from the

Table 3.7

x	y	$x_1{}^2 + y_1{}^2$	$(x_1{}^2 + y_1{}^2)^{\frac{3}{2}}$	$\dfrac{-0.955 \times 10^9}{(x_1{}^2 + y_1{}^2)^{\frac{3}{2}}} - 1$	Col. 5 times x_1	Col. 5 times y_1
4060	0	16.5×10^6	67.0×10^9	-1.014	4120	0
4060	500	16.8×10^6	68.2×10^9	-1.014	4120	507
4000	1000	17.0×10^6	70.0×10^9	-1.014	4050	1014
3880	1493	17.3×10^6	72.0×10^9	-1.013	3900	1516
3710	1972					
3480	2428					

computer on the first trial is shown below:

S = N//R
DOUBLE OPERATOR FOLLOWING THE SYMBOL N
X3 = 2.+X2+X1*(C/((X1*X1+Y1*Y1)**1.5) − 1.
PARENTHESES DO NOT BALANCE
Y3 = 2.+Y2+Y1*(C/((X1*X1+Y1*Y1)**1.5) − 1.
PARENTHESES DO NOT BALANCE
Y1 + Y2
ILLEGITIMATE FORTRAN STATEMENT

The first statement requires removing one of the slashes; the second and
third, adding a right parenthesis at the end of each statement; and the
fourth, replacing the plus sign by an equals sign.   After these errors have
been corrected, the program is run a second time, producing the following
results:

X-COORDINATE   Y-COORDINATE
4060.000000000   0.
4060.000000000   500.000000000
                              −0.55936401E 02 0.50200000E 03
                              −0.41705792E 04-0.29757232E 01
4170.579284668   −2.975723267
                              −0.36981201E 04-0.42231005E 04
                              0.52936389E 03-0.42180856E 04
529.363891602   −4218.085632324
  −0.95500000E 09
      00004   00002   00002

This time the program was executed, but the results are incorrect.   A check of the dump shows that $C = -0.955 \times 10^9$, $N = 4$, $R = 2$, and $S = 2$. Playing computer from the beginning of the program produces

$$L = 1$$
$$J = 1$$
$$X3 = 2. + 4060. + 4060. \left[\frac{-9.55 \times 10^8}{(4060^2 + 0^2)^{\frac{3}{2}}} - 1. \right]$$
$$= -58.0$$

Since this gives an answer which is about the same as that given by the computer, but different from that found in the hand calculations, there must be an error at this step. This error might be in the hand calculations. However, a careful check shows that 2. + X2 should be 2.*X2. Correcting this error and a similar error in the next statement and rerunning the program produce the following results:

```
X-COORDINATE Y-COORDINATE
4060.000000000        0.
4060.000000000        500.000000000
                                0.40020636E 04 0.09999999E 04
                                0.38874843E 04 0.14930242E 04
3887.484313965        1493.024276733
                                0.37184571E 04 0.19724435E 04
                                0.34980212E 04 0.24321189E 04
3498.021240234        2432.118927002
  -0.95500000E 09
                        00004      00002        00002
```

Since these results reasonably agree with those obtained by the hand calculations, the extra WRITE statement and the CALL DUMP statement can be removed and a CALL EXIT statement added at an appropriate point. The programmer can now have confidence that this program will solve other cases.

PROBLEMS

Section 3.4

Find the errors in the following statements:

◆3.1   READ (5,1) P, V, TEMPERATURE
   1   FORMAT (F10.3, E12.4, I7)

3.2   READ (5,2) A, B, C, D, E, F, G, H, I, J, K, L, M, N
   2   FORMAT (7F10.5)

3.3   READ (5 3) GAS, LIQUID, SOLID
   3   FORMAT (6E12.4)

3.4   READ (5,4) HEAT, COND., THERM, TEMP, PRESS, VOL, N,
$MM, 5M, BTU
4 FORMAT (5F10.3, 3I5, E15.3)

3.5   READ (5,5) M, N, PULSE, G*T, MASS, SOUNDV, WAVE,
$LENGTH, FREQ,
5 FORMAT (2I6, 5E12.4/(E20.12)

3.6   READ (6,5) ENERGY, PLANCKS, W1, W2, FREQ, ATOM, I J K L
6 FORMAT (3F15.7/4E15.7/4I8)

In the following problems:

$A = -46.193$ $B = 39.4 \times 10^{-17}$ $C = -14.65 \times 10^9$
$D = 157.38$ $I = 15$ $J = 3129$

In each problem sketch how the data should be punched on the data cards.

3.7   READ (5,7) A, C, I
7 FORMAT (F10.5, E12.6/I7)

♦3.8   READ (5,8) B, D, J
8 FORMAT (E12.6, F12.6, I10)

3.9   READ (5,9) A, C, I
9 FORMAT (2E15.7, I8)

3.10   READ (5,10) B, D, J
10 FORMAT (E15.7/E15.7, I6)

3.11   READ (5,11) A, B, C, D, I, J
11 FORMAT (F10.5, 3E12.4/(I9))

3.12   READ (5,12) I, J, A, B, C, D
12 FORMAT (2I10/(2E12.4))

♦3.13 Construct READ and FORMAT statements which read in nine integer variables with an I10 specification, three floating-point variables with an F12.7 specification, and six floating-point variables with an E12.7 specification.

3.14 Construct READ and FORMAT statements which read in 4 integer variables with an I8 specification, 15 floating-point variables with an F10.5 specification, and 2 floating-point variables with an E15.7 specification.

3.15 Construct READ and FORMAT statements which read in 12 integer variables with an I5 specification, 7 floating-point variables with an F8.3 specification, and 10 floating-point variables with an E10.5 specification.

3.16 Construct READ and FORMAT statements which read in your first name and age. Sketch how the data should be punched on the data card.

Section 3.5

Find the errors, if any, in the following arithmetic statements. Assume that no type statements appear in the program:

3.17 $A + B = C*DEF/M + G - XY***3$

♦3.18 LENGTH = SPEED*TIME

3.19 SPEED = LENGTH/TIME

3.20 R3 = 2R2 − R1*((C3/R1**2 − F)

3.21 P2 = (S(P1 + V**2/(2*G) + Z1))

3.22 V = (1/E0)(E**2./(2*N*H))

Write the following FORTRAN statements in an algebraic form:

3.23 W = −(1./E0**2)*AMASS*E**4/(8.*B*B + H**2)

3.24 A17 = 2.*HE**4 + 7.*AN − H

3.25 AJS = A*T**2*E**(−PHI/AK*T)

◆**3.26** PHI = AMM/(AL1/(H1*A1) + AL2/(U2*A2))

3.27 FORCE = (QN − QI)*QCL/(4.*PI*E0*R**2)

3.28 ANV = (4./3.)*A**3*G*(P − P1)/(6.*A*V)

Construct FORTRAN arithmetic statements for the following algebraic statements:

3.29 $q = \dfrac{4pa^3g(m - n)}{3E}$

3.30 $F = K\dfrac{q_1 q_2}{r^2}$

3.31 $y = 2a\left\{1 - \left[\dfrac{6.28(f_1 - f_2)t}{2}\right]^2\right\}\left\{1 - \left[\dfrac{6.28(f_1 + f_2)t}{2}\right]^2\right\}$

◆**3.32** $\rho = \dfrac{\Delta p}{V^2 A + \Delta X/AX}$

3.33 $H = \dfrac{6.28LK(t_a - t_b)}{1 - b/a}$

3.34 $t_f = \frac{9}{5}t_c + 32$

Compute the values of the variables on the left side of the following equations. Assume that no type statements appear in the program.

3.35 M = 7/3

3.36 Z = 7/3

3.37 M = 7./3.

3.38 Z = 7./3.

3.39 N = I/2*2   for I = 5

◆**3.40** N = J/2*2   for J = 4

◆**3.41** N = A/B*B   for A = 5.0 and B = 2.0

Section 3.6

In each problem, print results in an orderly manner, including titles, headings, etc.

3.42 The drag and lift forces on an airfoil or rocket can be approximated by the formulas

$D = C_D dA V^2$

$L = C_L dA V^2$

where

D = drag force

L = lift force

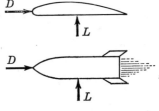

Fig. P3.42

d = density of air
A = cross-sectional area of airfoil or rocket
V = velocity of airfoil or rocket
C_D = experimentally determined drag coefficient
C_L = experimentally determined lift coefficient

Write a program which reads in the proper data and computes and prints out the values of the drag and lift forces.

♦**3.43**   Charles's and Boyle's gas laws may be stated by the one equation

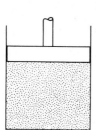

$$\frac{PV}{T} = \text{const}$$

where

P = pressure of gas
V = volume of gas
T = temperature of gas

Write a program which will read in initial values of P, V, and T and later values of V and T (when gas has changed its state), and compute and print out the later value of P.

Fig. P3.43 to P3.45

3.44   Same as Prob. 3.43, except, given later values of P and V, compute and print out T.

3.45   Same as Prob. 3.43, except, given later values of P and T, compute and print out V.

3.46   In the electrical circuit shown, the governing equations obtained from the laws of Ohm and Kirchhoff are

$$I_1 = \frac{V}{R_1}$$
$$I_2 = \frac{V}{R_2}$$
$$I = I_1 + I_2$$

where

V = voltage drop from A to B in volts
R_1, R_2 = resistance, ohms
I, I_1, I_2 = current, amp

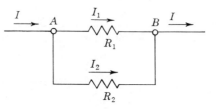

Fig. P3.46 and P3.47

Write a program which reads in values of V, R_1, and R_2 and computes and prints out the values of I_1, I_2, and I.

3.47 In Prob. 3.46, write a program which reads in values of I, R_1, and R_2 and computes and prints out the values of I_1, I_2, and V.

3.48 The equation given in the sample program (Sec. 3.1) is an empirical relation which predicts the flow of water in an open channel as illustrated. The

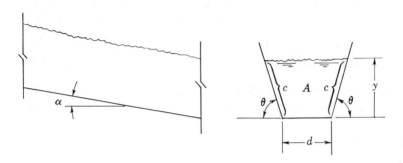

Fig. P3.48

symbols in the equation represent the following physical quantities:

Q = volume of water passing a given point in the channel per unit of time
A = cross-sectional area of channel, ft²
S = slope of channel = $\sin \alpha$
E = roughness coefficient (about 0.030 for canals and rivers)
P = "wetted perimeter," ft ($P = 2c + d$)

This empirical relation was first established in 1775 in a somewhat different form by Chezy and put in its present form by Manning in 1890. Thus, it is known as the *Chezy-Manning equation*.

Assume that the flow rate Q at several points in a certain channel is to be determined. Data are provided giving values to C, D, A, S, E, and $NAME$, where $NAME$ is an identification for each particular point in the channel. Two typical sets of data are shown in the accompanying table.

$NAME$	C	D	A	S	E
Station 1........	8.0	20.0	120.0	0.0001	0.030
Station 2........	7.6	21.5	115.0	0.0001	0.030

Write a program which reads in the given data, computes the flow rate, prints out the identification $NAME$ and the flow rate Q, and returns to the input statement for another set of data.

Section 3.7

3.49 In Prob. 3.43 Charles's and Boyle's law was given as

$$\frac{PV}{T} = \text{const}$$

Write a program that reads in initial values of P, V, and T and later values of any two of the quantitites and computes and prints the third quantity.

3.50 Write a program that will execute either of the programs for Probs. 3.46 and 3.47.

3.51 Write a program which reads in a temperature in either degrees Fahrenheit or degrees centigrade and computes and prints out the temperature in the opposite scale.

3.52 In the example programs on the satellite, would the following rearrangement of several of the statements alter the program in any way? If so, how?

```
X2 = X3
X1 = X2
Y2 = Y3
Y1 = Y2
```

3.53 In the electrical circuit shown, write a program that reads in the values of the resistances and the voltage drop from A to B and then computes and prints out all the currents. (See Prob. 3.46.)

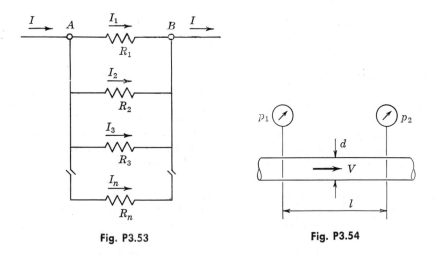

Fig. P3.53   Fig. P3.54

3.54 The equations governing the flow of a liquid in a horizontal pipe are

$$\frac{p_1}{S} - \frac{p_2}{S} = f\frac{l}{d}\frac{V^2}{2g}$$

$$Q = VA$$

where

p_1, p_2 = pressure at points 1 and 2, respectively

S = specific weight of liquid

l = length of pipe between points 1 and 2

d = diameter of pipe

V = velocity of liquid in pipe

g = acceleration due to gravity (about 32.17 ft/sec² on the earth)

f = experimentally determined "friction factor"

Q = flow rate (volume of liquid passing a given point per unit of time)

A = cross-sectional area of pipe

Assume that tests are made by varying the flow rate and measuring the pressures p_1 and p_2. Write a program which first reads in S, l, d, g, and A. Then form a DO which reads in n values of p_1, p_2, and Q and calculates an average value of f. The program should print out the values of n and the average f.

◆**3.55** An axial load P is applied to a steel bar, and the elongation e of the bar caused by the load is measured. Hooke's law states that

$$P = ke$$

where k is a constant. Assume that a test is made in which n different loads P are applied to the bar and in each case the elongation e is measured. Write a program which reads in the values of P and e and computes an average k. The program should print out the values of n and the average k.

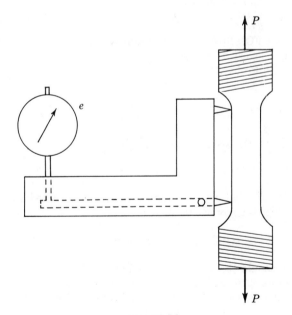

Fig. P3.55

3.56 Write a FORTRAN program corresponding to the flow chart given in Fig. 2.1.

3.57 Write a FORTRAN program corresponding to the flow chart given in Fig. 2.2, using the DO statement to form the loop.

3.58 Write a FORTRAN program corresponding to the flow chart given in Fig. 2.4, using the DO statement to form the loop.

Section 3.8

If the values of A and B are .TRUE. and .FALSE., respectively, determine the value of the given expression.

3.59 A.AND.B.AND..NOT.A

◆**3.60** A.AND.B.AND..NOT.B

3.61 A.AND.B.OR..NOT.A

3.62 A.AND.B.OR..NOT.B

3.63 A.OR.B.AND.A

3.64 A.OR.B.OR..NOT.B

If $A = 5.0$, $B = 3.0$, $C = -5.0$, and $D = -3.0$, determine the value of the given expression.

3.65 A.GT.B.AND.C.LT.D

3.66 A.GT.B.OR.C.LT.D

3.67 A.GE.$-$C.AND.B.LE.$-$D

3.68 A.GE.$-$C.OR.B.LE.$-$D

3.69 A.EQ.$-$C.AND.B.NE.$-$D

◆**3.70** A.EQ.$-$C.OR.B.NE.$-$D

3.71 Write a FORTRAN program corresponding to the flow chart given in Fig. 2.4, using a logical IF statement.

3.72–3.74 Do Probs. 3.49 to 3.51, using logical IF statements.

Section 3.9

3.75 In the DIMENSION statement

DIMENSION STEEL(6,8,5)

how many storage locations are reserved for the subscripted variable STEEL?

3.76 Consider the statement

READ (5,53) N, M, ((B(J,I), I = 1, N), J = 1, M)

In what order should the data be punched on the data cards?

3.77 In Prob. 3.76, $N = 2$, $M = 2$, $B_{11} = 1.0$, $B_{21} = 2.0$, $B_{12} = 3.0$, $B_{22} = 4.0$. Sketch the data cards for the format (2I5/(7F10.5)).

In Probs. 3.78 to 3.82 construct the programs as follows:

1. Read in values of all the variables, including subscripted variables.
2. Write statements for all the calculations, using DO's wherever possible.
3. Print out the desired results.

3.78 Write a FORTRAN program corresponding to the flow chart given in Fig. 2.2, using the x_i's as a subscripted variable.

3.79 Write a FORTRAN program corresponding to the flow chart given in Fig. 2.3, using the x_i's as subscripted variables.

3.80 Do Prob. 3.53, using subscripted variables for the resistances R_i and currents I_i.

3.81 Do Prob. 3.55, using subscripted variables for the loads P_i, elongations e_i, and individual constants k_i.

◆**3.82** In many matrix applications, including the solution of simultaneous equations, it is important to check to see whether or not the matrix is symmetric, i.e., whether or not each element a_{ij} is equal to its corresponding element a_{ji}, where the first subscript is the row number and the second subscript is the column number of the element. Write a FORTRAN program to check an $n \times n$ matrix for symmetry. The program should conserve computer time by checking each pair (a_{ij}, a_{ji}) only *once*. Include statements for reading the matrix into the computer and for printing the result of the symmetry check.

3.83 Another commonly occurring matrix form is one in which many of the elements are zero. Such a matrix is called *sparse*. In such cases it is convenient to have the computer initialize all elements to zero and then read in only the nonzero elements. Assuming that you are given the nonzero elements of an $n \times n$ matrix in the form i, j, a_{ij}, write a FORTRAN program to read the matrix into the computer, being careful, of course, to store the elements in the proper locations.

Section 3.10

Construct FORTRAN arithmetic statements for the following algebraic equations. All constants and variables should be of the floating-point type unless otherwise indicated.

3.84 $s = |a - b|$

3.85 $x = \left| \dfrac{b|c - d|}{|af/c - 3g|} \right|$

3.86 $y = a \sin \dfrac{m\pi x}{L} + b \cos \dfrac{m\pi x}{L}$

3.87 $w = c \sin \dfrac{m\pi x}{a} \sin \dfrac{n\pi y}{b}$

◆**3.88** $t = \sqrt{2gs}$

3.89 $\phi = \tan^{-1}(m\pi - x)$

3.90 $y = \cos x$, where x is in degrees

3.91 $z = \sqrt{\dfrac{a}{b}}\, e^{\alpha t - 1}$

3.92 $z = \sqrt{\dfrac{a}{b}}\, e^{\sqrt{\alpha t - 1}}$

3.93 $y =$ integer value of x

3.94 $y =$ odometer reading, where M is the total number of miles recorded and the largest possible odometer reading is 99,999.9 miles

3.95 $y =$ maximum of α, $3 - \beta$, and 2γ

3.96 $y =$ minimum of $\sqrt{\alpha}$, e^β, and γ

3.97 $y = 3\sqrt{\alpha}\,(\beta + 2I)$, where I is of the integer type

3.98 $y = \begin{cases} +1 \text{ if } z \text{ is negative} \\ -1 \text{ if } z \text{ is positive} \end{cases}$

◆**3.99** $y = \begin{cases} x_2 - x_1 \text{ if } x_2 > x_1 \\ 0 \text{ if } x_2 \leq x_1 \end{cases}$

Construct FORTRAN *arithmetic-statement functions* for the following functions (Probs. 3.100 to 3.108):

3.100 $f(x) = \tan x$

3.101 $f(x) = a_1 x^3 + a_2 x^2 + a_3 x + a_4$

3.102 $f(x, a_1, a_2, a_3, a_4) = a_1 x^3 + a_2 x^2 + a_3 x + a_4$

3.103 $f(x) = a_1 \sin^3 x + a_2 \sin^2 x + a_3 \sin x + a_4$

◆**3.104** $f(d, m, s) = \sin(d, m, s)$, where d, m, s represents an angle in the degree-minute-second form

3.105 $f(x) = \dfrac{ax^2 + bx + c}{dx^2 + ex + f}$

3.106 $f(a, b) = a^2 + b^2$

3.107 $f(a, b, c, d, x) =$ linearly interpolated value between c and d corresponding to x which is between a and b

3.108 $f(x) = 3(\text{function of Prob. 3.103}) - \sqrt{\text{function of Prob. 3.105}}$

3.109 Construct a FORTRAN FUNCTION which evaluates the largest root of the quadratic equation

$$ax^2 + bx + c = 0$$

If the discriminant is negative, a message should be printed stating this fact and the function set equal to $-b/2a$. Also, construct a main program which reads in values for a, b, and c, calls the above function in the arithmetic statement

$$d = 0.001375 e^{\text{largest root}}$$

and prints out the value of d.

3.110 Construct a FORTRAN FUNCTION which finds the largest of a group of n numbers. Also, construct a main program which reads in the n numbers, calls the above function in the arithmetic statement

$$r = 3.2896 \sqrt{\text{largest number}}$$

and prints out the value of r.

♦**3.111** Construct a FORTRAN FUNCTION which, if given values of the voltage and n resistances in a parallel circuit, determines the total current of the system. (See Prob. 3.53.) Also, construct a main program which reads in the values of the voltage and n resistances, calls the above function in an arithmetic statement which evaluates the equivalent resistance of the system, and prints out the value of the equivalent resistance.

3.112 Assume that the drag force D acting on a rocket fired vertically from the surface of the earth is related to the velocity V of the rocket according to the equation

$$D = C_D V^2$$

where C_D is an experimentally determined constant. Assume further that the constant C_D varies with the altitude H according to the given graph. Construct a FORTRAN FUNCTION which, if given values of H and V, computes the value of the drag force. Further, construct a main program which reads in values of the variables H, V, and t, calls the above function with the statement

$$f = T - W - D$$

and prints out the value of f. Assume that the values of T and W are evaluated by other given FORTRAN FUNCTIONS with an argument of t. In the above equation f represents the total force acting on the rocket, T the thrust, W the weight of the rocket and fuel, and t the time elapsed since lift-off.

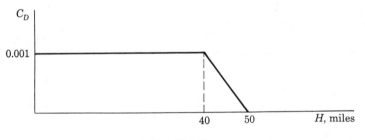

Fig. P3.112

3.113 Shown herewith is a table which gives the specific weight of the atmosphere as a function of the altitude. Construct a FORTRAN FUNCTION

Altitude, ft	Specific weight, lb/ft³	Altitude, ft	Specific weight, lb/ft³
0	0.0765	45,000	0.0147
5,000	0.0659	50,000	0.0116
10,000	0.0565	55,000	0.00914
15,000	0.0481	60,000	0.00721
20,000	0.0408	65,000	0.00567
25,000	0.0343	70,000	0.00447
30,000	0.0286	75,000	0.00354
35,000	0.0237	80,000	0.00277
40,000	0.0187		

which, if given the altitude, determines the specific weight by using linear interpolation of the tabulated values. Assume that the values of the specific weight are located in common storage. Also, construct an arithmetic statement which calls the function in accordance with the following algebraic equation:

$$f = T(t) - W(t) - C_D\gamma(h)V^2$$

Here, γ and h represent specific weight and altitude, respectively. (See Sec. 3.12 for explanation of common storage.)

◆3.114   Construct a SUBROUTINE which, when given the tangent of an angle, computes the angle in terms of degrees, minutes, and seconds. Also, construct a main program which reads in the two short sides of a right triangle, computes the hypotenuse, computes the interior angles by using the above SUBROUTINE, and prints out the values of all the sides and angles of the triangle.

3.115   The same as Prob. 3.114, with the exception that any two sides of a right triangle can be read in as input.

3.116   Construct a SUBROUTINE which sorts a group of n numbers into an increasing order. Then do Prob. 2.25, first using this SUBROUTINE to sort the trains.

3.117   Construct a SUBROUTINE which determines the quotient

$$\frac{a_1x^n + a_2x^{n-1} + \cdots + a_nx + a_{n+1}}{x - c}$$

Assume that the maximum value of n is 20. Also, construct a main program which reads in the values of the a_i's, c, and x, calls the above SUBROUTINE, evaluates the quotient for the given value of x, and prints out this value.

3.118   Construct a SUBROUTINE which evaluates the forces F_c and F_d in members c and d of the truss illustrated. Assume that the load L_1, force in

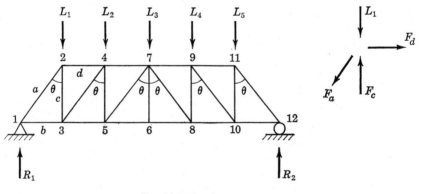

Fig. P3.118

member a, F_a, and angle θ are known. Equilibrium of the truss joint requires that

$$-F_a \sin\theta + F_d = 0$$

and

$$-F_a \cos \theta - L_1 + F_c = 0$$

Assume that similar SUBROUTINES have been constructed for all the other truss joints. Also, construct a main program which calls these SUBROUTINES to calculate the values of the forces in all the members of the truss. The order in which the SUBROUTINES must be called is given in the figure. The main program must read in the values of R_1, R_2, L_1 through L_5, and θ and print out the values of the forces in all the members of the truss. Place the values of these forces F_i in common storage. A COMMON statement must then appear in the main program and all the SUBROUTINES. (See Sec. 3.12 for explanation of common storage.)

4 NUMBER SYSTEMS AND MACHINE LANGUAGE

4.1 NUMBER SYSTEMS

The decimal number system is familiar to most persons today. In this system the number 10 (ten) is the *base*, or *radix*. Most digital computers use the binary system, in which the decimal number 2 (two) is the base. Other systems which are useful in the computer field are the octal system, the base of which is the decimal number 8 (eight), and the sexadecimal (more commonly called *hexadecimal*) system, the base of which is the decimal number 16 (sixteen). Table 4.1 presents a few integers from these systems. Notice that some new characters are needed to represent digits in the hexadecimal system. These are arbitrarily taken as *a*, *b*, *c*, *d*, *e*, and *f*.

In order to avoid confusion in using several number systems, some identification procedure must be established. In this text all numbers which are not decimal numbers will be subscripted by a decimal number which indicates the base being used. For example, the numbers

$$101_2 \qquad 23_8 \qquad 1c_{16} \qquad 29$$

represent 101 binary, 23 octal, 1c hexadecimal, and 29 decimal, respectively. The equivalents of these in the decimal system are 5, 19, 28, and 29, respectively.

4.1.1 Conversion of Integers

Table 4.1 can be utilized for converting some integers from one number system to another. However, to convert large numbers which do not appear in the table or to convert fractional parts of whole numbers, some method other than straight counting is needed.

Table 4.1

Decimal	Binary	Octal	Hexadecimal
0	0	0	0
1	1	1	1
2	10	2	2
3	11	3	3
4	100	4	4
5	101	5	5
6	110	6	6
7	111	7	7
8	1 000	10	8
9	1 001	11	9
10	1 010	12	a
11	1 011	13	b
12	1 100	14	c
13	1 101	15	d
14	1 110	16	e
15	1 111	17	f
16	10 000	20	10
17	10 001	21	11
18	10 010	22	12
19	10 011	23	13
20	10 100	24	14
21	10 101	25	15
22	10 110	26	16
23	10 111	27	17
24	11 000	30	18
25	11 001	31	19
26	11 010	32	1a
27	11 011	33	1b
28	11 100	34	1c
29	11 101	35	1d
30	11 110	36	1e
31	11 111	37	1f
32	100 000	40	20
33	100 001	41	21
34	100 010	42	22
35	100 011	43	23
.	.	.	.
.	.	.	.
.	.	.	.
40	101 000	50	28
.	.	.	.
.	.	.	.
.	.	.	.
100	1 100 100	144	64

Converting from the binary, octal, and hexadecimal systems to the decimal system is best accomplished by using *positional notation*. The positional notation of an integer number N in the base i system is

$$N = a_1 \times (10_i)^n + a_2 \times (10_i)^{n-1} + \cdots + a_n \times (10_i)^1 + a_{n+1} \times (10_i)^0 \tag{4.1}$$

where the a_i's represent integers. A few examples of numbers in their positional notation are

$$6283 = 6 \times 10^3 + 2 \times 10^2 + 8 \times 10 + 3$$
$$10110_2 = 1_2 \times 10_2{}^{100_2} + 1_2 \times 10_2{}^{10_2} + 1_2 \times 10_2$$
$$= 1_2 \times 2^4 + 1_2 \times 2^2 + 1_2 \times 2$$
$$437_8 = 4_8 \times 10_8{}^{2_8} + 3_8 \times 10_8 + 7_8 = 4_8 \times 8^2 + 3_8 \times 8 + 7_8$$
$$4c9_{16} = 4_{16} \times 10_{16}{}^{2_{16}} + c_{16} \times 10_{16} + 9_{16} = 4_{16} \times 16^2 + c_{16} \times 16 + 9_{16}$$

The conversion of the last three of these to the decimal system is illustrated below:

$$
\begin{array}{lll}
10110_2 = 1 \times 2^4 = 16 & 437_8 = 4 \times 8^2 = 256 & 4c9_{16} = 4 \times 16^2 = 1024 \\
 +0 \times 2^3 = 0 & +3 \times 8^1 = 24 & \phantom{4c9_{16} =} +12 \times 16^1 = 192 \\
 +1 \times 2^2 = 4 & +7 \times 8^0 = 7 & \phantom{4c9_{16} =} +9 \times 16^0 = 9 \\
 +1 \times 2^1 = 2 & \overline{287} & \phantom{4c9_{16} =} \overline{1225} \\
 +0 \times 2^0 = 0 & & \\
 \overline{22} & &
\end{array}
$$

The positional notation may also be utilized to convert from the decimal system to the binary, octal, and hexadecimal systems. For example, the decimal number 37 can be written as

$$37 = a_1 \times 2^n + a_2 \times 2^{n-1} + \cdots + a_n \times 2^1 + a_{n+1} \times 2^0$$

where each a_i is either a 0 or a 1. Dividing both sides by 2 gives

$$18 + \frac{1}{2} = (a_1 \times 2^{n-1} + a_2 \times 2^{n-2} + \cdots + a_n \times 2^0) + \frac{a_{n+1}}{2}$$

and since the term in parentheses represents an integer, it follows that

$$18 = (a_1 \times 2^{n-1} + a_2 \times 2^{n-2} + \cdots + a_n \times 2^0)$$
$$1 = a_{n+1}$$

Dividing the first of these by 2 gives

$$9 + \frac{0}{2} = (a_1 \times 2^{n-2} + a_2 \times 2^{n-3} + \cdots + a_{n-1} \times 2^0) + \frac{a_n}{2}$$
$$a_n = 0$$

Continuing the process,

$$4 + \frac{1}{2} = (a_1 \times 2^{n-3} + a_2 \times 2^{n-4} + \cdots + a_{n-2} \times 2^0) + \frac{a_{n-1}}{2}$$

$$a_{n-1} = 1$$

$$2 + \frac{0}{2} = (a_1 \times 2^{n-4} + a_2 \times 2^{n-5} + \cdots + a_{n-3} \times 2^0) + \frac{a_{n-2}}{2}$$

$$a_{n-2} = 0$$

$$1 + \frac{0}{2} = (a_1 \times 2^{n-5} + a_2 \times 2^{n-6} + \cdots + a_{n-4} \times 2^0) + \frac{a_{n-3}}{2}$$

$$a_{n-3} = 0$$

$$0 + \frac{1}{2} = 0 + \frac{a_{n-4}}{2}$$

$$a_{n-4} = a_1 = 1$$

The answer is then obtained by reading upward,

$$37 = 100101_2$$

Two more examples are given below for converting 200 to the octal and hexadecimal systems:

	Remainder		Remainder
$\frac{200}{8} = 25$	0	$\frac{200}{16} = 12$	8
$\frac{25}{8} = 3$	1	$\frac{12}{16} = 0$	12
$\frac{3}{8} = 0$	3		
$200 = 310_8$		$200 = c8_{16}$	

Converting between the binary, octal, and hexadecimal systems can be accomplished by the methods just described, but since the bases of these systems are multiples of each other, there is a much simpler method. From Table 4.1 it may be seen that there is a correspondence between a single octal digit and three binary digits, since $2^3 = 8$. For example,

```
010 110     binary
 2   6      octal
```

Thus, it is customary to write binary numbers in groups of three so that conversion between the octal and binary systems is easily performed. A few more examples are given below:

$$101\ 011\ 001\ 111_2 = 5317_8$$
$$264_8 = 010\ 110\ 100_2$$

Four binary digits correspond to one hexadecimal digit, since $2^4 = 16$, and binary numbers can be conveniently grouped in fours.   For example,

1010 1100 $1111_2 = acf_{16}$

$264_{16} = 0010\ 0110\ 0100_2$

One easy way of converting between octal and hexadecimal is by converting to binary first.   For example,

$264_8 = 010\ 110\ 100_2 = 1011\ 0100_2 = b4_{16}$

$264_{16} = 0010\ 0110\ 0100_2 = 001\ 001\ 100\ 100_2 = 1144_8$

$acf_{16} = 1010\ 1100\ 1111_2 = 101\ 011\ 001\ 111_2 = 5317_8$

4.1.2   Conversion of Fractions

The period in a decimal number such as 53.652 is called a *decimal point*. Numbers in the binary, octal, and hexadecimal systems may have binary, octal, and hexadecimal points, respectively.   Conversion of a fraction in any of these systems to the decimal system is easily accomplished by using positional notation, as illustrated below:

$0.101_2 = 1 \times 2^{-1} + 0 \times 2^{-2} + 1 \times 2^{-3} = \frac{1}{2} + \frac{1}{8} = \frac{5}{8} = 0.625$

$0.65_8 = 6 \times 8^{-1} + 5 \times 8^{-2} = \frac{6}{8} + \frac{5}{64} = \frac{53}{64} = 0.828125$

$0.2c_{16} = 2 \times 16^{-1} + 12 \times 16^{-2} = \frac{2}{16} + \frac{12}{256} = \frac{44}{256} = 0.171875$

To illustrate the conversion of a decimal fraction to another system, consider the conversion of 0.6 to the binary system.   Using positional notation,

$$0.6 = a_1 \times 2^{-1} + a_2 \times 2^{-2} + \cdots + a_n \times 2^{-n}$$

Multiplying by 2 gives

$$1.2 = a_1 + (a_2 \times 2^{-1} + \cdots + a_n \times 2^{-n+1})$$

Since the term in parentheses is a fraction,

$$1 = a_1$$

Continuing the process,

$$0.2 \times 2 = 0.4 \qquad a_2 = 0$$
$$0.4 \times 2 = 0.8 \qquad a_3 = 0$$
$$0.8 \times 2 = 1.6 \qquad a_4 = 1$$
$$0.6 \times 2 = 1.2 \qquad a_5 = 1$$
$$0.2 \times 2 = 0.4 \qquad a_6 = 0$$
$$\cdots \cdots \cdots \cdots$$

At this point one can easily see that the process continues ad infinitum, and the result is

$$0.6 = (0.1001\ 1001\ 1001\ \cdots)_2$$

A computer, of course, can process only finite binary numbers. Thus, slight errors, called *truncation errors*, often appear in binary numbers which have been converted from exact decimal fractions. The above binary number to six significant digits is

$0.100\ 110_2$

Examples showing the conversion of 0.6 to three significant digits in the octal and hexadecimal systems are given below:

$0.6 \times 8 = 4.8$	$a_1 = 4$	$0.6 \times 16 = 9.6$	$a_1 = 9$
$0.8 \times 8 = 6.4$	$a_2 = 6$	$0.6 \times 16 = 9.6$	$a_2 = 9$
$0.4 \times 8 = 3.2$	$a_3 = 3$	$0.6 \times 16 = 9.6$	$a_3 = 9$
	$0.6 = 0.463_8$		$0.6 = 0.999_{16}$

4.2   ARITHMETIC

The rules of addition, subtraction, multiplication, and division in the binary, octal, and hexadecimal systems are the same as in the decimal system. Tables 4.2 and 4.3 represent the addition tables for the binary and octal systems, respectively, while Tables 4.4 and 4.5 represent the multiplication tables for the binary and octal systems, respectively. The construction of the addition and multiplication tables for the hexadecimal system is left to the student as an exercise. Table 4.6 presents an example of each type of arithmetic operation in each of the number systems considered.

Table 4.2

	0	1
0	0	1
1	1	10

Table 4.3

	0	1	2	3	4	5	6	7
0	0	1	2	3	4	5	6	7
1	1	2	3	4	5	6	7	10
2	2	3	4	5	6	7	10	11
3	3	4	5	6	7	10	11	12
4	4	5	6	7	10	11	12	13
5	5	6	7	10	11	12	13	14
6	6	7	10	11	12	13	14	15
7	7	10	11	12	13	14	15	16

Table 4.4

	0	1
0	0	0
1	0	1

Table 4.5

	0	1	2	3	4	5	6	7
0	0	0	0	0	0	0	0	0
1	0	1	2	3	4	5	6	7
2	0	2	4	6	10	12	14	16
3	0	3	6	11	14	17	22	25
4	0	4	10	14	20	24	30	34
5	0	5	12	17	24	31	36	43
6	0	6	14	22	30	36	44	52
7	0	7	16	25	34	43	52	61

Table 4.6

Operation	Decimal	Binary	Octal	Hexadecimal
Addition	15	1 111	17	f
	13	1 101	15	d
	28	11 100	34	$1c$
Subtraction	29	11 101	35	$1d$
	15	01 111	17	f
	14	1 110	16	e
Multiplication	15	1 111	17	f
	13	1 101	15	d
	45	1 111	113	$c3$
	15	111 1	17	
	195	1 111	303	
		11 000 011		
Division	39	100 111	47	27
	4)156	100)10 011 100	4)234	4)9c
	12	10 0	20	8
	36	11 1	34	$1c$
	36	10 0	34	$1c$
	0	1 10	0	0
		1 00		
		100		
		100		
		0		

4.2.1   Complementing

By using the *complement* of the subtrahend, a problem in subtraction can be converted into a problem in addition.   It will shortly be shown that complementing is a very simple operation in the binary system and, for this reason, is a method often used to perform subtraction in digital computers.

To illustrate the principle of complementing, consider the subtraction

$$\begin{array}{r} 15 \\ -\ 7 \\ \hline 8 \end{array}$$

Replacing 7 by its base complement $(10 - 7 = 3)$ and adding produce

$$\begin{array}{r} 15 \\ +\ 3 \\ \hline 18 \end{array}$$

and dropping the first digit gives the correct answer, 8.   The subtraction $150 - 7$ is carried out as shown below:

$$\begin{array}{cc} 150 & 150 \\ -\ \ \ 7 \rightarrow +\ 993 \\ \hline & 1143 \rightarrow 143 \end{array}$$

In general terms, if $n - b$ is the base complement of b, then

$$a - b = a + (n - b) - n$$

This equation shows that the use of complementing is a valid procedure in any number system.   For example,

$$\begin{array}{cccc} 150_8 & 150_8 & 110_2 & 110_2 \\ -\ \ 7_8 \rightarrow +771_8 & -011_2 \rightarrow +101_2 \\ \hline 1141_8 \rightarrow 141_8 & 1011_2 \rightarrow 011_2 \end{array}$$

Another form of complementing is *base-minus-one complementing*. In the example $150 - 7$, the base-minus-one complement of 7 is $999 - 7 = 992$.   Subtraction is then performed as follows:

$$\begin{array}{r} 150 \\ +\ 992 \\ \hline 1142 \\ 1 \\ \hline 1143 \rightarrow 143 \end{array}$$

In general terms, if $m - b$ is the base-minus-one complement of b, then

$$a - b = a + (m - b) - (m + 1) + 1$$

This method is particularly convenient in the binary system, since the base-minus-one complement is obtained by simply replacing each digit by its inverse. For example,

$$
\begin{array}{r}
101\ 110_2 \\
-011\ 101_2
\end{array} \rightarrow
\begin{array}{r}
101\ 110_2 \\
+\ \ 100\ 010_2 \\
\hline
1\ 010\ 000_2 \\
+1_2 \\
\hline
1\ 010\ 001_2 \rightarrow 010\ 001_2
\end{array}
$$

The advantage of utilizing this technique in digital computers should be apparent. An electrical circuit which performs subtraction would have to include a circuit for altering each digit and an addition circuit. Since an addition circuit is necessary anyway, this circuit would be simpler than one which would have to perform the difficult task of borrowing.

4.3   THE DESIGN OF A DIGITAL COMPUTER

A FORTRAN program must be translated into a machine language by the compiler before execution. It is the purpose of the remainder of this chapter to present some of the general features of machine language. It is unlikely that the reader will find it necessary to learn such a language or to write a program in terms of it, but an understanding of at least the arithmetic that is used by the computer is very desirable.

In most computers both instructions and data are stored as strings of 0's and 1's, i.e., as numbers in the binary system. The reason why this system is used is that the 0 and 1 of the binary system can be represented by relatively simple electronic equipment. For example, a bistable magnetic element may be magnetized either in one direction or in the other, and if the direction can be sensed by some electrical circuitry, then the two states may be considered to be equivalent to 0 and 1, respectively. Many other devices have been used successfully. Some were mechanical, in the earliest computers, such as relays that were *open* or *closed*. Others were electronic, such as flip-flop circuits that were *set* or *reset* or cathode-ray tubes with *dark* or *illuminated* spots.

A step-by-step design of an imaginary digital computer, which we shall call the XYZ30, will now be considered. The steps fall into six categories:

1. The design of the *storage area* in which the data and the instructions of a program are stored

2. The design of an *arithmetic unit* in which all arithmetic operations are executed

3. The design of a *control unit* in which the flow of the program is controlled

4. The design of the *input-output mechanism* which provides for information to be placed into or taken out of the storage area

5. The design of a set of *instructions* which will cause the computer to perform certain operations

6. The design of the *electrical circuitry*

The details in the design of the XYZ30 following the six categories given above will now be considered.

4.4   STORAGE AREA

Data are stored in a computer in binary form.   Three types of data can generally be stored: integer numbers, floating-point numbers, and alphabetic characters.   Assume that in the XYZ30 one storage location contains 15 *on-off* electronic elements.   Such elements are often called *bits*, which is an abbreviation for *binary digits*.   A 15-bit binary number which is stored in a given location is called a *word*.

4.4.1   Integer Numbers

The first bit of a word which represents an integer number designates the sign of the number.   In our computer we shall arbitrarily let 0 be plus and 1 be minus.   Several examples are given in Table 4.7.   The magnitude of the largest integer which can be stored is fourteen 1's, or $2^{14} - 1 = 16,383$.

Table 4.7

Decimal number	Binary word	Octal representation
+7	000 000 000 000 111	00007
−7	100 000 000 000 111	40007
+100	000 000 001 100 100	00144
−100	100 000 001 100 100	40144

4.4.2   Floating-point Numbers

As explained in Chap. 3, most arithmetic in computers is done by use of floating-point numbers.   In FORTRAN such numbers are indicated by the inclusion of a decimal point. The general form of a floating-point number x is

$$x = (\text{fraction})(\text{base}^{\text{exponent}})$$

For example, 5.32 may be written in floating-point form as

0.532×10^1
0.0532×10^2
0.00000532×10^6
. . . .

If the first digit of the fraction is not zero, the number is said to be in *normalized form*, as in the first of the above examples.

Internally, floating-point numbers are stored in the form

(sign)(characteristic)(fraction)

where the *characteristic* is computed by

characteristic = exponent + constant

Adding a constant to the exponent eliminates the need for a negative characteristic, thus avoiding two signs in the number.

Let it be assumed that these three parts of a floating-point number are stored in the 15-bit word of the XYZ30 according to the following pattern:

Bit 1: The *sign;* a 0 here indicates a positive sign, and a 1 indicates a negative sign.

Bits 2–6: The *characteristic;* the binary number $10\,000_2$ here represents an exponent of 0. Thus, the characteristic is defined as

characteristic = exponent + $10\,000_2$

Bits 7–15: The *fraction;* nine significant binary digits, where the binary point is assumed to be in front of the first digit.

The binary floating-point number $0.101\,010\,111_2 \times 2^5$ would appear in a 15-bit binary word as follows:

0	10 101	101 010 111
sign	characteristic	fraction

Table 4.8 presents a few more examples. Since 15 binary digits is a rather large number to write, the octal form of the bit configuration is often utilized. Table 4.8 also shows the octal form for the given examples.

The range of numbers which can be held in this particular bit arrange-

Table 4.8

Decimal number	Binary number, normalized form	Bit configuration	Octal form of bit configuration
0.5	$0.1_2 \times 2^0$	010 000 100 000 000	20400_8
−0.5	$-0.1_2 \times 2^0$	110 000 100 000 000	60400_8
1.5	$0.11_2 \times 2^1$	010 001 110 000 000	21600_8
6.0	$0.11_2 \times 2^3$	010 011 110 000 000	23600_8
−6.0	$-0.11_2 \times 2^3$	110 011 110 000 000	63600_8
0.125	$0.1_2 \times 2^{-2}$	001 110 100 000 000	16400_8
−0.125	$-0.1_2 \times 2^{-2}$	101 110 100 000 000	56400_8

ment is from approximately $1 \times 2^{-16} \approx 1.5 \times 10^{-5}$ to $1 \times 2^{15} \approx 3 \times 10^4$. Nine significant binary digits is equivalent to three significant octal digits, which is almost equal to three significant decimal digits. Since this represents a rather limited range and a limited number of significant digits, most computers use words of much greater length than 15 bits.

4.4.3   Alphabetic Characters

In order to store alphabetic and other characters in a computer, a code must be established. Table 4.9 presents such a code which is the standard BCD (binary-coded decimal) code used by IBM. Two octal digits or six binary digits correspond to a single character. Thus, in a 15-bit word $2\frac{1}{2}$ characters could be stored. The table also shows the Hollerith[1] code often used in punching cards. The digits in the table indicate the row

[1] Named after Herman Hollerith, who designed a punched-card system for use in tabulating the census of 1890.

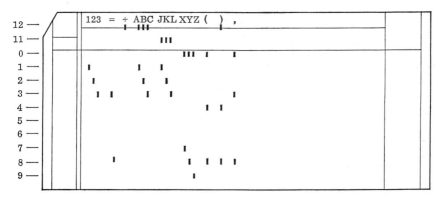

Fig. 4.1

or rows in which holes are punched in a card column. The rows on a standard IBM card are numbered as shown in Fig. 4.1. Several characters are punched on the card for illustration.

Table 4.9

Character	On card	In storage†	Character	On card	In storage†
0	0	00	J	11-1	41
1	1	01	K	11-2	42
2	2	02	L	11-3	43
3	3	03	M	11-4	44
4	4	04	N	11-5	45
5	5	05	O	11-6	46
6	6	06	P	11-7	47
7	7	07	Q	11-8	50
8	8	10	R	11-9	51
9	9	11	$	11-8-3	53
=	8-3	13	*	11-8-4	54
,	8-4	14	Blank		60
+	12	20	/	0-1	61
A	12-1	21	S	0-2	62
B	12-2	22	T	0-3	63
C	12-3	23	U	0-4	64
D	12-4	24	V	0-5	65
E	12-5	25	W	.0-6	66
F	12-6	26	X	0-7	67
G	12-7	27	Y	0-8	70
H	12-8	30	Z	0-9	71
I	12-9	31	,	0-8-3	73
.	12-8-3	33	(	0-8-4	74
)	12-8-4	34			
—	11	40			

† The *In storage* columns are given in octal.

4.4.4   Storage Area of XYZ30

In the storage area of our computer, let us place 512 fifteen-bit storage locations. Further, let us number these consecutively in the octal system from 000_8 to 777_8. The number of a particular storage location is called its *address*.

4.5   ARITHMETIC UNIT

The arithmetic unit of the XYZ30 shall consist of two registers, an *accumulator register*, abbreviated as the AC, and a *multiplier-quotient register*, abbreviated as the MQ.   The AC contains 16 bits.   The first bit is the sign bit, the next bit the overflow bit, and the remaining bits, which shall be numbered 2 to 15 to correspond to words in storage, contain a binary number.   A 1 will appear in the overflow bit if through an arithmetic calculation a number is obtained which is larger than the capacity of the computer register.   The primary use of the AC is in addition and subtraction.   The MQ is a 15-bit register which is used with the AC for the multiplication and division operations.

4.5.1   Addition

Integer addition is accomplished in two steps in the AC.   Suppose that the numbers $101\ 111_2$ and 010_2 are to be added.   The first step is to place one of the numbers in the AC.

AC

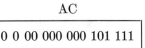

```
0 0 00 000 000 101 111
```

The next step is to add the second number.   The result is left in the AC.

AC

```
0 0 00 000 000 110 001
```

In the addition of floating-point numbers, the normalized form of one of the numbers must be altered so that the binary points line up.   Consider the addition of $0.100_2 \times 2^2$ and $0.101\ 111_2 \times 2^{-3}$.   Placing the first number in the AC gives

AC

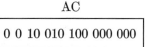

```
0 0 10 010 100 000 000
```

The second number must be changed to

$0.000\ 001\ 011\ 11_2 \times 2^2$

and adding the fractional part to the fraction already in the AC gives

AC

```
0 0 10 010 100 001 011
```

Notice that the last two digits of the second number are dropped as insignificant.

It is possible that an addition may cause an overflow from bit 8.   For example,

$$0.101_2 \times 2^2$$
$$0.110_2 \times 2^2$$
$$\overline{1.011_2 \times 2^2}$$

In such a case a shift to the right must occur, and a 1 is added to the characteristic.   The result would appear in the AC as

AC

| 0 0 10 011 101 100 000 |

4.5.2   Subtraction

In subtraction, the number to be subtracted is complemented and then the two numbers are added.   Consider the subtraction of 010_2 from $101\ 111_2$.   Placing $101\ 111_2$ in the AC,

AC

| 0 0 00 000 000 101 111 |

base-minus-one complementing of 010,

11 111 111 111 101

adding,

AC

| 0 1 00 000 000 101 100 |

placing a 0 in the second bit, and adding 1 finally give the correct answer in the AC.

AC

| 0 0 00 000 000 101 101 |

4.5.3   Multiplication

Integer multiplication is carried out by using the AC and the MQ together. Consider the multiplication of $001\ 011_2$ by 101_2.   The first step is to place one of the numbers, say $001\ 011_2$, into the MQ and to clear the AC.

AC   MQ

| 0 0 00 000 000 000 000 | 000 000 000 001 011 |

The computer checks the rightmost bit.   Since it is a 1, the second number, 101, is added to the AC.

AC   MQ

| 0 0 00 000 000 000 101 | 000 000 000 001 011 |

All the digits are now shifted to the right one place except for the sign bit in the MQ.

AC   MQ

| 0 0 00 000 000 000 010 | 010 000 000 000 101 |

Again a 1 is in the rightmost bit of the MQ; so 101_2 is added to the AC.

AC   MQ

| 0 0 00 000 000 000 111 | 010 000 000 000 101 |

Shift to the right.

AC   MQ

| 0 0 00 000 000 000 011 | 011 000 000 000 010 |

Since a zero is now in the rightmost bit of the MQ, only the shifting takes place.

AC   MQ

| 0 0 00 000 000 000 001 | 011 100 000 000 001 |

Adding,

AC   MQ

| 0 0 00 000 000 000 110 | 011 100 000 000 001 |

Shifting,

AC   MQ

| 0 0 00 000 000 000 011 | 001 110 000 000 000 |

The process continues until a total of 14 shifts occurs.   In this example, the remaining operations consist of only 10 more shifts.

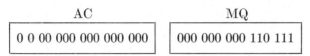

AC	MQ
0 0 00 000 000 000 000	000 000 000 110 111

The product appears in the MQ.   The signs of the two numbers must be checked and the correct digit placed in the first bit of the MQ.

With some thought, one can verify that the above operations are equivalent to the multiplication (in binary)

$$
\begin{array}{r}
101 \\
1\ 011 \\
\hline
101 \\
1\ 01 \\
101\ 0 \\
\hline
110\ 111
\end{array}
$$

Instead of saving the addition for the last step, the computer adds at each intermediate step.

If the product turns out to be more than 14 digits long, an overflow into the rightmost bits of the AC will occur.   Thus, after a multiplication operation, the AC should be checked.   If the binary number in the AC is not zero, an overflow has occurred.

Floating-point multiplication can be carried out by applying a similar process to the fractions of the two numbers.   Also, the exponents (not characteristics) must be added.   For example, multiplying $0.100_2 \times 2^2$ by $0.101\ 111_2 \times 2^{-3}$ gives $0.010\ 111\ 100_2 \times 2^{-1}$.   Putting this in normalized form gives $0.101\ 111_2 \times 2^{-2}$, and it would appear in the MQ as

MQ
001 110 101 111 000

4.5.4   Division

Integer division is also performed on the AC and the MQ together.   Suppose that $001\ 011_2$ is to be divided by 101_2.   The first step consists in placing $001\ 011_2$ in the MQ.

AC	MQ
0 0 00 000 000 000 000	000 000 000 001 011

Shift to the left one space, skipping the sign bit in the MQ.

AC	MQ
0 0 00 000 000 000 000	000 000 000 010 110

The number in the AC is now compared with the divisor. Since this number is less than the divisor, another shift to the left takes place. In this example, this process continues until a total of 13 shifts has occurred.

AC	MQ
0 0 00 000 000 000 101	010 000 000 000 000

When the number in the AC becomes equal to or greater than the divisor, the divisor is subtracted in the AC and a 1 is placed in the rightmost bit of the MQ.

AC	MQ
0 0 00 000 000 000 000	010 000 000 000 001

Shift to the left.

AC	MQ
0 0 00 000 000 000 001	000 000 000 000 010

The number in the AC is less than the divisor. Fourteen shifts have occurred, and so the operation is complete. The quotient 010_2 appears in the MQ and the remainder 001_2 in the AC. The signs of the dividend and the divisor are compared and the correct digit placed in the first bit of the MQ. One can verify that this process is equivalent to the division

$$
\begin{array}{r}
10 \quad \text{quotient} \\
101\overline{)1\ 011} \\
\underline{1\ 01} \\
1 \\
0 \\
\overline{1} \quad \text{remainder}
\end{array}
$$

Floating-point division can be carried out by applying a similar process to the fractions of the two numbers. Also, the exponent (not characteristic) of the divisor must be subtracted from the exponent of the dividend. For example, dividing $0.101\ 111_2 \times 2^{-3}$ by $0.100_2 \times 2^2$ gives

$$1.011\ 110 \times 2^{-5} = 0.101\ 111 \times 2^{-4}$$

This would appear in the MQ in the form

MQ

001 100 101 111 000

4.6   CONTROL UNIT

The control unit of our computer will consist of three registers: an *instruction location counter*, abbreviated by ILC, an *index register*, abbreviated by XR, and an *index limit register*, abbreviated by XL.   During the execution of a program, the storage location of the next instruction to be executed appears in the ILC.   This register need be only 9 bits long, since the largest address in our computer is 777_8.   In programming a loop, the XR and the XL contain the current values of the index and the upper limit on the index, respectively.   Let us restrict these numbers to be only positive and 9 bits long.

4.7   INPUT-OUTPUT MECHANISM

Figure 4.2 presents a schematic diagram of the input-output display unit for the XYZ30.   The placing of a piece of data or an instruction into a particular storage location is implemented in the following manner.   First,

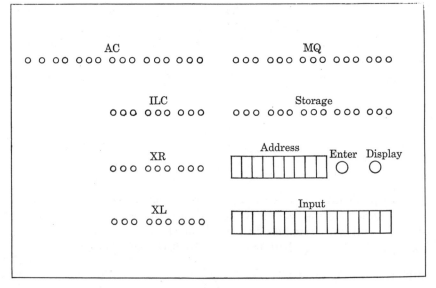

Fig. 4.2

the desired address (in binary) is keyed onto the *Address* keys.   Second, the binary digits corresponding to the input word are keyed onto the *Input* keys.   Depressing the *Enter* button then enters the binary digits into the proper storage location.   Displaying a number which is stored at a certain address is accomplished in a similar way.   Keying the desired address onto the *Address* keys and depressing the *Display* button cause the binary digits stored at the given address to be displayed on the *Storage* lights.

4.8   INSTRUCTIONS

The next item to be designed for the XYZ30 is a set of *instructions*.   An instruction must consist of two parts, an *operation* and an *address*.   It causes a certain operation to be performed on a number located at a certain address.

Instructions are stored in the storage area of a computer.   Thus, in the XYZ30 an instruction is stored by using 15 bits.   In order that every address in the storage area of our computer can be referenced, 9 bits must be reserved for the address portion of an instruction.   Let us reserve bits 7 through 15 for this purpose.

A code must be formed to specify the operation part of an instruction. Let us reserve bits 1 through 5 for this purpose.   The use of bit 6 will be explained in the next section.

Probably the most basic operation on a digital computer is the addition of two numbers.   Actually this consists of three separate operations: loading the first number into the AC, adding the second to the first in the AC, and finally storing the result at a specific address in the storage area. Let us choose the following binary codes for three instructions corresponding to these three operations:

00101   load the AC
00001   add to the number in the AC
00111   store

Suppose that the two numbers to be added are stored in locations 100_8 and 101_8 and that it is desired to store the result in location 102_8.   Assuming that a 0 should be placed in bit 6, the binary or machine-language program would be

Operation code	Operand address
001 010	001 000 000
000 010	001 000 001
001 110	001 000 010

The advantage of the octal system becomes apparent if this machine-language program is written in octal.

12 100
02 101
16 102

4.8.1 Assembly-language Programs

As a convenience to programming, alphabetic code names are given to machine instructions on most computers. If we assign the codes L for load, A for add, and ST for store, the above machine-language program in alphabetic code is

L 100_8
A 101_8
ST 102_8

A program written in an alphabetic code is called an *assembly-language program*. On most computers, the alphabetic code used in assembly languages is given a name. Let us call ours XAP (*X*YZ30 *A*ssembly *P*rogram).

In our computer we also need some other instructions which implement subtraction, multiplication, division, and control operations. Table 4.10 gives a list of these instructions. The table includes the description, the binary code, and the XAP code for each instruction.

Consider now the arithmetic operations in the equation

$$r = \frac{(a + b)c - d}{e}$$

Upon setting aside locations 101_8, 102_8, 103_8, 104_8, 105_8, and 106_8 for a, b, c, d, e, and r, respectively, we may write the following XAP program for performing the indicated operations:

Location	Instruction	
000_8	L	101_8
001_8	A	102_8
002_8	ST	106_8
003_8	LQ	106_8
004_8	M	103_8
005_8	STQ	106_8
006_8	L	106_8
007_8	S	104_8
010_8	ST	106_8
011_8	LQ	106_8
012_8	D	105_8
013_8	STQ	106_8
014_8	STOP	

Table 4.10

XAP code	Binary code	Description
STOP	00000	*Stop.* Stops the operation of the computer
A	00001	*Add.* Adds the integer number stored at the address to the integer number in the AC. Result appears in the AC
S	00010	*Subtract.* Subtracts the integer number stored at the address from the integer number in the AC. Result appears in the AC
M	00011	*Multiply.* Multiplies the integer number stored at the address by the integer number in the MQ. The AC is first cleared. Result appears in the MQ, overflowing into the AC if necessary
D	00100	*Divide.* Divides the integer number in the MQ by the integer number stored at the address. The AC is first cleared. Quotient appears in the MQ, remainder in the AC
L	00101	*Load* AC. Replaces the number in the AC by the number at the address
LQ	00110	*Load* MQ. Replaces the number in the MQ by the number at the address
ST	00111	*Store* AC. Replaces the number at the address by the number in the AC
STQ	01000	*Store* MQ. Replaces the number at the address by the number in the MQ
FA	01001	*Floating Add.* Adds the floating-point number at the address to the floating-point number in the AC. Result appears in the AC
FS	01010	*Floating Subtract.* Subtracts the floating-point number at the address from the floating-point number in the AC. Result appears in the AC
FM	01011	*Floating Multiply.* Multiplies the floating-point number at the address by the floating-point number in the MQ. The AC is first cleared. Result appears in the MQ. No overflow is possible into the AC
FD	01100	*Floating Divide.* Divides the floating-point number in the MQ by the floating-point number at the address. The AC is first cleared. Quotient appears in the MQ. Remainder appears in the AC
B	10000	*Branch.* Branch to the location designated by the address
BZ	10001	*Branch if Zero.* If the number in the AC is 0, branch to the location designated by the address. Otherwise proceed to the next sequential location
BM	10010	*Branch if Minus.* If there is a 1 in the first bit of the AC, branch to the location designated by the address. Otherwise proceed to the next sequential location
BXLE	10011	*Branch if Index Low or Equal.* If the number in the XR is less than or equal to the number in the XL, branch to the location designated by the address. Otherwise, proceed to the next sequential location
LX	10100	*Load Index.* Replace the integer number in the XR by the integer number at the address
LXL	10101	*Load Index Limit.* Replace the integer number in the XL by the integer number at the address
AX	10111	*Add to Index.* Add to the integer number in the XR the integer number stored at the address. Sum appears in the XR

Location (Octal)	Lights				
000	○ ○ ●	○ ● ○	○ ○ ●	○ ○ ○	○ ○ ●
001	○ ○ ○	○ ● ○	○ ○ ●	○ ○ ○	○ ● ○
002	○ ○ ●	● ● ○	○ ○ ●	○ ○ ○	● ● ○
003	○ ○ ●	● ○ ○	○ ○ ●	○ ○ ○	● ● ○
004	○ ○ ○	● ● ○	○ ○ ●	○ ○ ○	○ ● ●
005	○ ● ○	○ ○ ○	○ ○ ●	○ ○ ○	● ● ○
006	○ ○ ●	○ ● ○	○ ○ ●	○ ○ ○	● ● ○
007	○ ○ ○	● ○ ○	○ ○ ●	○ ○ ○	● ○ ○
010	○ ○ ●	● ● ○	○ ○ ●	○ ○ ○	● ● ○
011	○ ○ ●	● ○ ○	○ ○ ●	○ ○ ○	● ● ○
012	○ ○ ●	○ ○ ○	○ ○ ●	○ ○ ○	● ○ ●
013	○ ● ○	○ ○ ○	○ ○ ●	○ ○ ○	● ● ○
014	○ ○ ○	○ ○ ○	○ ○ ○	○ ○ ○	○ ○ ○
.					
.					
.					
101	○ ○ ○	○ ○ ○	○ ○ ○	○ ○ ○	○ ● ○
102	○ ○ ○	○ ○ ○	○ ○ ○	○ ○ ○	● ○ ○
103	○ ○ ○	○ ○ ○	○ ○ ○	○ ○ ○	○ ● ○
104	○ ○ ○	○ ○ ○	○ ○ ○	○ ○ ○	● ○ ○
105	○ ○ ○	○ ○ ○	○ ○ ○	○ ○ ○	○ ● ○
106	○ ○ ○	○ ○ ○	○ ○ ○	○ ○ ○	○ ○ ○

Fig. 4.3

The input switches can now be utilized to place the program and data in the computer. If $a = 2$, $b = 4$, $c = 2$, $d = 4$, and $e = 2$, the storage lights should appear as shown in Fig. 4.3 before the execution of the program. Placing 000 000 000 in the ILC and depressing the Start button begin execution of the program. At the end of the execution, the storage lights corresponding to location 106_8 would appear as

000 000 000 000 100

4.8.2   Address Modification

Consider the FORTRAN statements

        SUM = 0.
        DO 3 I = 1, N
    3   SUM = SUM + A(I)

The statement SUM = SUM + A(I) is a simple floating-point addition, and the corresponding XAP instruction would be FA. However, each time this step is executed, a number is added from a different storage location. Thus, the address of the FA instruction must be changed each time. On the XYZ30 let us accomplish this by establishing the following rule:

> If the sixth bit of an instruction contains a 1, the instruction is executed as if its address contained the stated address plus the integer number stored in the XR.

Use of this rule is called *address modification*. If the address of an instruction in an XAP program is to be modified, a comma and an X are placed after the instruction.

If we let the number 1 be stored in location 101_8, the number 0 in 102_8, N in 103_8, SUM in 104_8, and the array A starting in 105_8, the corresponding XAP program follows:

Location	Instruction	
000_8	LXL	103_8
001_8	LX	101_8
002_8	L	102_8
003_8	FA	104_8, X
004_8	AX	101_8
005_8	BXLE	003_8
006_8	ST	104_8
007_8	STOP	

4.9   ELECTRICAL CIRCUITRY

The electrical circuits needed to implement all the operations which have been discussed thus far would certainly be complex. No attempt will be made to design all these circuits, but for illustrative purposes an electrical

0

1

Fig. 4.4

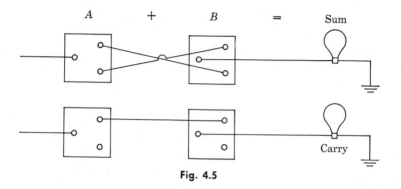

Fig. 4.5

circuit which performs the addition of two binary digits will now be presented.

First of all, let each of the two binary numbers 0 and 1 be represented by two switch settings as indicated in Fig. 4.4. The circuit shown in Fig. 4.5 then represents an addition circuit. If the sum of two binary numbers is 1, the upper light bulb is on. If there is a carry, the lower light bulb is on. Figure 4.6 shows all possible combinations. Obviously, a light bulb can be on only if the circuit is closed. If a circuit for addition can be designed, circuits for all the arithmetic operations can also be readily designed. Although the design of the complete circuitry of the XYZ30 appears prodigious, the circuitry of modern computers is many times more complex.

4.10   OTHER FEATURES OF COMPUTERS

The imaginary digital computer, the XYZ30, conceived in this chapter, is a great simplification of actual computers used commercially today. Many more bits are needed in a word to obtain a reasonable range of numbers and a large number of significant digits. Also, more machine instructions are needed to facilitate the programming of long, complicated programs. Many computers have several hundred machine instructions. Also, the input-output mechanisms of computers are much more sophisticated than the mechanism described in Sec. 4.7. Card readers, paper and magnetic-tape reels, printers, etc., are all employed for input-output operations.

The next step in the development of an overall computer system is the design of a compiler which can accept a statement-type language such as FORTRAN and translate it into machine language. The expressions *software* and *hardware* are common terms in computer technology. *Soft-*

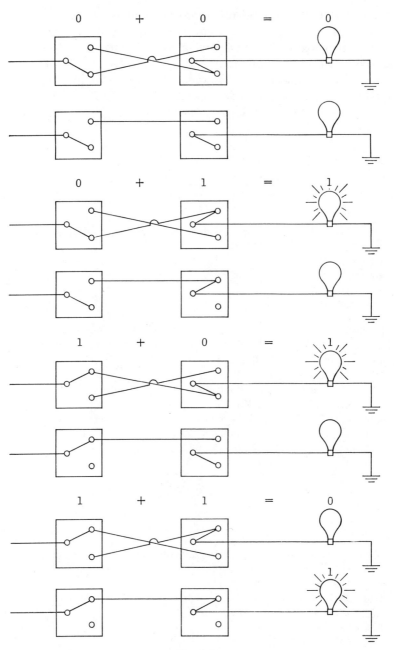

Fig. 4.6

ware includes all computer programs distributed by the computer man-
ufacturer, such as a program to translate a statement-type language into a
machine language.   The *hardware*, on the other hand, is the actual physical components of the computer.   The design of a compiler is complicated
and will not be considered.   However, with the introduction of the principles of machine language as presented here, one can realize the many
advantages of the FORTRAN language.

PROBLEMS

Section 4.1

In Probs. 4.1 to 4.5, convert the given numbers into positional notation:
4.1 56392
4.2 $111\ 001_2$
◆**4.3** 627_8
4.4 $d3e_{16}$
4.5 $100\ 111_2$

In Probs. 4.6 to 4.15, a number is given in each of the number systems, decimal,
binary, octal, and hexadecimal.   Convert each number to all the other systems.
◆**4.6**   78, $100\ 010\ 111_2$, 64_8, $3a5_{16}$
4.7   239, $011\ 111\ 001_2$, 241_8, $12f_{16}$
4.8   356, $001\ 001\ 100_2$, 121_8, $b33_{16}$
4.9   999, $110\ 110\ 110_2$, 421_8, $9cd_{16}$
4.10 681, $101\ 010\ 101_2$, 777_8, eee_{16}
4.11 0.2, $0.101\ 010\ 101_2$, 0.521_8, 0.521_{16}
4.12 0.33, $0.111\ 111\ 111_2$, 0.345_8, 32.2_{16}
4.13 0.414, $101.010\ 111_2$, 0.22_8, $0.fed_{16}$
◆**4.14** 0.99, $0.100\ 100\ 001_2$, 14.13_8, $0.fff_{16}$
4.15 62.3, $0.001\ 010\ 111_2$, 0.777_8, $0.3a7_{16}$

Section 4.2

In Probs. 4.16 to 4.39, perform the indicated arithmetic operation:
◆**4.16** $101\ 111\ 001_2 + 111\ 111\ 110_2$
4.17 $110\ 000\ 101_2 + 011\ 100\ 111_2$
4.18 $2467_8 + 312_8$
4.19 $4771_8 + 6226_8$
4.20 $3c6_{16} + fd5_{16}$
4.21 $a55_{16} + 3c9_{16}$
4.22 $101\ 111\ 011_2 - 100\ 000\ 100_2$
4.23 $100\ 010\ 110_2 - 100\ 100\ 111_2$

◆**4.24** $1627_8 - 777_8$
4.25 $451_8 - 1233_8$
4.26 $c39_{16} - 14e_{16}$
4.27 $32b_{16} - ddd_{16}$
4.28 $101\ 110_2 \times 010\ 111_2$
4.29 $110\ 001_2 \times 111\ 001_2$
4.30 $427_8 \times 623_8$
4.31 $333_8 \times 421_8$
◆**4.32** $333_{16} \times 421_{16}$
4.33 $3df_{16} \times 1dd_{16}$
4.34 $111\ 001_2 \div 101_2$
4.35 $100\ 010_2 \div 100\ 001_2$
4.36 $622_8 \div 41_8$
4.37 $427_8 \div 31_8$
4.38 $6c4_{16} \div f1_{16}$
◆**4.39** $d39_{16} \div 5a_{16}$

Perform the subtractions in Probs. 4.40 to 4.47 by complementing:
4.40 $2563 - 42$
4.41 $9621 - 4478$
4.42 $111\ 000_2 - 010\ 101_2$
◆**4.43** $101\ 101_2 - 100\ 001_2$
4.44 $632_8 - 421_8$
4.45 $4327_8 - 15_8$
4.46 $3c9_{16} - 11f_{16}$
4.47 $fd4_{16} - ee_{16}$

4.48–4.55 Do Probs. 4.40 to 4.47 by base-minus-one complementing.

Section 4.4

In Probs. 4.56 to 4.60, determine the decimal integer number corresponding to the binary or octal code given:
4.56 $000\ 000\ 001\ 111\ 100_2$
◆**4.57** $100\ 000\ 001\ 111\ 100_2$
4.58 00143_8
4.59 40032_8
4.60 60000_8

In Probs. 4.61 to 4.65, determine the decimal floating-point number corresponding to the binary or octal code given:
4.61 $010\ 011\ 101\ 000\ 000_2$
4.62 $101\ 000\ 110\ 000\ 000_2$
4.63 57610_8
◆**4.64** 24414_8
4.65 62400_8

In Probs. 4.66 to 4.70, determine the proper bit configuration in the XYZ30 for the given floating-point numbers:

4.66 23.5

4.67 −23.5

4.68 0.6

◆**4.69** −0.2

4.70 −14.7

4.71 What is the magnitude of the largest integer number which can be stored in a 36-bit word?

4.72 What are the magnitudes of the smallest and largest floating-point numbers which can be stored in a 36-bit word?   Assume that 8 bits are reserved for the exponent and 27 bits for the fraction.

In Probs. 4.73 to 4.77, give the *On Card* and *In Storage* codes for the given characters.   How many storage locations would be needed to store the characters in the XYZ30?   Sketch how the holes would be punched on a card.

4.73 XYZ30

4.74 DIGITAL COMPUTER

4.75 NUMERICAL METHODS

4.76 CHEMICAL PROCESS ∗1

4.77 F = MA

Section 4.5

4.78 Demonstrate how multiplication is performed on the XYZ30 by multiplying 6 by 12, showing sketches of the AC and the MQ after each step.

4.79 Demonstrate how division is performed on the XYZ30 by dividing 7 by 2, showing sketches of the AC and the MQ after each step.

Section 4.8

In Probs. ◆4.80, 4.81 to 4.83, ◆4.84, and 4.85, write an XAP program for the given flow chart.

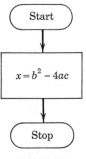

Fig. P4.80

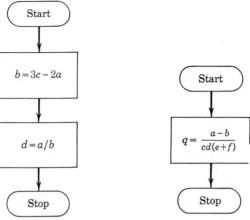

Fig. P4.81

Fig. P4.82

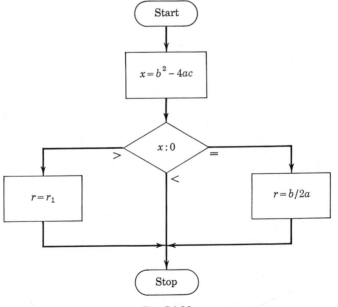

Fig. P4.83

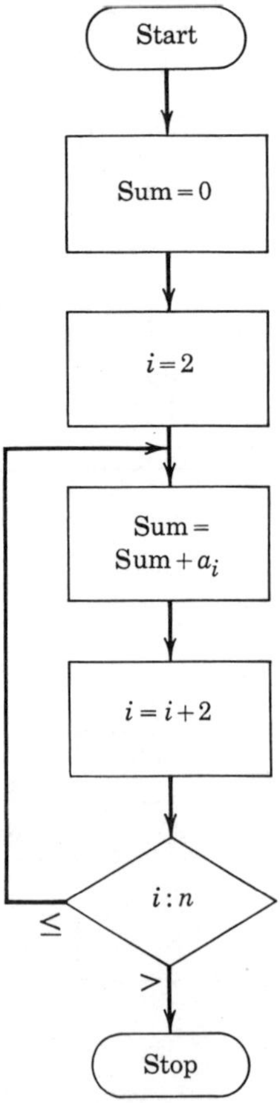

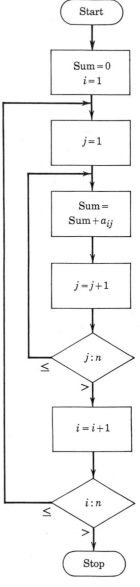

Fig. P4.84

Fig. P4.85

4.86 Write an XAP program to execute the following FORTRAN program:

NPRO = NUMB(2)
DO 100 I = 4, 12, 2
100 NPRO = NPRO*NUMB(I)

4.87 How does a computer determine whether a number stored in any given location is a piece of data or an instruction? What would happen if the computer mistook a piece of data as an instruction?

4.88 An XAP listing of a part of a larger program is given below. Using any FORTRAN statements you wish, write a FORTRAN program which will accomplish the same result as the XAP program.

Location	Instruction	
024_8	L	461_8
025_8	FA	460_8
026_8	ST	461_8
027_8	L	460_8
030_8	FS	457_8
031_8	ST	460_8
032_8	BM	034_8
033_8	B	024_8
034_8	L	461_8
035_8	ST	456_8

4.89 Before the program in Prob. 4.88 was executed, the addresses given below had the indicated numbers stored in them:

Address	Contents
461_8	00000_8
460_8	24500_8
457_8	21400_8
456_8	00000_8

What would the contents of these addresses be after the program had been executed? Express the numbers in octal form (i.e., as they would appear in the computer).

Section 4.9

4.90 Design an electrical circuit for the subtraction of one binary digit from another.

5 ROUNDING AND TRUNCATION ERRORS

5.1 INTRODUCTION

The numerical solution of an engineering problem is probably never exact, because of assumptions and approximations that have been made in obtaining the solution. Even such a simple problem as finding the circumference of a circle of which the diameter is known exactly can be solved only approximately in terms of a finite number of digits. In this case no assumptions are made in arriving at the mathematical formulation $C = \pi D$, but the value of π must be approximated in order to proceed to a numerical answer. In most problems the mathematical model of the physical situation also involves approximations.

In order to obtain an answer that is reliable and useful, the engineer must recognize when approximations are being made and then be able to estimate their effect on the answer. If we exclude accidental errors, such as misreading a scale or writing an incorrect FORTRAN statement, there are essentially three sources of approximation:

1. Simplifying assumptions made in setting up the mathematical model, such as the use of the ideal-gas law in the derivation of an equation for the frictional resistance due to the flow of a real gas
2. Inexactness in the data, because of limitations in measuring instruments
3. Approximations in the numerical solution of the mathematical model

We shall be concerned in this chapter mainly with the effect of the second and third of these approximations on calculated results. Some of the basic definitions and ideas relating to approximate calculations will be presented first, followed by a discussion of methods for estimating the accuracy of numerical results.

5.2   SIGNIFICANT FIGURES

The accuracy of a measured quantity is limited to the size of the smallest division of the measuring instrument.   For example, suppose that the weight of a carload of coal is 51.5 tons, as determined by a heavy-duty scale graduated to 0.1 ton, whereas the weight of a sample of the coal, taken for chemical analysis, is 2.3562 g, as found on a sensitive pan balance graduated to 0.0001 g.   Both measurements are assumed to be made carefully, with the reading taken to the nearest scale division.   Neither can be said to be *exact*, but the second value can be relied on to five figures and is therefore said *to be good to five significant figures*, while the first is good to only three figures.

Suppose now that the carload weight is converted from tons to grams by multiplying by 2000 lb/ton and 453.5924 g/lb, yielding 46,720,017.20000 g. Obviously, the measurement did not gain in accuracy by a conversion of units, and therefore the above number is misleading.   It should still be reported to only three significant figures, such as 4.67×10^7 g.

The decimal digits 1, 2, . . . , 9 are always considered to be significant. The digit 0 is also significant except when it is used merely to fix the position of the decimal point in a number.   For example, there are four significant figures in each of the following: 12.34, 2.350, 600.1, and 0.0006057.   It is not always clear whether or not a zero is significant, however, as in the number 12,370.   To avoid this difficulty, the number may be written as 1.237×10^4 or 1.2370×10^4, showing four or five significant figures, respectively.

5.3   ROUNDING AND TRUNCATION

In the above conversion of the carload weight, the factor of 453.5924 g/lb was used, good to seven significant figures.   Since the measured weight was good to only three significant figures, it was not necessary to use such an accurate value of the conversion factor to obtain the desired result. In fact, only three or four digits are needed.   Just using the leftmost digits as they stand is not the best procedure, however.   Instead, the number should be *rounded* according to the following rule: If the digits being discarded are 500 · · · or greater, the rightmost digit being retained should be increased by 1; otherwise it should not be changed.   This is equivalent to reading the value on a coarser scale, still to the nearest division.   Thus, the conversion factor rounded to four digits is 453.6 and to three digits is 454.   If the four-digit value is used in the calculation of the carload weight, the result is 46,720,800.00; when rounded to three significant figures, this gives the same answer as before, 4.67×10^7 g.

Truncation is the process of discarding unwanted digits without rounding the result.   The error, or difference between the approximate value and the exact value, can be considerably greater than with rounding.   For example, 1.9999 rounds to the integer 2 but truncates to the integer 1. On the average, the error in the last digit retained by truncation will be twice as large as by rounding.

The use of a finite number of terms from an infinite series expansion of a function is also a process of truncation.   Thus, the Taylor's series for $\cos x$ is

$$\cos x = 1 - \frac{x^2}{2!} + \frac{x^4}{4!} - \cdots + (-1)^{n-1} \frac{x^{2n-2}}{(2n-2)!}$$

and for $x \neq 0$ the exact sum of the series must be approximated by truncating the series to a finite number of terms.   Further truncation or rounding must then be applied in the calculation of each of the terms retained.   The overall accumulated error becomes difficult to determine.

In the conversion of a floating-point number to an integer in FORTRAN, the number is truncated, possibly causing an excessive error.   Rounding can be accomplished in this case by the addition of 0.5 to the floating-point value before the truncation, with due regard to the sign, i.e.,

I = SIGN(ABS(X)+0.5, X)

If the number is known to be positive, the simpler form

I = X + 0.5

may be used.

The conversion of decimal fractions to binary numbers is an important (and sometimes overlooked) source of truncation error.   For example, 0.2 (base 10) becomes a repeating fraction in binary,

0.001 100 110 011 001 100 110 011 $\cdots$

and all other decimal fractions except those which are combinations of 0.5, 0.25, 0.125, etc., behave similarly.   Thus, a constant in a FORTRAN statement or a piece of data may not be represented exactly in a binary machine.   Consequently, addition of fractions in a binary machine will not, in general, give the same results as in a decimal machine.   Furthermore, it is not possible to round to fractional decimal digits in a binary machine.

5.4   ABSOLUTE, RELATIVE, AND PERCENTAGE ERROR

Letting x^* be the exact value of a number and x the approximate value, the *absolute error* of x is defined as

$$e_a = x - x^*$$

and has the units of x.   The *relative error* of x is equal to the absolute error divided by x^*, provided that $x^* \neq 0$, or

$$e_r = \frac{e_a}{x^*} = \frac{x - x^*}{x^*}$$

and since it is a ratio, it has no units.   The *percentage error* is simply the relative error multiplied by 100.   All three methods of expressing error are commonly used; it is usually simple to convert from one form to another if desired.   In some instances the absolute error is of the greatest interest, as in a calculation of the likelihood of a rocket to the moon reaching its target.

As an example, let us calculate the error made by truncating the value of π to four digits.   Since we still must truncate π to get a value to be used as x^* in this calculation, the result is approximately

$$e_a = 3.141 - 3.1415927 = -0.0005927$$
$$e_r = \frac{-0.0005927}{3.1416} = -0.0001887$$

Per cent error $= -0.0001887 \times 100 = -0.01887$ per cent.

A difficulty with the use of the above formulas is that the exact value x^* either cannot be expressed, as in the above case, or is not known, as in the case of a measured quantity.   Returning to the example of the weight of the carload of coal, we can say only that the error is not greater than one-half the smallest scale division, assuming that the scale is perfectly accurate.   The error may be either positive or negative, and therefore $e_a = \pm 0.05$ ton.   The relative error must now be based on the measured weight instead of the exact weight, or

$$e_r \approx \frac{0.05}{51.5} = 0.000971 \qquad \text{which becomes rounded to } 0.001$$

Per cent error $= 0.001 \times 100 = 0.1$ per cent.

In general, the relative error and the percentage relative error should be expressed to no more significant figures than are given in the absolute error.   Thus, to say that the relative error in this measurement is 0.000971 would imply that the absolute error is known to three significant figures, which is not correct.

5.5   FUNCTIONS OF A SINGLE VARIABLE

Assume now that we have a measured value x, with an absolute error Δx, and that we are interested in knowing the error Δy in a quantity y which is a function of x.   In mathematical terms,

$$y = f(x) \tag{5.1}$$

and

$$y + \Delta y = f(x + \Delta x) \tag{5.2}$$

Differentiating Eq. (5.1) yields

$$\frac{dy}{dx} = f'(x)$$

and if Δx is sufficiently small, we may approximate the derivative by

$$\frac{\Delta y}{\Delta x} \approx \frac{dy}{dx} = f'(x)$$

giving

$$e_a = \Delta y = \Delta x \, f'(x) \tag{5.3}$$

Equation (5.3) can also be derived by expanding the right-hand side of Eq. (5.2) in a Taylor's series about x by the methods described in Chap. 10.   With the same assumption of sufficiently small values of Δx, only the first two terms of the series are needed, leading to the same final equation.

The relative error in y is

$$e_r = \frac{\Delta y}{y} = \Delta x \frac{f'(x)}{y} \tag{5.4}$$

As an example, assume that we wish to calculate the relative error in the function $y = x^2$ due to an error in x.   Direct application of Eq. (5.4) gives

$$\frac{\Delta y}{y} = 2 \frac{\Delta x}{x}$$

showing that the relative error in y is double that in x.   More generally, if $y = x^a$, then the relative error is

$$e_r = \frac{\Delta y}{y} = a \frac{\Delta x}{x} \tag{5.5}$$

5.6   FUNCTIONS OF SEVERAL VARIABLES

In the most general case, the calculated quantity y is a function of several independent variables x_i, each of which is subject to an error Δx_i. In functional notation,

$$y = f(x_1, x_2, \ldots , x_n) \tag{5.6}$$

and

$$y + \Delta y = f(x_1 + \Delta x_1, x_2 + \Delta x_2, \ldots , x_n + \Delta x_n) \tag{5.7}$$

Taking the total differential of y in Eq. (5.6), we obtain

$$dy = \frac{\partial f}{\partial x_1} dx_1 + \frac{\partial f}{\partial x_2} dx_2 + \cdots + \frac{\partial f}{\partial x_n} dx_n$$

where the notation $\partial f/\partial x_i$, called the *partial derivative* of f with respect to x_i, means that the derivative of the function is to be taken with respect to x_i alone, treating all the *other* x's as constants during the differentiation. Assuming again that Δx_i is sufficiently small, we may approximate the differential quantities, to get

$$e_a = \Delta y = \frac{\partial f}{\partial x_1} \Delta x_1 + \frac{\partial f}{\partial x_2} \Delta x_2 + \cdots + \frac{\partial f}{\partial x_n} \Delta x_n \tag{5.8}$$

The same result can be obtained by expanding the right-hand side of Eq. (5.7) in a Taylor's series about $(x_1, x_2, \ldots , x_n)$ and assuming that Δx_i is sufficiently small.

Equation (5.8) is sometimes referred to as a *general formula* for error estimation. Equation (5.3), which was derived for a single independent variable, is just a special case of Eq. (5.8).

If the function for y is given as a product of the form

$$y = \frac{\alpha x_1^a x_2^b}{x_3^c x_4^g} \tag{5.9}$$

the equation for dy is

$$dy = \alpha \left(\frac{a x_1^{a-1} x_2^b}{x_3^c x_4^g} dx_1 + \frac{b x_1^a x_2^{b-1}}{x_3^c x_4^g} dx_2 - \frac{c x_1^a x_2^b}{x_3^{c+1} x_4^g} dx_3 - \frac{g x_1^a x_2^b}{x_3^c x_4^{g+1}} dx_4 \right) \tag{5.10}$$

Substituting Eq. (5.9) into Eq. (5.10) yields

$$dy = \frac{ay}{x_1} dx_1 + \frac{by}{x_2} dx_2 - \frac{cy}{x_3} dx_3 - \frac{gy}{x_4} dx_4$$

or, approximately,

$$e_r = \frac{\Delta y}{y} = a \frac{\Delta x_1}{x_1} + b \frac{\Delta x_2}{x_2} - c \frac{\Delta x_3}{x_3} - g \frac{\Delta x_4}{x_4} \tag{5.11}$$

Thus, the relative error in y is the *algebraic sum* of the relative errors due to the several independent variables. Since the individual errors may be either positive or negative, the sum of the *absolute values* should be taken to give the maximum estimated error in y.

If a function contains an independent variable in more than one term, care must be taken to collect the coefficients of the error in that variable before calculating the total error. For example, if

$$y = ax_1 + bx_2 - cx_1{}^3$$
$$dy = a\,dx_1 + b\,dx_2 - 3cx_1{}^2\,dx_1$$
$$\Delta y \approx a\,\Delta x_1 + b\,\Delta x_2 - 3cx_1{}^2\,\Delta x_1$$

Although Δx_1 can be either positive or negative, it cannot be both at the same time. Therefore, the first and third terms should not be added in absolute values. By collecting terms first, this difficulty is avoided. Thus,

$$\Delta y = |(a - 3cx_1{}^2)\,\Delta x_1| + |b\,\Delta x_2|$$

5.7   PROPAGATION OF ERROR IN ARITHMETIC OPERATIONS

The basic arithmetic operations performed in a digital computer are addition, subtraction, multiplication, and division. The result of such operations will, in general, contain error because of error in the operands and rounding or truncation that occurs during the operation. If the operands are denoted as x_1 and x_2, with associated errors Δx_1 and Δx_2, a relation between these quantities and the error Δy in the result y may be derived by the application of Eq. (5.8).

For *addition* and *subtraction*

$$y = x_1 \pm x_2$$
$$dy = dx_1 \pm dx_2$$
$$e_a = \Delta y = \Delta x_1 \pm \Delta x_2$$

Again, the errors may be either positive or negative, and therefore the sum of the absolute values of the errors in x_1 and x_2 should be used in calculating Δy.

$$\Delta y = |\Delta x_1| + |\Delta x_2| \tag{5.12}$$

The relative error in y will depend on the relative magnitudes of x_1 and x_2 and whether the operation is addition or subtraction. For addition of two quantities that are nearly the same, the relative error is given by

$$e_r = \frac{\Delta y}{y} = \frac{|\Delta x_1| + |\Delta x_2|}{x_1 + x_2} \approx \frac{|\Delta x_1| + |\Delta x_2|}{2x_1}$$

which may be approximately the same as that of either x_1 or x_2. For subtraction, on the other hand, the relative error is

$$e_r = \frac{\Delta y}{y} = \frac{|\Delta x_1| + |\Delta x_2|}{x_1 - x_2}$$

If x_1 and x_2 are close together, the relative error in y will be much greater than that in x_1 and x_2.   As an example, suppose that a chemical compound containing carbon, hydrogen, and oxygen is analyzed for the percentages of two of those elements and the percentage of the third is determined by subtracting the sum of the two from 100.   If the analysis, with estimated errors, is

Carbon: 39.96 ± 0.05 per cent
Oxygen: 53.32 ± 0.05 per cent

the percentage of hydrogen by calculation is

Hydrogen: $100.00 - 93.28 = 6.72 \pm 0.10$ per cent

Although the percentage relative errors in the carbon and oxygen analyses are less than 0.15 per cent, that in the hydrogen is 1.5 per cent, an order of magnitude greater.   Practically, the analyses for the three elements would be made independently, rather than by the above calculation.   A comparison of the sum of the percentages with 100.00 would then provide an independent error check.

In the above example, the errors were assumed to be errors of measurement in the data.   The same kind of error analysis would apply, however, if the errors had come from round-off or truncation.   For example, in the calculation of

$$\log \frac{a}{b} = \log a - \log b$$

there may be a large error in the result even though a and b are known exactly.   Assuming that $a = 2.5000$ and $b = 2.4999$ and using five-place logarithms, we find

$$\log \frac{a}{b} = 2.39794 - 2.39792 = 0.00002$$

a result that is good to only one significant figure.

It is probably true that the subtraction of numbers that are close together is the most important source of error in computational processes. Wherever this is likely to occur, the programmer should try to avoid it by either a rearrangement of the order of the calculation or the use of a different algebraic form.   Unfortunately, there are no general rules for such procedures, and each problem has to be considered individually.   As an example, in the calculation of

$$y = \sqrt{a} - \sqrt{b}$$

where a and b are assumed to be exact and close to each other, there will be round-off or truncation errors in $\sqrt{a}$ and $\sqrt{b}$, leading to a larger error in y.   An equivalent form here is

$$y = \frac{a - b}{\sqrt{a} + \sqrt{b}}$$

which is less sensitive to the round-off error in the square roots.

For *multiplication*

$$y = x_1 x_2$$
$$dy = x_2 \, dx_1 + x_1 \, dx_2$$
$$e_a = \Delta y = |x_2 \, \Delta x_1| + |x_1 \, \Delta x_2|$$
$$e_r = \frac{\Delta y}{y} = \left| \frac{\Delta x_1}{x_1} \right| + \left| \frac{\Delta x_2}{x_2} \right| \tag{5.13}$$

Thus, the relative error in a product equals the *sum* of the relative errors of the quantities being multiplied.   If, for example, $x_1 = 58.6 \pm 0.05$ and $x_2 = 130.2 \pm 0.05$, the result is $y = 7630$, rounded to four figures.   The relative error is

$$e_r = \frac{0.05}{130.2} + \frac{0.05}{58.6} = 0.00038 + 0.00085 = 0.0012$$
$$\Delta y = e_r y = 0.0012 \times 7630 = 9.2$$

The product is, therefore, accurate to only three significant figures.

For *division*

$$y = \frac{x_1}{x_2}$$
$$dy = \frac{x_2 \, dx_1 - x_1 \, dx_2}{x_2{}^2}$$
$$e_a = \Delta y = \left| \frac{\Delta x_1}{x_2} \right| + \left| \frac{x_1 \, \Delta x_2}{x_2{}^2} \right|$$
$$e_r = \frac{\Delta y}{y} = \left| \frac{\Delta x_1}{x_1} \right| + \left| \frac{\Delta x_2}{x_2} \right| \tag{5.14}$$

As might have been suspected, the relative error for division is the same as for multiplication.   With the data of the previous example, the calculation becomes

$$y = \frac{130.2}{58.6} = 2.222 \qquad \text{rounded to four figures}$$
$$e_r = 0.0012$$
$$\Delta y = e_r y = 0.0012 \times 2.222 = 0.0027$$

Again the result is seen to be good to only three figures.   Hence, it should be written as 2.22.

In summary, and as an aid to memory, the *absolute error* in an *addition* or *subtraction* is the sum of the *absolute errors* of the operands, and the *relative error* in a *multiplication* or *division* is the sum of the *relative errors* of the operands.

5.8 ACCUMULATION OF ROUNDING ERROR

The methods described in the previous section may be used to estimate the error due to a single arithmetic operation. In practice, a program may involve thousands of operations, and the problem of estimating the error accumulated in the final result has no simple, or even guaranteed, solution. To take a relatively straightforward example, consider the addition of 10,000 decimal numbers, each of which has been rounded in the last digit. Since the maximum error in each number is 0.5 with respect to the last digit, the accumulated *maximum error* is 5000 units, indicating that the last four digits are now no longer significant. This is a very pessimistic estimate, however, for it assumes that the rounding errors are all maximal and are all in the same direction. Normally, of course, the errors tend to cancel, and the *probable error* would be much lower, as shown in Sec. 5.8.2.

5.8.1 Range Arithmetic

It is obviously desirable to have the computer do the work of keeping account of the accumulation of error, if possible. One method that has been used involves the storing and use of a pair of *range numbers*, the high and low values, for each operand. For addition and subtraction, the pair is ordered algebraically; for multiplication and division, it is ordered in terms of absolute values. Letting $u = 57.2 \pm 0.04$ and $v = -23.4 \pm 0.05$ and letting the subscripts H and L symbolize high and low values, respectively, the results of the arithmetic operations on two variables u and v are

Addition:
$$y_H = u_H + v_H = 57.24 + (-23.35) = 33.89$$
$$y_L = u_L + v_L = 57.16 + (-23.45) = 33.71$$
Subtraction:
$$y_H = u_H - v_L = 57.24 - (-23.45) = 80.79$$
$$y_L = u_L - v_H = 57.16 - (-23.35) = 80.51$$
Multiplication:
$$y_H = u_H \times v_H = 57.24 \times (-23.45) = -1342$$
$$y_L = u_L \times v_L = 57.16 \times (-23.35) = -1335$$
Division:
$$y_H = u_H \div v_L = 57.24 \div (-23.35) = -2.452$$
$$y_L = u_L \div v_H = 57.16 \div (-23.45) = -2.437$$

Note that there will be an accumulation of rounding error in the calculation of the ranges, unless high values (in magnitude) are always rounded upward and low values are truncated. Range numbers are of limited utility in estimating an overall error because, as mentioned above, the probable error may be much lower than that indicated in their calculation.

As an example of a difficulty in the use of range numbers, the error will be calculated for the solution of the simultaneous equations

$$x + y = 3.00 \pm 0.01$$
$$2x + y = 4.00 \pm 0.02$$

The coefficients of the x and y terms are assumed to be exact; in general they would not be exact.

Solving the first equation for y yields

$$y = (3.01, 2.99) - x$$

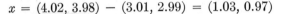

and substituting into the second equation gives

$$x = (4.02, 3.98) - (3.01, 2.99) = (1.03, 0.97)$$

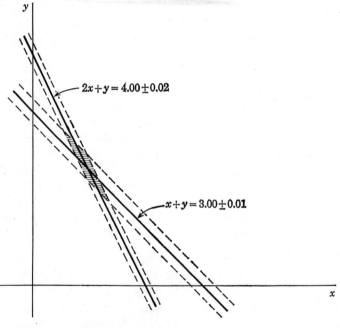

Fig. 5.1

From the first equation

$$y = (3.01, 2.99) - (1.03, 0.97) = (2.04, 1.96)$$

Although these results indicate the range in which the values of x and y must lie, they do not present the entire picture. A graph of the above equations in Fig. 5.1 shows, as expected, that the intersection of the lines is not a point. Instead, the *area* of intersection is a diamond, showing that the range for the two variables is not a simple combination of the ranges of the single variable.   For example, the maximum positive error in x is not associated with the maximum positive error in y.

5.8.2   Distribution of Rounding Error

The material presented here is intended to give only an intuitive idea of the possible statistical distribution of accumulated round-off error.   It is not complete, but it should demonstrate that the maximum possible error is very unlikely to occur.

Consider the addition of a group of positive numbers in a binary machine.   Assume that they have a round-off error that is either $+1$ or -1 in the last digit position.   These two states may be diagramed as

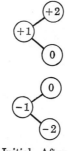

If, now, two numbers with the same round-off characteristics are added together, there are four possible combinations.   If the initial state of the first number was $+1$, the resulting state may be $+2$ or 0; if it was -1, the result is -2 or 0.   In diagram form,

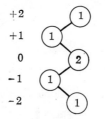

Initial   After
state   addition

There are, in reality, only three possible second states, with two ways to get to zero.   If the diagram is redrawn to show this, we have

$+2$ (1)
$+1$ (1)
0 (2)
-1 (1)
-2 (1)

where the states have been written outside at the left and the number in

each circle indicates how many paths there are to that state.   Carrying
the process two steps further produces

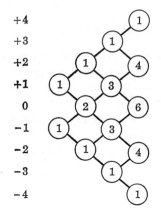

It is obvious that there is a pattern to the numbers in the circles.   In
this case, they are just the coefficients of the terms in the binomial expan-

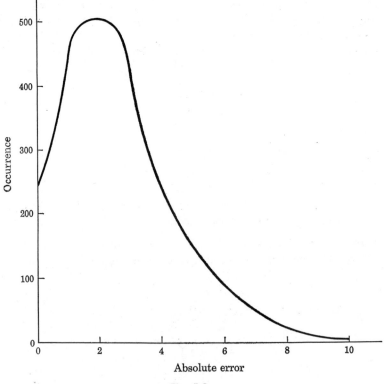

Fig. 5.2

sion of $(a + b)^n$, for $n = 1, 2, 3, \ldots$. Taking n as 10 yields the numbers 1, 10, 45, 120, 210, 252, 210, 120, 45, 10, 1. If the symmetrical pairs are combined, so as to present a distribution of absolute error, the result is that shown in Fig. 5.2. Although the maximum absolute error is 10 units, the probability of its occurrence is very small.

In Fig. 5.3 the data for 8, 12, and 16 steps have been recalculated by dividing by the respective total, putting the curves on the same basis of probability vs. absolute error. It can be seen that the probability of the rounding errors completely canceling decreases as the number of steps increases. The probable error increases, but not in direct proportion to the number of steps.

The process described above is that of a "random walk," in which each step is made either backward or forward on a straight-line path, without regard to the direction of a previous step. More complex processes can be considered, such as possible rounding errors of $+1$, 0, or -1 at each stage.

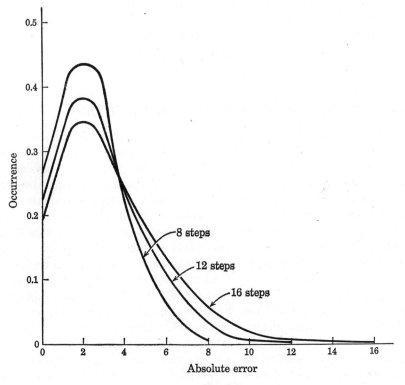

Fig. 5.3

A diagram for four stages of that process is as follows:

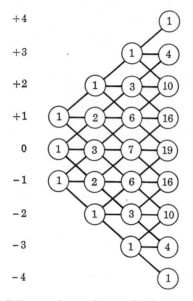

The coefficients are different from those which were derived above, but they confirm the previous qualitative conclusions.

5.9   RECOMMENDED CALCULATIONAL PROCEDURE

Two sources of approximation have been discussed in this chapter: errors arising from inexactness of measured values and errors arising from inexactness of numerical methods and calculations used in processing the data. Essentially, the two kinds of error affect results in the same way, and methods have been given by which an estimate can be made of the accumulated error. In many engineering calculations, however, the calculational error can be reduced to the point where it is not significant, making the overall analysis considerably simpler. If the total number of calculations is in the order of a few thousand, rounding and truncation errors will usually affect only the low-order two or three decimal digits of the answer, unless numbers that are nearly alike are being subtracted or numbers that are different by several orders of magnitude are being added. With single-precision arithmetic using eight digits, this still leaves a result that is sufficiently accurate for most purposes. If the total number of calculations is in the order of millions, however, it is probable that at least double-precision arithmetic will be needed.

In summary, then, the recommended procedure is to look for sensitive steps in the calculation and to consider alternative numerical methods where such exist.   If there is a question as to the effect of rounding and truncation errors in the calculation, a comparison of the results from runs made in single- and double-precision arithmetic will often be of help.

5.10   APPLICATIONS

The solution to virtually every engineering problem should include an estimate of the reliability of the results.   Thus, the possible applications of error analysis are very large in number.   Two typical examples are considered in this section: a problem in surveying and a problem in heat-exchanger design.

5.10.1   Area of a Polygon

The polygon $ABCD$ shown in Fig. 5.4 and the data of Table 5.1 represent a surveyed traverse; from these data it is desired to determine the area enclosed by the polygon and also to estimate the error in the result.

The north-south component of a given side is called its *latitude*.   The east-west component is called its *departure*.   If a north-south line through the most westerly point (point A) is taken as a reference line, the *double-*

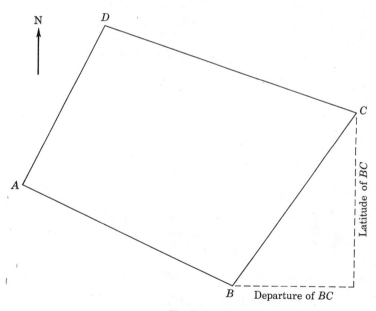

Fig. 5.4

meridian distance (DMD) of a side is defined as twice the distance from the reference line to the midpoint of the side. The area A of an n-sided

Table 5.1

Side	Bearing	Length
AB	S67°59′E	168.51
BC	N33°04′E	153.58
CD	N74°57′W	188.83
DA	S26°38′W	129.19

polygon is then given by the relation

$$A = \tfrac{1}{2} \sum_{i=1}^{n} (\text{DMD})_i (\text{latitude})_i \tag{5.15}$$

Positive directions are taken to be north for latitudes and east for departures. The signs can be taken care of automatically by changing the bearings to angles measured clockwise from north. Thus, if the bearing is northeast, there is no change; if it is southeast, it should be subtracted from 180°; if it is southwest, it should be added to 180°; and if it is northwest, it should be subtracted from 360°.

It will be assumed that the bearings are exact but that the measured lengths have a maximum error of 0.005 ft each. Letting subscripts 1, 2, 3, and 4 refer to sides AB, BC, CD, and DA, respectively, the following equations may be derived in terms of the corrected bearings:

Latitudes: $\quad \text{lat}_i = l_i \cos \alpha_i \qquad i = 1, 2, 3, 4$ $\hfill (5.16)$

Departures: $\quad d_i = l_i \sin \alpha_i \qquad i = 1, 2, 3, 4$ $\hfill (5.17)$

DMD's: $\quad \begin{aligned}[t] &\text{DMD}_1 = d_1 \\ &\text{DMD}_2 = \text{DMD}_1 + d_1 + d_2 \\ &\text{DMD}_3 = \text{DMD}_2 + d_2 + d_3 \\ &\text{DMD}_4 = \text{DMD}_3 + d_3 + d_4 \end{aligned}$ $\hfill (5.18)$

Substituting Eqs. (5.16) to (5.18) into Eq. (5.15) gives

$$\begin{aligned} A = \tfrac{1}{2}[& l_1{}^2 \sin \alpha_1 \cos \alpha_1 + (2l_1 \sin \alpha_1 + l_2 \sin \alpha_2)l_2 \cos \alpha_2 \\ & + (2l_1 \sin \alpha_1 + 2l_2 \sin \alpha_2 + l_3 \sin \alpha_3)l_3 \cos \alpha_3 \\ & + (2l_1 \sin \alpha_1 + 2l_2 \sin \alpha_2 + 2l_3 \sin \alpha_3 + l_4 \sin \alpha_4)l_4 \cos \alpha_4] \end{aligned}$$

Letting $c_{ij} = \sin \alpha_i \cos \alpha_j$ and rearranging, we have

$$\begin{aligned} A = \tfrac{1}{2}(& l_1{}^2 c_{11} + 2l_1 l_2 c_{12} + 2l_1 l_3 c_{13} + 2l_1 l_4 c_{14} + l_2{}^2 c_{22} \\ & + 2l_2 l_3 c_{23} + 2l_2 l_4 c_{24} + l_3{}^2 c_{33} + 2l_3 l_4 c_{34} + l_4{}^2 c_{44}) \end{aligned}$$

By applying Eq. (5.8), we get

$$\Delta A = c_{11}l_1\, \Delta l_1 + c_{12}(l_2\, \Delta l_1 + l_1\, \Delta l_2) + c_{13}(l_3\, \Delta l_1 + l_1\, \Delta l_3)$$
$$+ c_{14}(l_4\, \Delta l_1 + l_1\, \Delta l_4) + c_{22}l_2\, \Delta l_2 + c_{23}(l_3\, \Delta l_2 + l_2\, \Delta l_3)$$
$$+ c_{24}(l_4\, \Delta l_2 + l_2\, \Delta l_4) + c_{33}l_3\, \Delta l_3 + c_{34}(l_4\, \Delta l_3 + l_3\, \Delta l_4)$$
$$+ c_{44}l_4\, \Delta l_4$$

The coefficients of each error term must be collected. The result may be put in symmetric form as

$$\Delta A = (c_{11}l_1 + c_{12}l_2 + c_{13}l_3 + c_{14}l_4)\, \Delta l_1$$
$$+ (c_{12}l_1 + c_{22}l_2 + c_{23}l_3 + c_{24}l_4)\, \Delta l_2$$
$$+ (c_{13}l_1 + c_{23}l_2 + c_{33}l_3 + c_{34}l_4)\, \Delta l_3$$
$$+ (c_{14}l_1 + c_{24}l_2 + c_{34}l_3 + c_{44}l_4)\, \Delta l_4 \quad (5.19)$$

Because of the form of the equations for latitudes, departures, and DMD's, a computer program can readily be written that will calculate the area and the error in the area for an n-sided polygon.

5.10.2 Design of a Heat Exchanger

Figure 5.5 presents a diagram of a simple counterflow heat exchanger consisting of two concentric pipes. Such equipment is widely used in petroleum refineries and elsewhere to transfer heat from one fluid stream into another fluid stream. In the diagram, fluid A enters the inner pipe at $T_1°F$ and is cooled to $T_2°F$, while fluid B is heated from $t_2°F$ to $t_1°F$ in the annulus between the two pipes. The heat is transferred through the inner pipe wall; there is no mixing of the fluids.

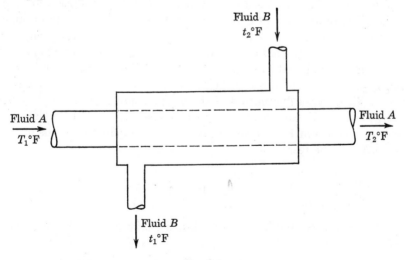

Fig. 5.5

If the flow rates and physical properties of the fluids are given, the *heat load* of the exchanger can be calculated from either of the two equations

$$q = w_A c_A (T_1 - T_2) = w_B c_B (t_1 - t_2) \tag{5.20}$$

where

q = heat load, Btu/hr
w = flow rate, lb/hr
c = heat capacity, Btu/(lb)(°F)

A part of the design of an exchanger is the calculation of the length that is needed to handle the given heat load. In the simplest equations that are used, the rate at which heat is transferred in the exchanger is considered to be proportional to the length of the exchanger and to the logarithmic mean temperature difference between the fluids. Thus,

$$q = U \pi D_i L \frac{\Delta_1 - \Delta_2}{\ln (\Delta_1/\Delta_2)} \tag{5.21}$$

where

U = coefficient of heat transfer, Btu/(hr)(ft²)(°F)
D_i = inner diameter of inner pipe, ft
L = length of exchanger, ft
Δ_1 = temperature difference at hot end $(T_1 - t_1)$
Δ_2 = temperature difference at cold end $(T_2 - t_2)$

In practice, the coefficient U depends on the geometry and material of construction of the exchanger, the flow rates of the fluids, and the physical properties of the fluids. For given conditions, it can be estimated from empirical formulas and has a usual reliability of 10 to 25 per cent.

The effect of an error in U on the calculated length is easily obtained, either by the use of Eq. (5.8) or by inspection of Eq. (5.21). Since the product UL must be held constant, the maximum length corresponds to the minimum predicted value of U. Other possibilities can be considered, such as the effect of an error in the measurement of the inlet temperature of fluid B (or just a fluctuation of the temperature with time).

If q, T_1, and T_2 are considered to be constant, the independent variables are t_1, t_2, and U. However, t_1 can be eliminated in terms of t_2 by Eq. (5.20). The result is

$$L = \frac{q \ln \dfrac{T_1 - (q/w_B c_B) + t_2}{T_2 - t_2}}{U \pi D_i [T_1 - (q/w_B c_B) - T_2]}$$

Applying Eq. (5.8) and letting α denote the argument of the ln function give

$$\Delta L = \frac{q}{U \pi D_i [T_1 - (q/w_B c_B) - T_2]} \left[- \ln \alpha \frac{\Delta U}{U} + \frac{(1 + \alpha) \Delta t_2}{\alpha (T_2 - t_2)} \right]$$

It may be observed that the error in U causes a like relative error in L. The effect of the error in t_2 depends on the magnitude of the difference $T_2 - t_2$. If that quantity is small, the second term can become very important.

PROBLEMS

5.1 Calculate the value of $\sqrt{102} - \sqrt{101}$, correct to five significant figures.

◆**5.2** The following quantities are in error, as indicated:

$u = 10.0 \pm 0.05$ $x = 12{,}300 \pm 100$

$v = 0.00252 \pm 0.00002$ $y = 62{,}000 \pm 500$

$w = 100 \pm 1$ $z = 7.61 \pm 0.03$

Determine the maximum values of the absolute and relative errors in the result of each of the following operations:

(a) $u + v + x + y$

(b) $u + 5x - y$

(c) y^3

(d) $\dfrac{u}{w}$

(e) uvw

(f) $\log{(u + z)}$

(g) $\log{(u - z)}$

◆**5.3** The velocity of a liquid through an orifice is given by the equation

$$V = C_o \sqrt{\frac{2g\,\Delta p}{\rho}}$$

where

V = velocity, ft/sec
C_o = orifice coefficient
g = gravitational constant, (ft)(lb mass)/(sec²)(lb force)
ρ = density of liquid, lb mass/ft³
Δp = pressure drop across orifice, lb force/ft²

The pressure drop Δp is usually calculated from a manometer reading by the formula

$$\Delta p = \frac{R}{12}(\rho_M - \rho)$$

where

R = manometer reading, in.
ρ_M = density of manometer fluid, lb mass/ft³

Calculate V and ΔV for the following conditions:

$C_o = 0.60 \pm 0.02$
$g = 32.2$
$\rho = 62.4$
$\rho_M = 848.6$
$R = 20.0 \pm 0.1$

5.4 Solve the following system of simultaneous equations.   Assuming that the numerical values have an error equal to one-half of the low-order digit, find the limits of the answer.

$1.00x + 2.00y = 6.00$
$1.00x - 1.00y = 0.00$

5.5 Repeat Prob. 5.4 for the equations

$1.00x + 2.00y = 6.00$
$1.25x + 1.00y = 3.25$

♦**5.6** The quadratic equation

$x^2 - 4x + 4 = 0$

is to be solved for two values of x.   Using range arithmetic and assuming that the above coefficients are 4.0 ± 0.05, calculate the ranges of the answers.

5.7 Referring to Sec. 5.8.2, derive the coefficients for the fifth step of a process in which the possible rounding errors are $+1$, 0, and -1.

5.8 Write a FORTRAN program to calculate the area of an n-sided polygon, given the bearings and lengths of the sides.   The program should also include a calculation of the error in the area.   Execute the program, using the data given in the example of Sec. 5.10.1.

♦**5.9** It is desired to determine the height h of the tower shown.   Surveyors make the following measurements: $h_1 = 12.28$ ft, $d = 473.22$ ft, and $\alpha = 32°44'$.

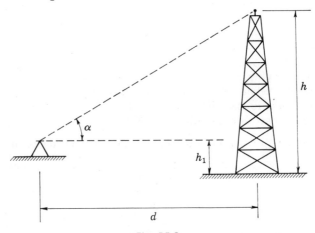

Fig. P5.9

Assuming that the tower is perfectly vertical, calculate h and the error in this calculation. Assume that the angle α is measured to the nearest minute and that the distances are measured to the nearest 0.01 ft.

5.10 A bridge is to be constructed across a river from A to C as shown. Surveyors measure a base line AB as 300.00 ft, angle BAC as 90°00′, and angle ABC as 72°24′. Determine the length AC of the bridge and the error in this length. (*Note:* Angle BAC may not be exactly a right angle.) Assume that the angles are measured to the nearest minute and that the distance AB is measured to the nearest 0.01 ft.

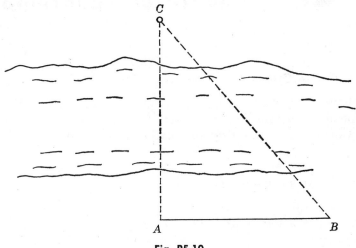

Fig. P5.10

5.11 A simple counterflow heat exchanger of the type shown in Fig. 5.5 is being operated under steady-state conditions. Measurements of the flow rates and temperatures are made, and these data, along with the heat capacities of the fluids A and B, permit a calculation of the heat-transfer coefficient U by means of Eq. (5.21). Derive an expression for the absolute error in U that is caused by errors in the temperature measurements. Assume that the other quantities entering into the calculation are exact.

ROOTS OF EQUATIONS

6.1 INTRODUCTION

The solving of an engineering problem may, in part, require the determination of the roots of some equation. *Roots* of an equation are defined as the values of x which satisfy an equation of the form

$$f(x) = 0 \qquad (6.1)$$

Examples of such equations are

$$x^2 + 3x + 2 = 0$$
$$x^6 - 2x^5 - 3x^4 + 16x^2 - 2x + 12 = 0$$
$$x - \tan x = 0$$

The first two equations are examples of *polynomial equations*, and the third is an example of a *transcendental equation*.

The *zeros* of a function $f(x)$ are equal to the roots of Eq. (6.1). A *function* is said to have *zeros*, and an *equation* is said to have *roots*. In general,

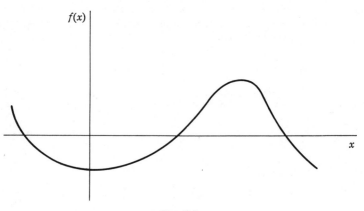

Fig. 6.1

equations can have roots which take on either real or imaginary values. The real roots of an equation can be considered in graphical terms. If a graph of the function $f(x)$ is plotted as shown in Fig. 6.1, the real roots are represented by the points where the curve crosses the x axis.

This chapter is concerned with the determination of roots of equations. One method of obtaining real roots is by actually plotting a graph of the function. Another very simple but often very effective method is trial and error. Other methods generally start with trial and error and then calculate successive trials by some analytical scheme. Sometimes the scheme converges to a root, and at other times it diverges from all the roots.

6.2 PROPERTIES OF POLYNOMIALS

An nth-order polynomial is an algebraic expression of the form

$$f(x) = a_1 x^n + a_2 x^{n-1} + \cdots + a_n x + a_{n+1} \tag{6.2}$$

where $a_1 \neq 0$. If the polynomial is equated to zero, there are n roots which satisfy the equation. The roots may be either *real* or *complex* as illustrated below:

$$x^2 + x - 6 = 0 \qquad \text{has roots } x = 2, \, x = -3$$
$$x^2 + 1 = 0 \qquad \text{has roots } x = i, \, x = -i \qquad \text{where } i = \sqrt{-1}$$

If the coefficients a_i are restricted to be real only, the complex roots must occur in pairs of the form

$$x = a + bi \qquad x = a - bi$$

Such complex numbers are known as *complex conjugates*. Also, the roots may be either *distinct* or *nondistinct*. The previous two examples illustrate distinct roots, while the following example illustrates nondistinct roots:

$$x^2 - 2x + 1 = 0 \qquad \text{has roots } x = 1, \, x = 1$$

The root $x = 1$ is said to be a root of multiplicity 2. If the roots of a polynomial are represented by r_i, the polynomial can be written in the factored form

$$f(x) = a_1(x - r_1)(x - r_2)(\cdot \cdot \cdot)(x - r_i)(\cdot \cdot \cdot)(x - r_n) \tag{6.3}$$

6.2.1 Descartes's Rule of Signs

This rule can sometimes be used to obtain the number of positive or negative roots of a polynomial. Descartes's rule is stated as follows:

> The number of positive roots of a polynomial is equal to, or less than by an even integer, the number of sign changes in the coefficients.
> The number of negative roots of a polynomial is equal to, or less than by an even integer, the number of sign changes in the coefficients, if x is replaced by $-x$.

The proof of this rule can be found in many algebra books.

As an illustration, consider the equation

$$x^3 - 23x^2 + 62x - 40 = 0$$

Since there are three sign changes $(+ - + -)$, the number of positive roots is either three or one. Replacing x by $-x$ gives

$$-x^3 - 23x^2 - 62x - 40 = 0$$

indicating no negative roots. Thus, the equation has either three real positive roots or one real positive root and two imaginary roots.

6.2.2   Newton's Relations

Another property of a polynomial may be derived from the equations which relate the roots of a polynomial to its coefficients. These equations are called *Newton's relations*. They are useful in determining an upper bound for the largest root (absolute value) of a polynomial if it is known that all the roots are real.

Suppose that an nth-order polynomial $f(x)$ is expressed in the form of Eq. (6.3),

$$f(x) = a_1(x - r_1)(x - r_2)(\cdot \cdot \cdot)(x - r_n)$$

Consider the multiplication of the first four factors, excluding a_1,

$$(x - r_1)(x - r_2) = x^2 - (r_1 + r_2)x + r_1 r_2$$
$$(x - r_1)(x - r_2)(x - r_3)$$
$$= x^3 - (r_1 + r_2 + r_3)x^2 + (r_1 r_2 + r_1 r_3 + r_2 r_3)x - r_1 r_2 r_3$$
$$(x - r_1)(x - r_2)(x - r_3)(x - r_4) = x^4 - (r_1 + r_2 + r_3 + r_4)x^3$$
$$+ (r_1 r_2 + r_1 r_3 + r_1 r_4 + r_2 r_3$$
$$+ r_2 r_4 + r_3 r_4)x^2 - (r_1 r_2 r_3 + r_1 r_2 r_4 + r_2 r_3 r_4)x$$
$$+ r_1 r_2 r_3 r_4$$

By using inductive reasoning, Newton's relations for the coefficients of the polynomial

$$f(x) = a_1 x^n + a_2 x^{n-1} + \cdot \cdot \cdot + a_n x + a_{n+1}$$

are

$$\frac{a_2}{a_1} = -\text{sum of the roots}$$

$$\frac{a_3}{a_1} = \text{sum of the products of the roots two at a time}$$

$$\frac{a_4}{a_1} = -\text{sum of the products of the roots three at a time} \tag{6.4}$$

$$\cdots \cdots \cdots \cdots \cdots \cdots \cdots \cdots \cdots$$

$$\frac{a_{n+1}}{a_1} = (-1)^n \,(\text{product of the roots})$$

To determine an upper bound of the roots (assuming that all are real), consider the following:

$$r_1^2 + r_2^2 + \cdots + r_n^2 = (r_1 + r_2 + \cdots + r_n)^2$$
$$- 2(r_1 r_2 + r_1 r_3 + r_1 r_4 + \cdots + r_1 r_n$$
$$+ r_2 r_3 + r_2 r_4 + \cdots + r_2 r_n$$
$$+ r_3 r_4 + \cdots + r_3 r_n$$
$$\cdots \cdots \cdots \cdots \cdots \cdots$$
$$+ r_{n-1} r_n)$$

Using Newton's relations,

$$r_1^2 + r_2^2 + \cdots + r_n^2 = \left(\frac{a_2}{a_1}\right)^2 - 2\left(\frac{a_3}{a_1}\right)$$

and it follows that

$$r_{\max}^2 \le r_1^2 + r_2^2 + \cdots + r_n^2$$

$$|r|_{\max} \le \sqrt{\left(\frac{a_2}{a_1}\right)^2 - 2\frac{a_3}{a_1}} \tag{6.5}$$

Consider again the equation

$$x^3 - 23x^2 + 62x - 40 = 0$$

The upper bound is

$$|r|_{\max} \le \sqrt{(-23)^2 - 2(62)}$$
$$\le 20.12$$

6.2.3   Reducing a Polynomial

If one root r_1 of a polynomial is known, a reduced polynomial is obtained from the polynomial division

$$\frac{f(x)}{x - r_1}$$

This operation is useful in determining all the roots of a polynomial equation. After one root is determined, the polynomial can be reduced and a second root determined from this reduced polynomial. This process of determining a root, reducing the polynomial, etc., can be continued until the reduced polynomial becomes a quadratic of the form

$ax^2 + bx + c$

The quadratic formula

$$x = \frac{-b \pm \sqrt{b^2 - 4ac}}{2a}$$

can then be utilized to obtain the last two roots.

6.3 PROPERTIES OF TRANSCENDENTAL FUNCTIONS

A *transcendental function* is defined as a function whose value cannot be determined for any given argument by a finite number of additions, subtractions, multiplications, divisions, and extractions of roots. Trigonometric, logarithmic, exponential, and hyperbolic functions are examples.

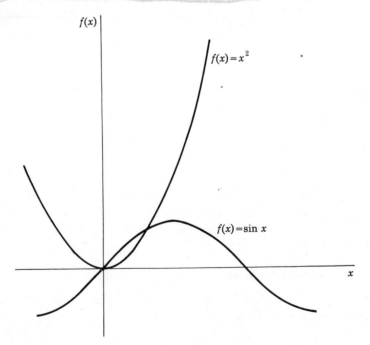

Fig. 6.2

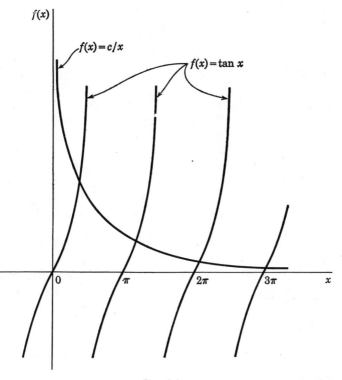

Fig. 6.3

An equation containing transcendental functions is called a *transcendental equation*.   Examples are

$x - \cos x = 0$
$x^2 - \sin x = 0$
$x \tan x = c$

Since transcendental equations can take many forms, few general properties can be stated.   They can possess both real and imaginary roots. There may be a finite number of real roots, as in the case of $x^2 - \sin x = 0$, illustrated in Fig. 6.2, or an infinite number of real roots, as in the case of $x \tan x = c$, illustrated in Fig. 6.3.   Graphing is probably the best method of determining the number and approximate locations of the real roots.

A transcendental equation can be approximated by a polynomial equation by replacing the transcendental function with an approximating polynomial.   Methods of accomplishing this are given in Chaps. 8 and 10. However, the methods of the present chapter generally apply to both polynomial and transcendental equations, and thus replacing a transcendental function with an approximating polynomial is seldom feasible.

6.4   SEARCHING

A good first step in determining the roots of an equation is to make a search for the number and approximate location of the real roots. The first consideration in making the search is to choose an interval along the x axis. If a polynomial possessing all real zeros is considered, the upper bound of all the zeros, given by Eq. (6.5), can be utilized in choosing an interval. Even if it is not known that all the zeros are real, this still may be a convenient procedure. In other types of equations, an interval can be chosen from a graph of the equation.

After an interval is chosen, the equation is evaluated at incremented values of x until two successive points with functional values of opposite sign are found. One root must lie between these two points.

For example, it has already been determined that the equation

$$x^3 - 23x^2 + 62x - 40 = 0$$

has either one or three positive roots and no negative roots. Further, if it has three real roots,

$$|r|_{max} \leq 20.12$$

With this information, a reasonable interval might be $0 \leq x \leq 21$, with an increment $\Delta x = 3$. Table 6.1 gives functional values obtained by this search. It is noticed that a root lies between $x = 18$ and $x = 21$.

Table 6.1

x	$f(x)$
0	$- 40$
3	$- 34$
6	-280
9	-616
12	-880
15	-910
18	-544
21	$+380$

As a second example, consider the equation

$$x^2 - \sin x = 0$$

One root is obviously $x = 0$. Examining Fig. 6.2 shows a second root near $x = 1$. The results of a search between $x = 0$ and $x = 1$ with an

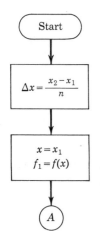

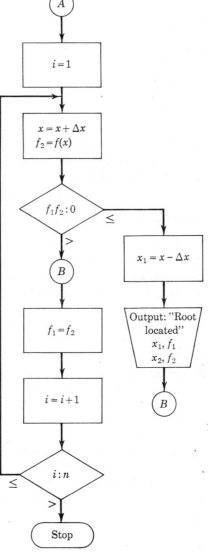

Fig. 6.4

increment of $\Delta x = 0.2$ is given in Table 6.2. Since $\sin x$ can never be greater than 1, the search need not be conducted past $x = 1$.

A flow chart for searching in an interval $x_1 \le x \le x_2$, which is divided into n equal increments, is presented in Fig. 6.4. Notice that the sign of the product of two successive functional values can be used to indicate

Table 6.2

x	$f(x)$
0.0	0
0.2	-0.159
0.4	-0.229
0.6	-0.205
0.8	-0.077
1.0	$+0.159$

whether or not a root lies between the two points. If this product is negative or zero, a root lies within that particular subinterval and a message is printed out to that effect.

6.4.1   Half-interval Method of Determining a Root

Once a subinterval containing a root of an equation has been found by a searching technique, the subinterval can be further subdivided to locate the root more closely. This process can be continued until the subinterval is so small that, for practical purposes, the root is determined.

One method of further subdividing an interval is called the *half-interval,* or *binary search, method.* This consists in evaluating the function at the midpoint of the interval and then retaining the half-interval containing the root. The method can be repeated until the root is obtained with the desired accuracy.

To illustrate, consider the equation

$$x^3 - 23x^2 + 62x - 40 = 0$$

It has been determined from Table 6.1 that a root lies between $x = 18$ and $x = 21$. At $x = 19.5$, the value of the function is -162 to three significant figures. Thus, the root lies between $x = 19.5$ and $x = 21.0$. The completion of the problem, which produces the root $r_1 = 20$, is given in Table 6.3.

Table 6.3

x_1	x_{midpt}	x_2	$f(x_1)$	$f(x_{midpt})$	$f(x_2)$
18	19.5	21	-544	-162	$+380$
19.5	20.2	21	-162	$+70$	$+380$
19.5	19.8	20.2	-162	-67	$+70$
19.8	20.0	20.2	-67	0	$+70$

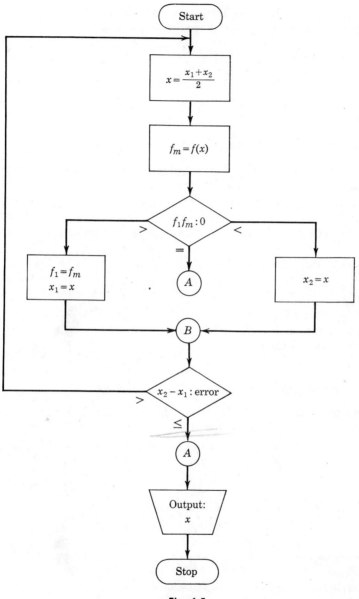

Fig. 6.5

To obtain the other two roots, the polynomial can first be reduced. This is easily done by means of synthetic division.

$$
\begin{array}{r r r r}
20) \quad 1 & -23 & 62 & -40 \\
 & 20 & -60 & 40 \\
\hline
1 & -\ 3 & 2 & 0
\end{array}
$$

The reduced equation is then

$$x^2 - 3x + 2 = 0 \quad \cdot$$

from which the remaining roots can be obtained by factoring or using the quadratic formula.   Their values are $r_2 = 2$ and $r_3 = 1$.

An alternative procedure of determining the second and third roots consists in continued use of the searching technique and the half-interval method.   The search given in Table 6.1 does not indicate any other real roots.   However, if the increment is reduced to $\Delta x = 1.5$, the search would indicate roots between 0 and 1.5, and between 1.5 and 3.   The half-interval method could then be applied to obtain the roots $r_2 = 2$ and $r_3 = 1$.

A flow chart of the half-interval method is given in Fig. 6.5.   Two initial points x_1 and x_2 along with the corresponding functional value f_1 are needed to initiate the flow.

6.4.2   Sources of Divergence

Two problems arise in connection with a search.   The first is that the interval may not be large enough to contain all or any of the roots. Second, the search may fail if two roots are close together and the increment Δx is too large.   The second case is illustrated graphically in Fig. 6.6 and numerically in Table 6.1.

Suppose that a search is made from $-\sqrt{(a_2/a_1)^2 - 2(a_3/a_1)}$ to $+\sqrt{(a_2/a_1)^2 - 2(a_3/a_1)}$ on an nth-order polynomial and that $n-4$ real zeros are counted.   There are then two possibilities:

1. The increment used in the search was not small enough to detect the other zeros.

2. The polynomial possesses at least two imaginary zeros.   The other two zeros could be either real or imaginary.   If they are real, they could be outside the interval used in the search, since the bounds used in the above search apply only in the case of all real zeros.

Consider, now, the technique of determining successive zeros of a polynomial from reduced polynomials.   The first zero to be determined pos-

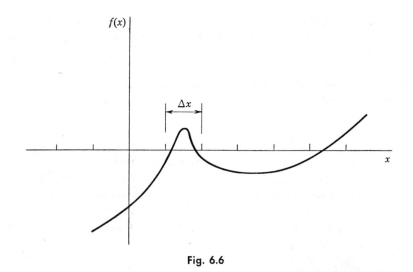

Fig. 6.6

sesses a small error, since the zero is generally determined to within only a tolerable interval. This error is propagated through division to the coefficients of the reduced polynomial. A similar error occurs in determining the second zero, in addition to the error caused by the inaccurate coefficients. Thus, an accumulated error can cause the zeros which are found last to contain a much larger error than that contained in the zero which is found first. Sometimes it may be necessary to improve the accuracy of the last few zeros. This can be done by using the calculated zeros as first trials in the half-interval method applied to the original polynomial.

This accumulated error occurs, not only in the half-interval method, but in any method which utilizes polynomial reduction in determining the zeros of a polynomial. As an illustration of the effect of round-off error, consider the polynomial

$$x^3 - 23x^2 + 62x - 40 = 0$$

which has the three real zeros 1, 2, and 20. If the initial estimate is given by Eq. (6.5), the above method will converge to the largest zero, 20. Assume, however, that the iteration is stopped at the approximate value 20.02, which is in error by 0.1 per cent. Reduction of the polynomial by synthetic division (with no further rounding error) gives the polynomial

$$x^2 - 2.98x + 2.3404 = 0$$

The quadratic formula may be used to find the remaining zeros, which are $1.49 \pm i$. In this case, a relatively small error in the first zero developed into a much larger error in the other zeros and even caused them to become complex. If a value of 19.98 had been used, the other zeros would have been 0.725 and 2.295.

Greater accuracy can be obtained in the case of the above polynomial and in similar cases where the zeros cover a large range in magnitude by making the substitution

$$w = \frac{1}{x}$$

The new equation is rearranged into polynomial form and solved for a zero. If Eq. (6.5) is again used to find the starting value, the iteration will converge to the largest zero in w, which will be the smallest in x. That value of x may be used to reduce the *original* polynomial by synthetic division, and then the process may be repeated on the reduced polynomial.

If this method is applied to the previous example, the first zero is $w = 1$. If we take it to be 1.001 so as to have the same relative error as before, the value of x is 0.999. The reduced polynomial is

$$x^2 - 22.001x + 40.021 = 0$$

which has zeros of 2.000 and 20.000. Obviously, the round-off error has been practically eliminated by the extraction of the zeros in the order of increasing magnitude.

6.5   SECANT METHOD

The method of determining a real zero of a function $f(x)$ using the *secant method* is best illustrated graphically. In Fig. 6.7, x_1 and x_2 represent two arbitrary trials, and $f(x_1)$ and $f(x_2)$ represent the functional values at x_1 and x_2, respectively. Point C on the straight line ABC is taken as a third trial x_3. A fourth trial x_4 is then determined by using points x_2 and x_3 in the same manner, a fifth point x_5 by using points x_3 and x_4, etc., until the root r_1 is obtained within the desired accuracy.

To illustrate the secant method, consider the equation

$$f(x) = x^3 - 23x^2 + 62x - 40 = 0$$

Using $x_1 = 21$ as a first trial gives the functional value $f(21) = 380$. Similarly, a second trial of $x_2 = 20.5$ gives $f(20.5) = 180$. Referring to Fig. 6.7, the value of e can be determined from similar triangles.

$$\frac{e}{180} = \frac{e + 0.5}{380}$$

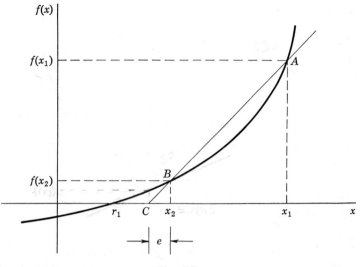

Fig. 6.7

From this

$$e = 0.45$$

and the third trial is obtained.

$$x_3 = 20.5 - 0.45 = 20.05$$

The process is continued until the root $r_1 = 20$ is reached, as illustrated in Table 6.4.   The second and third roots can be determined by the same procedure used in the example on the half-interval method in Sec. 6.4.1.

Table 6.4

x	$f(x)$	e
21.0	380	
20.5	180	0.45
20.05	17.14	0.04
20.01	3.422	0.01
20.00	0	

6.5.1   Interpolation, or False Position

In this method the first two trials are chosen in such a manner that the two functional values are of opposite sign.   The root of the equation

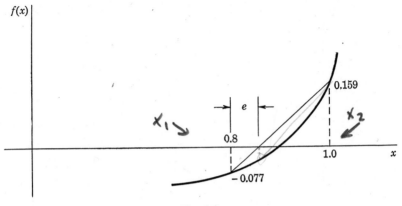

Fig. 6.8

lying between these two points can then be approximately obtained by
linear interpolation. Successive trials are made by always using two
points whose functional values are of opposite sign. This procedure is
also called the *method of false position*. As an example, consider the
equation

$$x^2 - \sin x = 0$$

Table 6.2 shows that a root lies between $x = 0.8$ and $x = 1.0$. Linear
interpolation between these two points is illustrated graphically in
Fig. 6.8. The value of e is obtained from

$$\frac{e}{0.077} = \frac{0.2 - e}{0.159}$$

which gives $e = 0.065$. The next trial is then

$$x = 0.800 + 0.065 = 0.865$$

The completion of the problem is given in Table 6.5, where the root
$r = 0.877$ is obtained.

Table 6.5

x	$f(x)$	e
0.8	−0.077	
1.0	+0.159	0.065
0.865	−0.126	0.010
0.875	−0.0018	0.002
0.877	0.0000	

If we let f_1 and f_2 be the functional values at successive points x_1 and x_2, respectively, a general equation for linear interpolation can be constructed from Fig. 6.8.

$$\frac{e}{|f_1|} = \frac{x_2 - x_1 - e}{|f_2|}$$

$$e = \frac{(x_2 - x_1)|f_1|}{|f_1| + |f_2|}$$

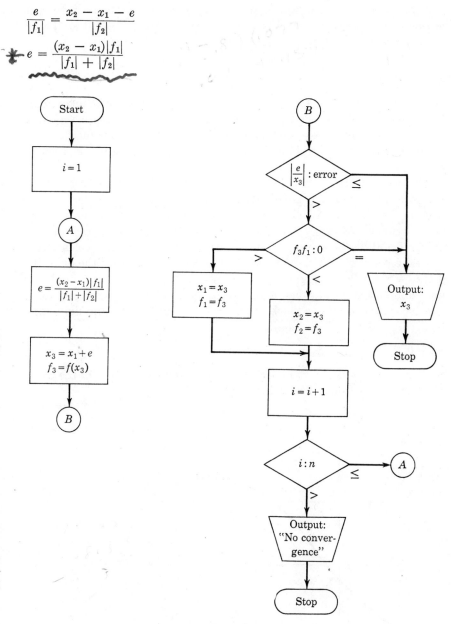

Fig. 6.9

A third trial x_3 is then obtained from

$$x_3 = x_1 + e$$

Figure 6.9 presents a flow chart for applying the method. The root is assumed to be accurately obtained when the change e is small compared with the approximation x_3. The quantity

$$\left| \frac{e}{x_3} \right|$$

is the relative error. A maximum number of iterations n is introduced to prevent an infinite loop in case of a program error or in case the relative error never becomes less than the prescribed amount. This might occur because of round-off errors or if the root happened to be in the vicinity of $x = 0$.

The use of relative error as defined above as a test for convergence is undesirable if there is a possibility that the denominator of the fraction may become zero. An alternative procedure for checking the relative error between two values x_1 and x_2 is based on the fact that a fixed number of significant digits is retained in the result of each computer calculation. For example, if x_1 and x_2 differ only in the last decimal digit, the quantity $0.1(x_1 - x_2)$ will be insignificant with respect to either x_1 or x_2 and will not change that value when added to it. Consequently, checking for a relative error of 10^{-7} in a machine that retains eight significant digits may be accomplished by checking to see if

$$x_2 + 0.1(x_1 - x_2) = x_2$$

Similarly, if it is desired to check for a relative error of 10^{-5} in the same machine, the factor to be used is 0.001.

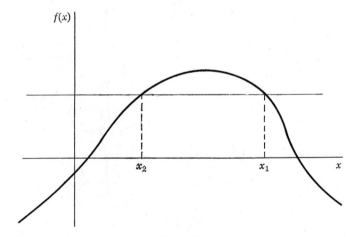

Fig. 6.10

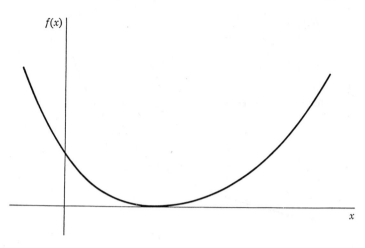

Fig. 6.11

6.5.2   Sources of Divergence

A special instance when the secant method does not converge is given in Fig. 6.10. Here, $f(x_1) \approx f(x_2)$, which produces a third trial near infinity. Using the method of false position eliminates this possibility but contains another source of divergence, as illustrated in Fig. 6.11. The figure shows a plot of a function which is tangent to the x axis. A polynomial with a zero of multiplicity n, where n is even, is an example of such a function. Quite obviously, the method of false position could not be used to determine such roots.

6.6   NEWTON'S METHOD

Newton's method of determining a root of an equation is illustrated graphically in Fig. 6.12. If x_1 represents an arbitrary first trial, a value for the second trial x_2 is obtained by constructing a line tangent to the curve representing $f(x)$ at x_1. The intersection of this tangent line with the x axis, as illustrated, is the second trial x_2.

By definition, the slope m of the tangent line is

$$m = \frac{f(x_1)}{x_1 - x_2}$$

from which

$$x_2 = x_1 - \frac{f(x_1)}{m}$$

From calculus it is known that the slope m is the derivative of the function

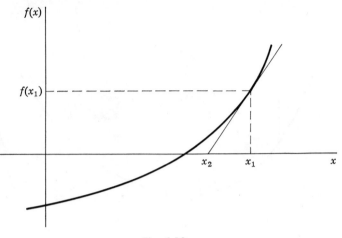

Fig. 6.12

at point x_1, which is symbolized by $f'(x_1)$.   Thus, Newton's equation for the second trial becomes

$$x_2 = x_1 - \frac{f(x_1)}{f'(x_1)} \tag{6.6}$$

Similarly to the previous methods, a third trial is gotten from the second, a fourth from a third, etc., until the root is obtained.

As an example, again consider the equation

$$f(x) = x^3 - 23x^2 + 62x - 40 = 0$$

The derivative of $f(x)$ is

$$f'(x) = 3x^2 - 46x + 62$$

Taking a first trial $x_1 = 21$,

$$f(21) = 380$$
$$f'(21) = 419$$
$$x_2 = 21 - \tfrac{380}{419} = 20.10$$

The fraction is rounded, since only four significant digits are retained. The complete process in determining the root $r_1 = 20$ is given in Table 6.6.

Table 6.6

x	$f(x)$	$f'(x)$	$f(x)/f'(x)$
21.0	380	419	0.9i
20.09	32.2	349	0.09
20.0	0		

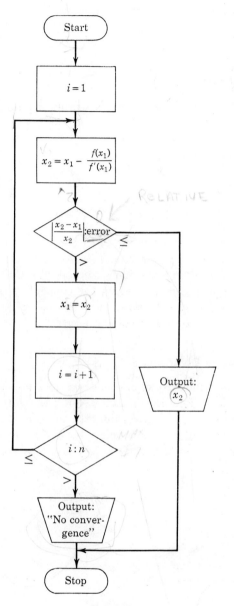

Fig. 6.13

As a second example, consider the equation

$$f(x) = x^2 - \sin x = 0$$

The derivative is

$$f'(x) = 2x - \cos x$$

and Newton's method is used as demonstrated in Table 6.7.

Table 6.7

x	$f(x)$	$f'(x)$	$f(x)/f'(x)$
1.000	0.159	1.460	0.109
0.891	0.017	1.154	0.014
0.877	0.000		

The method of applying Newton's method to a function $f(x)$ is given in the flow chart of Fig. 6.13. Again the concepts of relative error and limiting the number of iterations, as presented in the flow chart for linear interpolation (Fig. 6.9), are incorporated.

6.6.1   Sources of Divergence

Two instances in which Newton's method does not converge are illustrated in Fig. 6.14. The first instance shows that, if the slope $f'(x_1)$ is close to zero, the succeeding trial value will be near infinity. The second instance illustrates an infinite looping between two trials x_1 and x_2. Both these instances can generally be corrected by starting with a new trial x_n.

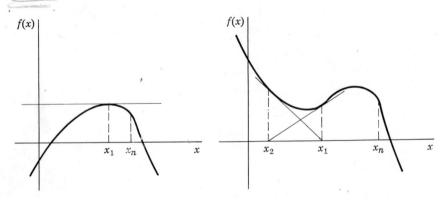

Fig. 6.14

6.7   APPLICATION OF SYNTHETIC DIVISION

In the example presented on the half-interval method, synthetic division was used to reduce the polynomial after one root had been determined. Synthetic division can be utilized with polynomials to a greater extent in both the method of linear interpolation and Newton's method. This is because of two important properties of polynomials. The first is that synthetic division can be used as an alternative method of determining functional values. Consider a polynomial $f(x)$ of degree n divided by a linear factor $x - x_1$.

$$\frac{f(x)}{x - x_1} = f_1(x) + \frac{R}{x - x_1}$$

The expression $f_1(x)$ is a polynomial of degree $n - 1$, and R is the remainder resulting from the division. Multiplying the equation by $x - x_1$ produces

$$f(x) = (x - x_1)f_1(x) + R$$

and it follows that

$$f(x_1) = R \tag{6.7}$$

Thus, the remainder R is equal to the functional value $f(x_1)$.

The second property is concerned with the derivative of the polynomial $f'(x)$. In the above derivation consider the division of $f_1(x)$ by $x - x_1$,

$$\frac{f_1(x)}{x - x_1} = f_2(x) + \frac{R_1}{x - x_1}$$
$$f_1(x) = (x - x_1)f_2(x) + R_1$$

The expression $f_2(x)$ is a polynomial of degree $n - 2$, and R_1 is the remainder resulting from the division. Substitution of the above expression for $f_1(x)$ into the expression for $f(x)$ gives

$$f(x) = (x - x_1)[(x - x_1)f_2(x) + R_1] + R$$

Taking the derivative of $f(x)$ gives

$$f'(x) = (x - x_1)[(x - x_1)f_2'(x) + f_2(x)] + [(x - x_1)f_2(x) + R_1]$$

and it follows that

$$f'(x_1) = R_1 \tag{6.8}$$

Thus, the remainder R_1 equals the derivative of the function at x_1, $f'(x_1)$.

The example presented in the section on Newton's Method will now be solved by applying synthetic division. This particular method is called the *Birge-Vieta method*. Using a first trial of $x_1 = 21$ on the equation

$$x^3 - 23x^2 + 62x - 40 = 0$$

synthetic division is applied in the following way:

$$
\begin{array}{r|rrrr}
21) & 1 & -23 & +62 & -40 \\
   &   & +21 & -42 & 420 \\
\hline
   & 1 & -2 & 20 & 380 = R = f(x_1)
\end{array}
$$

Applying synthetic division again,

$$
\begin{array}{r|rrr}
21) & 1 & -2 & 20 \\
   &   & 21 & 399 \\
\hline
   & 1 & 19 & 419 = R_1 = f'(x_1)
\end{array}
$$

The second trial is

$$x_2 = 21 - \tfrac{380}{419} = 20.1$$

The process is now continued,

$$
\begin{array}{r|rrrr}
20.1) & 1 & -23 & 62 & -40 \\
     &   & 20.1 & -58.29 & 74.6 \\
\hline
20.1) & 1 & -2.9 & 3.71 & 34.6 = f(x_2) \\
     &   & 20.1 & 345 & \\
\hline
     & 1 & 17.2 & 349 = f'(x_2) &
\end{array}
$$

$$x_3 = 20.1 - \frac{34.6}{349} = 20.0$$

Since division must be used anyway to reduce the polynomial after one zero has been determined, it seems plausible to utilize it to its fullest extent.

6.7.1 Flow Chart for Synthetic Division

Figure 6.15 represents a flow chart for performing the division

$$
\frac{a_1 x^n + a_2 x^{n-1} + \cdots + a_n x + a_{n+1}}{x - c}
$$

$$
= b_1 x^{n-1} + b_2 x^{n-2} + \cdots + b_{n-1} x + b_n + \frac{b_{n+1}}{x - c}
$$

Multiplying by the factor $x - c$ gives

$$
a_1 x^n + a_2 x^{n-1} + \cdots + a_n x + a_{n+1} = (x - c)
$$
$$
\times (b_1 x^{n-1} + b_2 x^{n-2} + \cdots + b_{n-1} x + b_n) + b_{n+1}
$$

Start

$b_1 = a_1$

$i = 2$

$b_i = a_i + c b_{i-1}$

$i = i + 1$

$i : n+1$

$\leq$

$>$

Stop

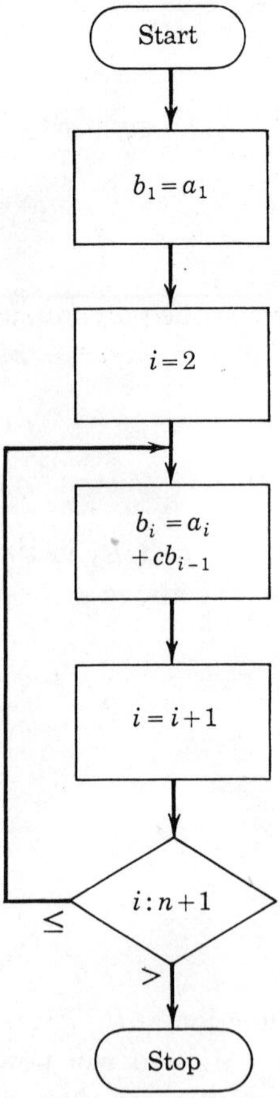

Fig. 6.15

and equating coefficients of like powers of x produces a sequence of expressions for the b_i's.

$b_1 = a_1$
$b_2 = a_2 + cb_1$
. . . .
$b_i = a_i + cb_{i-1}$
. . . .
$b_{n+1} = a_{n+1} + cb_n$

This method of obtaining the coefficients b_i is known as *Horner's method*.

6.8 ITERATION

In the previous methods the function $f(x)$ has always been considered in the form of Eq. (6.1),

$$f(x) = 0 \qquad (6.1)$$

The first step in an iteration scheme is to put the function in the form

$$x = F(x) \qquad (6.9)$$

After a first trial x_1, successive trials are obtained by substitution into $F(x)$. If we let x_2 be the second trial, iteration is represented algebraically as

$$x_2 = F(x_1)$$

As an example, consider again

$$x^3 - 23x^2 + 62x - 40 = 0$$

which can be rearranged to

$$x = 23 - \frac{62}{x} + \frac{40}{x^2} = F(x)$$

$x^3 = 23x^2 - 62x - 40$

$x = 23x^2 - \frac{62x}{x^2} - \frac{40}{x^2}$

With a first trial of $x = 21$, the following calculations illustrate the iteration scheme:

$x_1 = 21.0 \qquad F(21.0) = 23 - \dfrac{62}{21} + \dfrac{40}{21^2} = 20.1$

$x_2 = 20.1 \qquad F(20.1) = 23 - \dfrac{62}{20.1} + \dfrac{40}{20.1^2} = 20.0$

$x_3 = 20.0 \qquad F(20.0) = 23 - \dfrac{62}{20.0} + \dfrac{40}{20.0^2} = 20.0$

Root $= 20.0$

The above equation can also be put into the form

$$x = \frac{x^2}{23} + \frac{62}{23} - \frac{40}{23x}$$

$23x^2 = x^3 + 62x^2 - 40$

and assuming a first trial of $x_1 = 21$ produces the following results:

$$x_1 = 21.0 \qquad F(21.0) = \frac{21.0^2}{23} + \frac{62}{23} - \frac{40}{(23)(21.0)} = 21.8$$

$$x_2 = 21.8 \qquad F(21.8) = \frac{21.8^2}{23} + \frac{62}{23} - \frac{40}{(23)(21.8)} = 23.2$$

$$x_3 = 23.2 \qquad F(23.2) = \frac{23.2^2}{23} + \frac{62}{23} - \frac{40}{(23)(23.2)} = 26.1$$

$$x_4 = 26.1 \qquad F(26.1) = \frac{26.1^2}{23} + \frac{62}{23} - \frac{40}{(23)(26.1)} = 32.3$$

It is apparent that the process does not converge. However, if a first trial of $x_1 = 18$ is taken, the process does converge, as illustrated in Table 6.8. Note that two distinct roots have been obtained by using two separate iteration schemes.

Table 6.8

x	$F(x)$
18.0	16.7
16.7	14.7
14.7	12.0
12.0	8.78
8.78	5.85
5.85	3.89
3.89	2.90
2.90	2.46
2.46	2.25
2.25	2.14
2.14	2.08
2.08	2.05
2.05	2.03
2.03	2.01
2.01	2.00
2.00	2.00

As another example, consider again the equation

$$x^2 - \sin x = 0$$

Upon rearranging to

$$x = F(x) = \frac{\sin x}{x}$$

and taking $x = 1$ as a first trial, we see that the iteration quickly converges.

$$x_1 = 1.000 \qquad F(1.000) = \frac{\sin 1.000}{1.000} = 0.841$$

$$x_2 = 0.841 \qquad F(0.841) = \frac{\sin 0.841}{0.841} = 0.886$$

$$x_3 = 0.886 \qquad F(0.886) = \frac{\sin 0.886}{0.886} = 0.874$$

$$x_4 = 0.874 \qquad F(0.874) = \frac{\sin 0.874}{0.874} = 0.877$$

$$x_5 = 0.877 \qquad F(0.877) = \frac{\sin 0.877}{0.877} = 0.877$$

Root = 0.877

An alternative iteration scheme involves taking the square root of both sides of the equation,

$$x = F(x) = \sqrt{\sin x}$$

$$x_1 = 1.000 \qquad F(1.000) = \sqrt{\sin 1.000} = 0.917$$
$$x_2 = 0.917 \qquad F(0.917) = \sqrt{\sin 0.917} = 0.891$$
$$x_3 = 0.891 \qquad F(0.891) = \sqrt{\sin 0.891} = 0.882$$
$$x_4 = 0.882 \qquad F(0.882) = \sqrt{\sin 0.882} = 0.879$$
$$x_5 = 0.879 \qquad F(0.879) = \sqrt{\sin 0.879} = 0.877$$
$$x_6 = 0.877 \qquad F(0.877) = \sqrt{\sin 0.877} = 0.877$$

Root = 0.877

A flow chart for determining a root of an equation by iteration is given in Fig. 6.16.

6.8.1 Sources of Divergence

The criterion for convergence of an iteration scheme can be obtained by studying the graphs of $y = x$ and $y = F(x)$. Points of intersection of the two graphs represent the roots of the equation. In Fig. 6.17, the zigzag patterns represent graphically the iteration scheme. In both cases first trials x_1 are made on both sides of the root. In the first case there is convergence, and in the second case there is divergence. In the first case the slope $F'(x) < 1$ in the neighborhood of the root and in the second case $F'(x) > 1$ in the neighborhood of the root. This is also true if $F'(x)$ is negative; so it follows that the convergence criterion for an iteration on the equation $x = F(x)$ is that

$$|F'(x)| < 1 \tag{6.10}$$

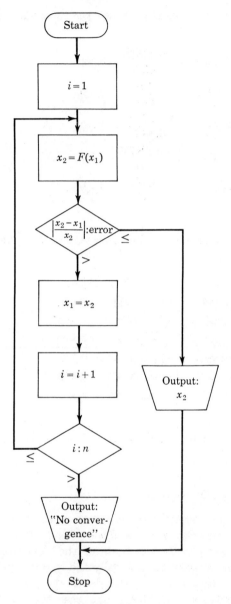

Fig. 6.16

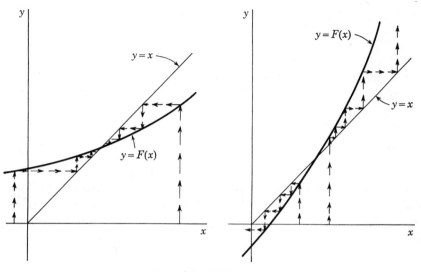

Fig. 6.17

in the neighborhood of the root. Of course, an iteration scheme may diverge from one root to converge on another root. This was illustrated in the last example.

The graph shown in Fig. 6.18 illustrates the convergence and divergence of the iteration method of the polynomial example just considered. Note that, starting with $x = 21$, the iteration scheme follows a diverging zigzag pattern, while, starting with $x = 18$, the scheme follows a zigzag pattern which converges to the root $x = 2$. Examination of the graph also shows that $x = 2$ is the only root to which this particular iteration scheme converges.

Failure to converge because of infinite looping between two points, as illustrated with Newton's method, is also possible in an iteration scheme. Such a case is illustrated in Fig. 6.19. Note that $|F'(x)| < 1$ at A and in the immediate vicinity of the root, although, at B, $|F'(x)| > 1$. A first trial x_1 closer to the root would probably start a converging sequence.

6.8.2 Comparison of Methods

All the methods of finding roots presented in this chapter are, in fact, iteration methods. Consider, first, Newton's method, whereby

$$F(x) = x - \frac{f(x)}{f'(x)}$$

$$F'(x) = \frac{f(x)f''(x)}{[f'(x)]^2}$$

and it follows that the convergence criterion is

$$\left| \frac{f(x)f''(x)}{[f'(x)]^2} \right| < 1$$

If x is sufficiently close to the root, then

$$f(x) \approx 0$$

and provided that

$$f'(x) \neq 0$$
$$f''(x) \neq \infty$$

the convergence criterion is satisfied.   Generally speaking, then, Newton's method is more likely to converge than the standard iteration technique.

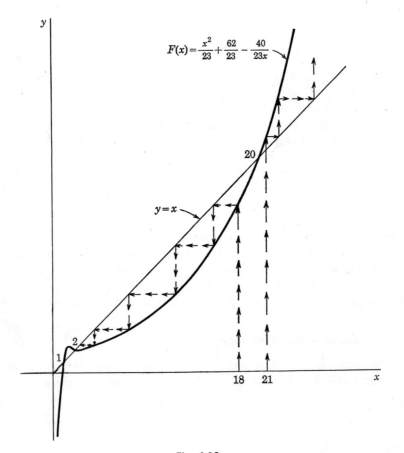

Fig. 6.18

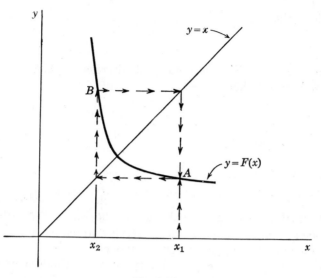

Fig. 6.19

Another criterion used in comparing methods is that of examining the *rate of convergence*. In Chap. 10 it is shown that a function can be expressed in its Taylor's-series form as

$$f(x) = f(x_1) + (x - x_1)f'(x_1) + (x - x_1)^2 \frac{f''(\xi)}{2}$$

where ξ lies within the (x,x_1) interval. Letting r be a root of the equation

$$f(r) = 0$$

gives

$$0 = f(x_1) + (r - x_1)f'(x_1) + (r - x_1)^2 \frac{f''(\xi)}{2}$$

Dividing by $f'(x_1)$ and rearranging produce

$$r = x_1 - \frac{f(x_1)}{f'(x_1)} - (r - x_1)^2 \frac{f''(\xi)}{2f'(x_1)}$$

Comparing this with Newton's equation shows that the error is

$$\text{Error} = -(r - x_1)^2 \frac{f''(\xi)}{2f'(x_1)}$$

If we let e_1 be the error in the trial x_1 and e_2 that in x_2, then

$$e_1 = r - x_1$$

and it follows that

$$e_2 \propto e_1{}^2$$

Thus, the error in Newton's method decreases *quadratically.* For example, if $e_2 \approx \frac{1}{2}e_1$, then

$$e_3 \approx \frac{e_1}{8}$$ ⟵ 4 power

$$e_4 \approx \frac{e_1}{128}$$

$$e_5 \approx \frac{e_1}{32,768}$$

This is, of course, neglecting the changes in the factor $f''(\xi)/2f'(x_1)$. In many problems, as borne out in the examples presented in Sec. 6.6, the error decreases in the manner described above.

The iteration of the secant and interpolation methods depends upon two previous trials, and thus the convergence criterion given by Eq. (6.10) does not apply. Considering the discussion in Secs. 6.5.2 and 6.6.1, it can generally be stated that the secant method has the same type of convergence difficulties as Newton's method. The interpolation method, on the other hand, will always converge.

To analyze the rate of convergence of these latter two methods, consider the equation $p(x)$ of the straight line passing through the two points $[x_1,f(x_1)]$ and $[x_2,f(x_2)]$.

$$p(x) = f(x_1) + \frac{f(x_1) - f(x_2)}{x_1 - x_2}(x - x_1)$$

The error in replacing a function $f(x)$ by a straight line is derived in Chap. 8 as

$$\frac{(x - x_1)(x - x_2)f''(\xi)}{2}$$

where ξ is a point in the interval containing x, x_1, and x_2. Thus,

$$f(x) = f(x_1) + \frac{f(x_1) - f(x_2)}{x_1 - x_2}(x - x_1) + \frac{(x - x_1)(x - x_2)f''(\xi)}{2}$$

At a root r of $f(x)$,

$$0 = f(x_1) + \frac{f(x_1) - f(x_2)}{x_1 - x_2}(r - x_1) + \frac{(r - x_1)(r - x_2)f''(\xi)}{2}$$

Multiplying this equation by

$$\frac{x_1 - x_2}{f(x_1) - f(x_2)}$$

and rearranging give

$$r = x_1 - \frac{x_1 - x_2}{f(x_1) - f(x_2)} f(x_1) + \frac{x_1 - x_2}{f(x_1) - f(x_2)} \frac{(r - x_1)(r - x_2)f''(\xi)}{2}$$

The first two terms on the right-hand side of the equation are equivalent to the secant and linear-interpolation methods given in Secs. 6.5 and 6.5.1. Thus, the error in these methods is

$$\frac{x_1 - x_2}{f(x_1) - f(x_2)} \frac{(r - x_1)(r - x_2)f''(\xi)}{2}$$

Using the mean-value theorem illustrated in Fig. 9.14,

$$f'(\xi_1) = \frac{f(x_1) - f(x_2)}{x_1 - x_2}$$

where ξ_1 is in the (x_1, x_2) interval, and it follows that the error is

$$\frac{(r - x_1)(r - x_2)f''(\xi)}{2f'(\xi_1)}$$

Thus, the error in a trial x_3 using the secant method is expressed in terms of the two previous errors,

$$e_3 \propto e_2 e_1$$

For comparison, assume that $e_2 \approx e_1$ (since the first two trials should statistically have the same error) and $e_3 \approx \frac{1}{2}e_1$. Then it follows that

$$e_4 \approx \frac{e_1}{4}$$

$$e_5 \approx \frac{e_1}{16}$$

$$e_6 \approx \frac{e_1}{128}$$

and the rate of convergence is slower than that of Newton's method.

As illustrated in Table 6.5, one of the points, say x_1, usually remains constant during a linear-interpolation iteration. In such a case,

$$e_3 \propto e_2$$

and by use of the same initial assumptions as in the secant method,

$$e_4 \approx \frac{e_1}{4}$$

$$e_5 \approx \frac{e_1}{8}$$

$$e_6 \approx \frac{e_1}{16}$$

Such a rate of convergence is called *linear convergence*.

Linear interpolation has the disadvantage that roots of an even multiplicity cannot be determined. Newton's method also has difficulties here, since $f'(x) = 0$. The convergence criterion

$$\left| \frac{f(x)f''(x)}{[f'(x)]^2} \right| < 1$$

is in an indeterminate form. Applying L'Hospital's rule twice gives

$$\left| \frac{f(x)f^{iv}(x) + 2f'(x)f'''(x) + [f''(x)]^2}{2\{f'(x)f'''(x) + [f''(x)]^2\}} \right| < 1$$

Near a double root, $f(x) \approx f'(x) \approx 0$, and on the assumption that the given higher derivatives are bounded in the region of the root, the above function reduces to $\frac{1}{2}$ and the convergence criterion is satisfied. It can be shown that Newton's method also converges for higher-ordered multiple roots.

Consider, now, the rate of convergence in such a case. The Taylor's series for $f'(x)$ is

$$f'(x) = f'(x_1) + (x - x_1)f''(\xi_1)$$

where ξ_1 is in the (x,x_1) interval. At $x = r$, the derivative is zero.

$$0 = f'(x_1) + (r - x_1)f''(\xi_1)$$
$$f'(x_1) = -(r - x_1)f''(\xi_1)$$

Substituting this last equation into the expression for the error in Newton's method gives

$$\text{Error} = (r - x_1)\frac{f''(\xi)}{2f''(\xi_1)}$$

which shows a *linear rate of convergence* for multiple roots.

In summing up, Newton's method converges faster than interpolation and somewhat faster than the secant method. Newton's method also has an advantage over the linear-interpolation method in that even-ordered multiple roots can be determined, although at a slower rate of convergence. On the other hand, for all other types of roots, linear interpolation does converge, while the secant method and Newton's method might diverge, as shown in Figs. 6.10 and 6.14. The half-interval method is similar to the linear-interpolation method but generally converges slower. This can be illustrated by solving the problem given in Sec. 6.5.1 by the half-interval method and comparing the number of iterations with the number used in Table 6.5. Sometimes it is desired to obtain a zero of a function whose derivative is not known. In such a case it would be necessary to approximate the derivative numerically in order to use Newton's method.

Considering the Taylor's series of $f(x)$ suggests another method of finding roots. By retaining another term in Taylor's series, a higher-ordered Newton's equation can be derived. The rate of convergence is cubic in this case. The derivation of the equation is left as an exercise for the student. Keeping additional terms in Taylor's series is useless, since the derivation of the iteration formula itself requires the determination of the root of a polynomial equation.

The rounding and truncation errors in all the methods presented here for finding a single root are minimal, since they reflect only the errors in the final iteration.

6.9 COMPLEX ROOTS

The methods presented so far have been concerned with determining real roots of an equation. Attention will now be focused upon determining complex or imaginary zeros of polynomials. Complex roots of transcendental equations are not considered in this text. All the methods of determining real roots can be utilized to determine imaginary roots, provided that complex arithmetic is used. As described earlier, complex roots occur in conjugate pairs of the form

$$a + bi \qquad a - bi$$

It then follows that after one root, $a + bi$, is determined, a second root is simply $a - bi$.

In order to avoid complex arithmetic, many methods of determining complex roots involve finding quadratic factors of the form

$$x^2 + px + q$$

An nth-order polynomial $f(x)$ with the zeros $a + bi$ and $a - bi$ can be expressed in the form

$$f(x) = (x - a - bi)(x - a + bi)f_1(x)$$

where $f_1(x)$ is a polynomial of degree $n - 2$. Multiplication of the two factors puts $f(x)$ in the form

$$f(x) = (x^2 + px + q)f_1(x)$$

where

$$p = -2a$$
$$q = a^2 + b^2$$

$(x - a - bi)(x - a + bi)$

Note that both p and q are real. After the quadratic factor is determined, the two zeros can be obtained upon application of the quadratic formula.

6.9.1   Existence of Complex Roots

Before presenting some methods of determining imaginary zeros, the problem of whether or not a given polynomial has imaginary zeros should be considered. Graphing a polynomial is an excellent way of locating approximately the real zeros. Suppose that the graph of a fourth-order polynomial $f(x)$ with four real zeros is drawn as shown in Fig. 6.20. The derivative $f'(x)$ is a third-order polynomial and therefore has three zeros. Between every two successive real zeros of $f(x)$ lies at least one of the real zeros of $f'(x)$. This fact, a special case of Rolle's theorem, is easily ascertained from the figure. It follows that the three zeros of $f'(x)$ are all real. Now, suppose that the graph of a fourth-order polynomial appears as shown by the dashed line in Fig. 6.20. Notice that, between x_1 and x_2, the derivative $f'(x)$ has three zeros. Since $f'(x)$ is a third-order polynomial, it has no other zeros. Use of Rolle's theorem then proves that x_1 and x_2 are the only real zeros of $f(x)$. Generally speaking, it follows that, if there is more than one zero of the derivative function $f'(x)$ between two real zeros of $f(x)$, there must be at least one pair of complex conjugate zeros. If a graph of a polynomial is constructed, the appearance of a configuration similar to that of the dashed curve in Fig. 6.20 indicates the presence of imaginary zeros.

Of course, if there is no "hump" in the graph, the polynomial can still have imaginary zeros. For example, assume that the graph of a fourth-

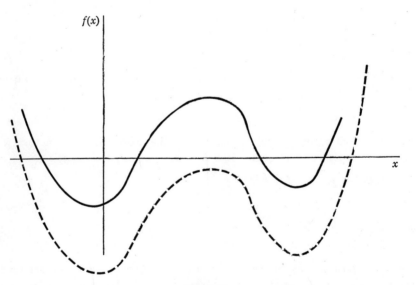

Fig. 6.20

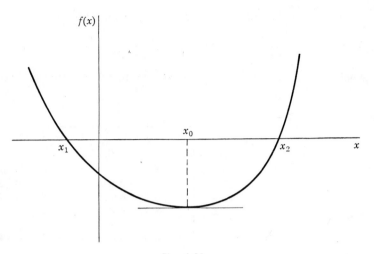

Fig. 6.21

order polynomial appears as shown in Fig. 6.21. On the assumption that there are no real zeros of $f(x)$ or $f'(x)$ outside the interval between x_1 and x_2, there exist two possibilities. The derivative $f'(x)$ either has a triple zero at x_0 or has two imaginary zeros. In either case the function $f(x)$ has two imaginary zeros.

6.9.2 Example Illustrating Use of Newton's Method

Newton's method will be utilized to determine one complex root of the equation

$$f(x) = x^4 - 6x^3 + 18x^2 - 24x + 16 = 0$$

The derivative of this polynomial is

$$f'(x) = 4x^3 - 18x^2 + 36x - 24$$

Sometimes a good first trial can be obtained by considering the first three terms of the polynomial as a quadratic equation. In this case

$$x^2 - 6x + 18 = 0$$

$$x = \frac{6 \pm \sqrt{36 - 72}}{2} = 3 \pm 3i$$

Upon taking $3 + 3i$ as a first trial, $f(x)$ and $f'(x)$ can be determined by direct substitution and use of complex arithmetic.

$$f(3 + 3i) = -56 - 72i$$
$$f'(3 + 3i) = -132$$

The second trial is then determined by Newton's method, from Eq. (6.6).

$$x_2 = 3 + 3i - \frac{-56 - 72i}{-132} = 2.58 + 2.45i$$

The process is continued as shown in Table 6.9. Thus, two roots of the

Table 6.9

x	$f(x)$	$f'(x)$	$f(x)/f'(x)$
$3 + 3i$	$-56 - 72i$	-132	$0.42 + 0.55i$
$2.58 + 2.45i$	$-17.6 - 20.3i$	$-60.1 - 2.98i$	$0.31 + 0.32i$
$2.27 + 2.13i$	$-5.25 - 4.79i$	$-30.2 - 4.51i$	$0.20 + 0.13i$
$2.07 + 2.00i$	$-1.27 - 0.58i$	$-18.7 - 6.18i$	0.07
$2.00 + 2.00i$	0		

equation are $2 + 2i$ and $2 - 2i$. The polynomial can now be put in the form

$$f(x) = (x - 2 - 2i)(x - 2 + 2i)f_1(x)$$
$$= (x^2 - 4x + 8)f_1(x)$$

where the polynomial $f_1(x)$ is determined from synthetic division.

```
4  -8)  1  -6     18   -24     16
              - 8   +16   -16
           4  - 8   + 8
         ─────────────────────────
         1  -2     2     0      0
```

Thus,

$$f_1(x) = x^2 - 2x + 2$$

from which the final two roots, $1 + i$ and $1 - i$, are determined upon use of the quadratic formula.

The flow chart for Newton's method given in Fig. 6.13 also applies to determining complex roots provided that complex arithmetic is used. The relative error in both the real and imaginary parts of a trial x_2 should be considered.

6.9.3 Lin's Method

Lin's method employs synthetic division by dividing the polynomial $f(x)$ by the trial factor $x^2 + p_1 x + q_1$. Algebraically, this is represented by

$$\frac{f(x)}{x^2 + p_1 x + q_1} = f_1(x) + \frac{R}{x^2 + p_1 x + q_1}$$

The division is stopped after $f_1(x)$ is determined, and the trial values p_1 and q_1 are replaced by two variables p_2 and q_2, respectively. These variables represent the coefficients of a new trial quadratic $x^2 + p_2x + q_2$ and are determined by setting the remainder R in the above division equal to zero. The process is continued until successive values of p_1, p_2 and q_1, q_2 show no significant change.

To demonstrate Lin's method of determining complex roots, consider the equation

$$x^4 - 6x^3 + 18x^2 - 24x + 16 = 0$$

Let the factor $x^2 - 1.33x + 0.89$ be a first trial. Application of the method using synthetic division gives

```
1.33  −0.89)    1    −6      +18       −24        +16
                           − 0.9     + 4.2      −10.9q₂
                    +1.33   − 6.2    −10.9p₂
                    1   −4.67   +10.9        0          0
```

$$-24 + 4.2 - 10.9p_2 = 0 \qquad p_2 = -1.82$$
$$16 - 10.9q_2 = 0 \qquad q_2 = 1.47$$

Now the second-trial quadratic factor is $x^2 - 1.82x + 1.47$. Again applying synthetic division results in

```
1.82  −1.47)    1    −6      +18       −24        +16
                           − 1.47    + 6.2      − 8.92q₃
                    +1.82   − 7.61    − 8.92p₃
                    1   −4.18   + 8.92        0          0
```

with new trial values determined as follows:

$$-24 + 6.2 - 8.92p_3 = 0 \qquad p_3 = 2.00$$
$$16 - 8.92q_3 = 0 \qquad q_3 = 1.79$$

This process is continued until there is no significant change in the p_i's and q_i's. Table 6.10 shows the completion of the problem. Thus, the

Table 6.10

p	q	Dividend
−2.00	1.79	$x^2 - 4.00x + 8.21$
−2.04	1.95	$x^2 - 3.96x + 7.96$
−2.05	2.01	$x^2 - 3.95x + 7.89$
−2.04	2.03	$x^2 - 3.96x + 7.97$
−2.01	2.01	$x^2 - 3.99x + 7.98$
−2.00	2.00	$x^2 - 4.00x + 8.00$

polynomial has been factored into the form

$$(x^2 - 2x + 2)(x^2 - 4x + 8) = 0$$

from which the roots are calculated from the quadratic formula to be

$$1 \pm i \qquad 2 \pm 2i$$

Often good first trial values are $p = 0$ and $q = 0$. Lin's method then produces

$$
\begin{array}{ccccccc}
0 & 0) & 1 & -6 & +18 & -24 & +16 \\
 & & & & 0 & 0 & -18q_1 \\
 & & & 0 & 0 & -18p_1 & \\
\hline
 & & 1 & -6 & +18 & 0 & 0
\end{array}
$$

$$-24 - 18p_1 = 0 \qquad p_1 = -1.33$$
$$16 - 18q_1 = 0 \qquad q_1 = 0.89$$

and the method would proceed as before.

6.9.4   Another Iteration Method

Lin's method is an example of an iteration scheme. To illustrate another iteration method, consider the equation

$$x^4 - 2x^3 + 3x^2 - 2x + 2 = 0$$

It is desired to obtain quadratic factors of the form $x^2 + Px + Q$ and $x^2 + px + q$. Thus, it follows that

$$
\begin{aligned}
x^4 - 2x^3 + 3x^2 - 2x + 2 &= (x^2 + Px + Q)(x^2 + px + q) \\
&= x^4 + (P + p)x^3 + (Q + Pp + q)x^2 \\
&\qquad\qquad + (Pq + Qp)x + Qq
\end{aligned}
$$

Since this is an identity in x, the corresponding coefficients must be the same on both sides of the equation. Therefore,

$$P + p = -2$$
$$Q + Pp + q = 3$$
$$Pq + Qp = -2$$
$$Qq = 2$$

The solution of these four simultaneous nonlinear algebraic equations can sometimes be obtained through iteration. Let $q = 4$ be a first trial. Then from the fourth equation $Q = \frac{1}{2}$, and the first and third equations become

$$
\left. \begin{aligned}
P + p &= -2 \\
4P + \tfrac{1}{2}p &= -2
\end{aligned} \right\} \qquad P = -\tfrac{2}{7}, \, p = -\tfrac{12}{7}
$$

The second equation can be written as

$$q = 3 - Pp - Q$$

and upon substitution,

$$q = 3 - (-\tfrac{2}{7})(-\tfrac{12}{7}) - \tfrac{1}{2} \approx 2$$

which differs from the assumed value of q. Taking 2 to be a new trial and repeating the process give

$$Q = 1$$
$$\left.\begin{array}{l} P + p = -2 \\ 2P + p = -2 \end{array}\right\} \quad P = 0, p = -2$$
$$q = 3 - 0 - 1 = 2$$

which checks. The equation has now been factored into the form

$$(x^2 + 1)(x^2 - 2x + 2) = 0$$

from which the roots $\pm i$ and $1 \pm i$ are readily determined.

Note that these two methods work equally well if the roots are real. Thus, both methods could be utilized to extract the zeros of any polynomial of even order. If the order of a polynomial is odd, there must be at least one real zero. In such a case a procedure such as Newton's method could be used to find one real zero, and then one of the methods discussed in this section could be used to obtain the rest of the zeros, be they real or imaginary.

As with the methods of finding real roots, there is no assurance that these methods of finding imaginary roots will always converge.

6.10   FLOW CHART FOR ROOTS OF A POLYNOMIAL

The general form of a polynomial equation can be written as

$$a_1 x^n + a_2 x^{n-1} + \cdots + a_n x + a_{n+1} = 0$$

It will be assumed that all the roots are real. In applying the Birge-Vieta method, two subroutines considered previously are utilized. The first is SUBROUTINE ROOTS (Sec. 3.10.4), which determines the two roots r_1 and r_2 of the quadratic equation

$$ax^2 + bx + c = 0$$

The arguments of the subroutine are a, b, c, r_1, and r_2. The second subprogram is called SUBROUTINE DIV (Fig. 6.15), which performs the

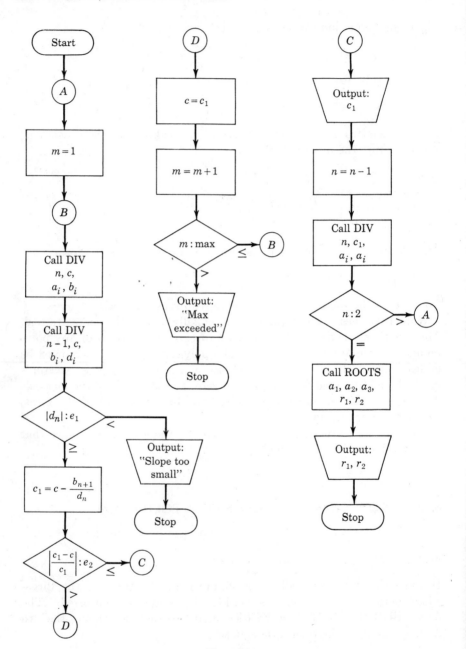

Fig. 6.22

division

$$\frac{a_1 x^n + a_2 x^{n-1} + \cdots + a_n x + a_{n+1}}{x - c} = b_1 x^{n-1} + b_2 x^{n-2}$$

$$+ \cdots + b_{n-1} x + b_n + \frac{b_{n+1}}{x - c}$$

The term b_{n+1} is the remainder obtained in the division. The arguments of the subroutine are n, c, the a_i's, and the b_i's. The division is utilized in determining the functional values of the polynomial and its derivatives, as well as in reducing the equation.

The flow chart for the solution of the problem is given in Fig. 6.22. An initial value of c must be provided. The purpose of e_1 is to prevent division by zero. As pointed out in Sec. 6.6.1, the method diverges when the slope is close to zero. The variable e_2 specifies the desired accuracy of the root. The variable *max* prevents the possibility of an infinite loop in the event that the method does' not converge to the desired root. When all but two of the roots have been determined, the quadratic formula (SUBROUTINE ROOTS) is used to obtain the last two roots. A computer program can be easily formed from the flow chart.

6.11   APPLICATIONS

Four engineering applications which require the determination of roots of equations will be discussed. The first is concerned with the flow of water in an open channel. The second analyzes an electrical circuit containing a varistor. The third considers the angular rotation of a heavy lever attached to the end of a circular bar. The final application considers the deflection of a flexible cable due to its own weight.

6.11.1   Flow of Water in an Open Channel

As an engineering application which requires the determination of a root of a polynomial, consider the following problem from the field of hydraulics: The figure for Prob. 3.48 illustrates the flow of water in an open channel. An empirical relation for the flow rate is the Chezy-Manning equation

$$Q = \frac{1.49}{n} A R^{\frac{2}{3}} S^{\frac{1}{2}} \tag{6.11}$$

where

Q = flow rate, ft³/sec
n = experimentally determined roughness coefficient, which is between
     0.025 and 0.035 for most canals and rivers
A = cross-sectional area of channel
R = "hydraulic radius," which is defined as area A divided by wetted
     perimeter P
S = slope of channel

Consider a rectangular channel, and assume that the quantities Q, n, S, and d are given. It is then desired to calculate y, the depth of flow. Substitution into Eq. (6.11) gives

$$Q = \frac{1.49dy}{n}\left(\frac{dy}{d+2y}\right)^{\frac{2}{3}}S^{\frac{1}{2}}$$

Cubing both sides of the equation, multiplying through by $(d+2y)^2$, and collecting terms produce

$$\left[\left(\frac{1.49}{n}\right)^3 d^5 S^{\frac{3}{2}}\right]y^5 - [4Q^3]y^2 - [4Q^3d]y - [Q^3d^2] = 0$$

which is a fifth-order polynomial in y. Since all the bracketed terms are positive, a check of Descartes's rule shows that the equation possesses only one positive root.

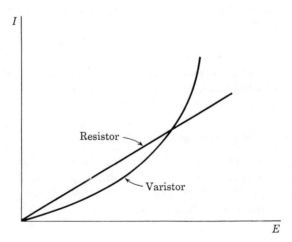

Fig. 6.23

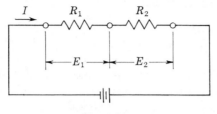

Fig. 6.24

6.11.2   Varistors

A second example makes use of Ohm's law, which is stated as

$$E = IR \tag{6.12}$$

where E, I, and R represent voltage, current, and resistance, respectively. If a graph is constructed by plotting I versus E, it is a straight line, as shown in Fig. 6.23. Several types of resistors, however, do not possess this linear property. Thus, the current may vary with voltage in a manner illustrated by the curve in Fig. 6.23. Such a resistor is called a nonlinear resistor, or a *varistor*. Most vacuum tubes are varistors. Generally, the relation between the current and the voltage for a varistor can be approximated by a polynomial of the form

$$I = a_1 E + a_2 E^2 + a_3 E^3 + \cdots + a_n E^n = \sum_{i=1}^{n} a_i E^i \tag{6.13}$$

Consider now a circuit consisting of a linear resistor R_1 and a varistor R_2, as shown in Fig. 6.24. The voltage applied to the circuit is E. The equations pertaining to such a circuit are

$$I = \frac{E_1}{R_1} = \sum_{i=1}^{n} a_i E_2{}^i$$
$$E = E_1 + E_2$$

Solving the first equation for E_1 and substituting into the second equation give

$$R_1 \sum_{i=1}^{n} a_i E_2{}^i + E_2 = E \tag{6.14}$$

which is a polynomial equation for E_2. After E_2 is determined, I and E_1 can be obtained from the first two equations.

6.11.3   Heavy Lever Attached to a Circular Bar

As an example of a transcendental equation derived from a physical situation, consider the problem illustrated in Fig. 6.25 of a heavy lever attached to the end of a circular bar. The weight of the lever causes a torque $T = Wr \cos \theta$, which is applied to the end of the circular bar. The end of the bar, along with the lever, rotates through the angle θ. If the circular bar is made of an elastic material, the torque is related to the angle θ by the equation

$$T = K\theta$$

where K is a proportionality constant which depends upon the dimensions and physical properties of the circular bar. Substituting the value of the torque into this equation produces

$$Wr \cos \theta = K\theta \tag{6.15}$$

which represents a transcendental equation to be solved for the angle θ.

6.11.4   Flexible Cables

Figure 6.26 illustrates a uniform cable, such as a transmission line, hanging from two supports and sagging because of its own weight. The shape

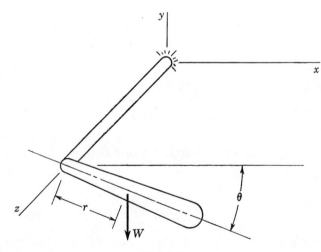

Fig. 6.25

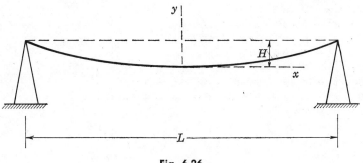

Fig. 6.26

of such a cable is a catenary, the equation of which is

$$y = \frac{T_0}{\mu}\left(\cosh\frac{\mu x}{T_0} - 1\right) \tag{6.16}$$

where T_0 is the tension in the cable at $x = 0$, μ is the weight per unit length of the cable, and cosh represents the hyperbolic cosine function. At $x = L/2$, $y = H$, and

$$H = \frac{T_0}{\mu}\left(\cosh\frac{\mu L}{2T_0} - 1\right) \tag{6.17}$$

Given H, μ, and L, Eq. (6.17) is a transcendental equation for the tension T_0 at the center of the cable. The length S of the cable is given by

$$S = \frac{2T_0}{\mu}\sinh\frac{\mu L}{2T_0} \tag{6.18}$$

where sinh represents the hyperbolic sine function.

PROBLEMS

The equations given in Probs. 6.1 to 6.15 possess only real roots. Determine the roots of these equations, using any of the methods described in the text.

  6.1 $x^2 + 3x - 18 = 0$

♦**6.2** $x^2 + 2.4x - 4.32 = 0$

6.3 $x^2 - 10.0x - 1.4 = 0$

◆**6.4** $x^3 - 20.6x^2 - 15.3x + 17.6 = 0$

6.5 $x^3 + 1.40x^2 - 6.72x + 4.32 = 0$

◆**6.6** $x^3 - 7x + 6 = 0$

6.7 $x^4 - 10x^3 + 35x^2 - 50x + 24 = 0$

◆**6.8** $x^4 - 95.4x^3 - 4800x^2 + 13,500x = 0$

6.9 $x^4 + 1.38x^3 - 8.98x^2 - 0.25x + 7.74 = 0$

◆**6.10** $x^2 = \cos x$

6.11 $x - \tan x = 0$

◆**6.12** $x \tan x = 1$

6.13 $xe^x = 1$

◆**6.14** $x \ln x = 1$

6.15 $x^2 = \sinh x$

6.16   Construct a convergence-divergence graph for the first iteration example given in Sec. 6.8.

6.17   Construct a third iteration scheme (the first two are given in Sec. 6.8) for the equation

$$x^3 - 23x^2 + 62x - 40 = 0$$

Which root is obtained by starting the iteration with a trial value $x = 21$? Construct a convergence-divergence graph for this scheme.

6.18   The equation

$$x^3 - 23x^2 + 62x - 40 = 0$$

can be rearranged as follows:

$$x^2 = 23x - 62 + \frac{40}{x}$$

$$x = \sqrt{23x - 62 + \frac{40}{x}}$$

Starting with a trial value of $x = 21$, iterate this equation until a root is reached.

The equations given in Probs. 6.19 to 6.24 contain at least one pair of complex roots.   Determine all roots of these equations, using any suitable method.

6.19 $x^3 - 7x^2 + 16x - 10 = 0$

◆**6.20** $x^3 + x^2 + 4.41x + 4.41 = 0$

6.21 $x^3 - 1.4x^2 + 0.48x + 2.88 = 0$

6.22 $x^4 - 12x^3 + 72x^2 - 240x + 400 = 0$

◆**6.23** $x^4 - 10.1x^3 + 37.0x^2 - 42.5x + 13.3 = 0$

6.24 $x^4 - 2x^3 + 2x - 1 = 0$

6.25   Find a complex root of

$$x^4 - 2x^3 + 3x^2 - 2x + 2 = 0$$

using Newton's method and complex arithmetic.   Then determine the rest of the roots.

6.26 Attempt to solve

$$x^4 - 2x^3 + 3x^2 - 2x + 2 = 0$$

using Lin's method. What happens?

6.27 Starting with a value of $q = 5$, determine the roots of

$$x^4 - 6x^3 + 18x^2 - 24x + 16 = 0$$

using the iteration method discussed in Sec. 6.9.4.

6.28 Construct a computer program which, if given the real roots, determines the coefficients of the corresponding polynomial. Assume that $a_1 = 1$.

6.29 Referring to Fig. 6.4, construct a computer program which searches for the real roots of a given equation.

6.30 Referring to Fig. 6.5, construct a computer program which determines a real root of a given function by using the half-interval method.

6.31 Referring to Fig. 6.9, construct a computer program which determines a real root of a given function by using the method of false position.

6.32 Referring to Fig. 6.13, construct a computer program which determines a real root of a given function by using Newton's method.

6.33 Revise the program in Prob. 6.32 so that a complex root could also be determined.

6.34 Referring to Fig. 6.16, construct a computer program which determines a real root of a given function by using iteration.

6.35 Construct a flow chart and computer program which determines a pair of complex roots of a polynomial by using Lin's method.

6.36 Referring to Fig. 6.22, construct a computer program which determines all the roots of a polynomial by using Newton's method.

6.37 Construct a computer program which determines the roots of a cubic polynomial of the form

$$x^3 + ax^2 + bx + c = 0$$

6.38 Determine the polynomial equation in y governing the flow in the channel illustrated in Prob. 3.48.

6.39 Given a linear resistor and a varistor in parallel, determine the equation for the voltage drop across the circuit in terms of the current I, resistance R_1, and the characteristics of the varistor R_2.

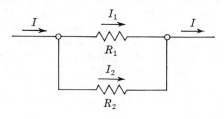

Fig. P6.39

◆**6.40**   The bar-and-lever system has a horizontal force P applied as shown. Determine the governing equation for the angle θ.   The physical constant of the bar relating the torque T to the angle θ is K.

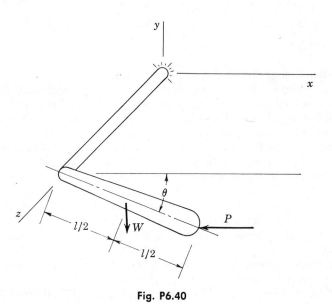

Fig. P6.40

6.41   Use the data given below to determine the depth of flow y in a rectangular channel of width $d = 25$ ft:

$Q = 500$ ft^3/sec
$S = 0.0001$
$n = 0.025$

◆**6.42**   Assume that the relation between the current I and the voltage E of a varistor is

$$I = 8.0 \times 10^{-9}E^3 + 2 \times 10^{-3}E$$

Referring to Fig. 6.24, determine the current I in the system if $E = 2000$ volts, $R_1 = 20$ ohms, and R_2 has the characteristics of the above equation.

6.43   In Prob. 6.40 determine the angle θ if $W = 5000$ lb, $P = 0$, and $l = 24$ in. A torque of 875 lb-in. is required to rotate the bar through $1°$.

6.44   In Prob. 6.40 determine the angle θ if $W = 5000$ lb, $P = 1000$ lb, and $l = 24$ in.   A torque of 875 lb-in. is required to rotate the bar through $1°$.

6.45   A transmission cable suspended from two towers weighs 25 lb/ft.   If the distance L between the towers is 2000 ft and the sag H is 300 ft, determine the tension T_0 at the center of the cable.   Also, calculate the total length S of the cable.

6.46 A telephone cable suspended from two poles weighs 5 lb/ft. If the distance between poles is 90 ft and the length of the cable is 100 ft, determine the sag H.

6.47 Derive a polynomial equation in x for the simultaneous equations

$$x^3 + y^2 = 9$$
$$xy^2 + yx^2 = 6$$

Then determine a solution for x and y.

6.48 Letting Δa_i be the error in the coefficient a_i, derive an expression for the error Δr in a root r of the equation

$$a_1 x^3 + a_2 x^2 + a_3 x + a_4 = 0$$

Assume that the errors are small enough that their products and second or higher powers can be neglected.   Apply the results to Prob. 6.5.

6.49 Assuming three initial trials, derive a parabolic interpolating formula for determining roots of equations.   Apply the result to determining a root of Prob. 6.5.

6.50 By retaining one more term in Taylor's series, derive a higher-ordered Newton's method for determining roots of equations.   Apply the result to determining a root of Prob. 6.5.

6.51 Derive an expression for the error in the higher-ordered Newton's method given in Prob. 6.50.

6.52 Determine a convergence criterion for the higher-ordered Newton's method given in Prob. 6.50.

7   SIMULTANEOUS LINEAR EQUATIONS

7.1   INTRODUCTION

The stress analysis of a frame, the analysis of a complicated electrical circuit, the solution of a chemical mixing problem, and the vibration analysis of a mechanical system are all examples of engineering problems which require the solving of a system of simultaneous linear equations. Such a system is usually given in the form

$$
\begin{aligned}
a_{11}x_1 + a_{12}x_2 + \cdots + a_{1n}x_n &= c_1 \\
a_{21}x_1 + a_{22}x_2 + \cdots + a_{2n}x_n &= c_2 \\
\cdots \cdots \cdots \cdots \cdots \cdots \cdots \cdots \cdots \\
a_{n1}x_1 + a_{n2}x_2 + \cdots + a_{nn}x_n &= c_n
\end{aligned}
\tag{7.1}
$$

where the a_{ij}'s are known coefficients, the c_i's are known constants, and the x_i's are the unknowns for which the equations are to be solved. It is assumed that all the coefficients and constants are real. Generally speaking, then, there is a set of real x_i's which satisfies the system of equations. Under certain circumstances, however, there is no set of x_i's which satisfies the system, while under other circumstances there is an infinite number of sets of x_i's which satisfy the system.

There are three general methods for solving a system of linear equations: elimination, iteration, and the use of determinants. All three methods will be considered in later sections of this chapter.

7.2   TWO EQUATIONS

The general theory of simultaneous equations can be presented most simply by examining two simultaneous equations of the form

$$
\begin{aligned}
a_{11}x_1 + a_{12}x_2 &= c_1 \\
a_{21}x_1 + a_{22}x_2 &= c_2
\end{aligned}
\tag{7.2}
$$

One method of determining the solution is as follows: Solve the first equation for x_1 in terms of x_2, substitute this into the second equation, calculate the value of x_2 from the second equation, substitute this value back into the first equation, and finally calculate the value of x_1 from the first equation. This procedure is illustrated below:

$$x_1 = \frac{1}{a_{11}}(c_1 - a_{12}x_2)$$

$$\frac{a_{21}}{a_{11}}(c_1 - a_{12}x_2) + a_{22}x_2 = c_2$$

$$x_2 = \frac{c_2 - a_{21}c_1/a_{11}}{a_{22} - a_{21}a_{12}/a_{11}} = \frac{a_{11}c_2 - a_{21}c_1}{a_{11}a_{22} - a_{21}a_{12}}$$

$$x_1 = \frac{c_1 - a_{12}[(a_{11}c_2 - a_{21}c_1)/(a_{11}a_{22} - a_{21}a_{12})]}{a_{11}} = \frac{a_{22}c_1 - a_{12}c_2}{a_{11}a_{22} - a_{21}a_{12}}$$

$$(7.3)$$

Another method of solution is to multiply the first equation by a_{21}, the second by a_{11}, and then subtract the second from the first. This leads to the same result, as shown below:

$$
\begin{array}{rl}
a_{21}a_{11}x_1 + a_{21}a_{12}x_2 = & a_{21}c_1 \\
-a_{11}a_{21}x_1 - a_{11}a_{22}x_2 = & -a_{11}c_2 \\
\hline
(a_{21}a_{12} - a_{11}a_{22})x_2 = & a_{21}c_1 - a_{11}c_2
\end{array}
$$

$$x_2 = \frac{a_{21}c_1 - a_{11}c_2}{a_{21}a_{12} - a_{11}a_{22}} = \frac{a_{11}c_2 - a_{21}c_1}{a_{11}a_{22} - a_{21}a_{12}}$$

$$x_1 = \frac{a_{22}c_1 - a_{12}c_2}{a_{11}a_{22} - a_{21}a_{12}}$$

Generally, Eqs. (7.3) give the solution to the system of two simultaneous equations. Notice that the denominators in the two equations are the same. Consider now some special cases. First of all, if c_1 and c_2 are both zero, the equations are said to be *homogeneous*. Provided that the denominator is not zero, the only possible solution is

$$x_1 = 0$$
$$x_2 = 0$$

If the denominator, $a_{11}a_{22} - a_{21}a_{12}$, is zero, the equations are said to be *singular*. Provided that the numerators are not zero in such a case, there are no solutions to the system of equations. If *both* the numerators and denominator are zero, there is an infinite number of solutions. In this case the equations are said to be *dependent;* i.e., one equation can be obtained from the other by some algebraic operation.

To illustrate, consider the homogeneous equations

$$x_1 + 2x_2 = 0$$
$$2x_1 + 4x_2 = 0$$

Clearly, both the numerators and the denominator are zero. Several solutions of the system are given in Table 7.1.

Table 7.1

x_1	x_2
0	0
1	$-\frac{1}{2}$
-1	$+\frac{1}{2}$
10	-5

If the above equations are altered to

$$x_1 + 2x_2 = 3$$
$$2x_1 + 4x_2 = 5$$

there are no solutions. However, the dependent equations

$$x_1 + 2x_2 = 3$$
$$2x_1 + 4x_2 = 6$$

have an infinite number of solutions, some of which are given in Table 7.2. Notice that the second equation can be obtained from the first by multiplying the first by 2.

Table 7.2

x_1	x_2
0	1.5
3	0
1	1

A final example illustrates a system of *nonhomogeneous, nonsingular, independent* equations.

$$x_1 + 2x_2 = 3$$
$$2x_1 - 4x_2 = 6$$

The solution of these equations is

$$x_1 = 3$$
$$x_2 = 0$$

The results presented here for two equations can be generalized for any number of equations. A system of n equations can be made homogeneous by setting the c_i's to zero. It will be shown in Sec. 7.6.2 that the solution of a system of equations is a set of x_i's of the form

$$x_i = \frac{f_i(a_{ij}\text{'s, } c_i\text{'s})}{g(a_{ij}\text{'s})} \tag{7.4}$$

Further, the function $g(a_{ij}\text{'s})$ is the same for each x_i, and it follows that, if this denominator is zero, the system of equations is singular. Finally, if all of the numerators $f_i(a_{ij}\text{'s}, c_i\text{'s})$ are zero as well as the denominator, the equations are not all independent.

Before analyzing these special cases, methods will now be presented for solving nonhomogeneous, nonsingular, independent, simultaneous linear equations.

7.3   ELIMINATION METHODS

A simple and straightforward method of solution is the elimination method that was used in Sec. 7.2 for solving two simultaneous equations. In principle, if there are more than two equations, the first equation may be divided by a_{11} and solved for x_1 and the resulting equation may be used to eliminate x_1 from all the subsequent equations. This produces a new system of $n - 1$ equations in the $n - 1$ unknowns $x_2, x_3, \ldots, x_n$. Repeating the procedure on these equations eliminates x_2 and produces a new system of $n - 2$ equations in the $n - 2$ unknowns $x_3, x_4, \ldots, x_n$. Finally, after $n - 1$ such eliminations, the result is a single equation involving only the one unknown x_n. Substitution of the value of x_n into the system of two equations involving x_n and x_{n-1} gives the value of x_{n-1}; together the values of x_n and x_{n-1} may be used to find the value of x_{n-2}, etc., until the first equation is again reached, giving the value of x_1.

In practice the method can be simplified and condensed, by using the technique attributed to Gauss. First let us list the operations that may be performed on the system of equations without changing the solution. They are:

1. The positions of any two equations in the system may be exchanged. Clearly, the order in which the equations are written is arbitrary. The order in which they are solved, however, may affect the result because of rounding and truncation errors accumulated during the calculations.

2. Any equation may be multiplied or divided throughout by a non-zero constant.

3. Any equation may be added to any other equation, and the resulting equation may then replace either of the two original equations in the system. Also, rules 2 and 3 may be combined, so that an equation is first multiplied by a constant and then added to another equation.

Second, we may omit writing the symbols x_i in the equations, as well as the equals signs, since we are simply doing arithmetic on the coefficients and the constant terms. The array will then appear as

$$
\begin{array}{ccccc}
a_{11} & a_{12} & \cdots & a_{1n} & c_1 \\
a_{21} & a_{22} & \cdots & a_{2n} & c_2 \\
\cdots & \cdots & \cdots & \cdots & \cdots \\
a_{n1} & a_{n2} & \cdots & a_{nn} & c_n
\end{array}
$$

The a_{ij} array is frequently called the *coefficient matrix*, and the c_i column is called the *constant vector*. To simplify subscripting, c_1 may be renamed as $a_{1,n+1}$, c_2 as $a_{2,n+1}$, etc. The a_{ij} array is now called the *augmented matrix*. In accordance with standard practice a matrix is enclosed in parentheses ( ) or brackets [ ]. The latter will be used here. The augmented matrix is, then,

$$
\begin{bmatrix}
a_{11} & a_{12} & \cdots & a_{1n} & a_{1,n+1} \\
a_{21} & a_{22} & \cdots & a_{2n} & a_{2,n+1} \\
\cdots & \cdots & \cdots & \cdots & \cdots \\
a_{n1} & a_{n2} & \cdots & a_{nn} & a_{n,n+1}
\end{bmatrix}
\tag{7.5}
$$

It will be convenient to speak of the matrix as made up of rows R_i and columns C_j. The position of an element in the matrix is given by the row and column numbers, that is, a_{ij}.

7.3.1 Gaussian Elimination

As an example of the use of Gaussian elimination, let us solve the equations

$$
\begin{aligned}
x_1 + x_2 + x_3 + x_4 &= 10 \\
2x_1 + x_2 + 3x_3 + 2x_4 &= 21 \\
x_1 + 3x_2 + 2x_3 + x_4 &= 17 \\
3x_1 + 2x_2 + x_3 + x_4 &= 14
\end{aligned}
\tag{7.6}
$$

The augmented matrix is

$$
\begin{bmatrix}
1 & 1 & 1 & 1 & 10 \\
2 & 1 & 3 & 2 & 21 \\
1 & 3 & 2 & 1 & 17 \\
3 & 2 & 1 & 1 & 14
\end{bmatrix}
$$

The steps in the elimination are shown in the following diagrams. In each step the matrix is transformed by the indicated operations on the rows. Note that the row being used to modify the other rows, often called the *pivot row*, is first divided by its diagonal element. The resulting row, shown in the matrix at the right, is marked by a prime. In the elimination of the coefficients of x_2, for example, R_3' is formed by subtracting $2R_2'$ from R_3.

Elimination of coefficients of x_1

$$\begin{bmatrix} 1 & 1 & 1 & 1 & 10 \\ 2 & 1 & 3 & 2 & 21 \\ 1 & 3 & 2 & 1 & 17 \\ 3 & 2 & 1 & 1 & 14 \end{bmatrix} \quad \begin{matrix} R_1/1 \\ R_2 - 2R_1' \\ R_3 - R_1' \\ R_4 - 3R_1' \end{matrix} \quad \begin{bmatrix} 1 & 1 & 1 & 1 & 10 \\ 0 & -1 & 1 & 0 & 1 \\ 0 & 2 & 1 & 0 & 7 \\ 0 & -1 & -2 & -2 & -16 \end{bmatrix}$$

Elimination of coefficients of x_2

$$\begin{bmatrix} 1 & 1 & 1 & 1 & 10 \\ 0 & -1 & 1 & 0 & 1 \\ 0 & 2 & 1 & 0 & 7 \\ 0 & -1 & -2 & -2 & -16 \end{bmatrix} \quad \begin{matrix} R_2/(-1) \\ R_3 - 2R_2' \\ R_4 - (-1)R_2' \end{matrix} \quad \begin{bmatrix} 1 & 1 & 1 & 1 & 10 \\ 0 & 1 & -1 & 0 & -1 \\ 0 & 0 & 3 & 0 & 9 \\ 0 & 0 & -3 & -2 & -17 \end{bmatrix}$$

Elimination of coefficients of x_3

$$\begin{bmatrix} 1 & 1 & 1 & 1 & 10 \\ 0 & 1 & -1 & 0 & -1 \\ 0 & 0 & 3 & 0 & 9 \\ 0 & 0 & -3 & -2 & -17 \end{bmatrix} \quad \begin{matrix} R_3/3 \\ R_4 - (-3)R_3' \end{matrix} \quad \begin{bmatrix} 1 & 1 & 1 & 1 & 10 \\ 0 & 1 & -1 & 0 & -1 \\ 0 & 0 & 1 & 0 & 3 \\ 0 & 0 & 0 & -2 & -8 \end{bmatrix}$$

Solution for x_4

$$\begin{bmatrix} 1 & 1 & 1 & 1 & 10 \\ 0 & 1 & -1 & 0 & -1 \\ 0 & 0 & 1 & 0 & 3 \\ 0 & 0 & 0 & -2 & -8 \end{bmatrix} \quad \begin{matrix} \\ \\ \\ R_4/(-2) \end{matrix} \quad \begin{bmatrix} 1 & 1 & 1 & 1 & 10 \\ 0 & 1 & -1 & 0 & -1 \\ 0 & 0 & 1 & 0 & 3 \\ 0 & 0 & 0 & 1 & 4 \end{bmatrix}$$

The forward elimination is now complete, and the equations corresponding to the matrix form are

$$\begin{aligned} x_1 + x_2 + x_3 + x_4 &= 10 \\ x_2 - x_3 &= -1 \\ x_3 &= 3 \\ x_4 &= 4 \end{aligned}$$

Backward substitution may now be used to solve for the x_i's in reverse order. Hence,

$$\begin{aligned} x_4 &= 4 \\ x_3 &= 3 \\ x_2 &= -1 + x_3 = 2 \\ x_1 &= 10 - x_2 - x_3 - x_4 = 1 \end{aligned}$$

An alternative procedure is to continue using the elimination technique to convert the coefficient matrix to a form in which the only nonzero terms are on the diagonal. The constant vector will then contain the solution. Continuing the above elimination, for example,

Elimination of coefficients of x_4

$$
\begin{bmatrix}
1 & 1 & 1 & 1 & 10 \\
0 & 1 & -1 & 0 & -1 \\
0 & 0 & 1 & 0 & 3 \\
0 & 0 & 0 & 1 & 4
\end{bmatrix}
\begin{matrix}
R_1 - R_4 \\
R_2 - 0R_4 \\
R_3 - 0R_4 \\
\\
\end{matrix}
\begin{bmatrix}
1 & 1 & 1 & 0 & 6 \\
0 & 1 & -1 & 0 & -1 \\
0 & 0 & 1 & 0 & 3 \\
0 & 0 & 0 & 1 & 4
\end{bmatrix}
$$

Elimination of coefficients of x_3

$$
\begin{bmatrix}
1 & 1 & 1 & 0 & 6 \\
0 & 1 & -1 & 0 & -1 \\
0 & 0 & 1 & 0 & 3 \\
0 & 0 & 0 & 1 & 4
\end{bmatrix}
\begin{matrix}
R_1 - R_3 \\
R_2 - (-1)R_3 \\
\\
\\
\end{matrix}
\begin{bmatrix}
1 & 1 & 0 & 0 & 3 \\
0 & 1 & 0 & 0 & 2 \\
0 & 0 & 1 & 0 & 3 \\
0 & 0 & 0 & 1 & 4
\end{bmatrix}
$$

Elimination of coefficients of x_2

$$
\begin{bmatrix}
1 & 1 & 0 & 0 & 3 \\
0 & 1 & 0 & 0 & 2 \\
0 & 0 & 1 & 0 & 3 \\
0 & 0 & 0 & 1 & 4
\end{bmatrix}
\begin{matrix}
R_1 - R_2 \\
\\
\\
\\
\end{matrix}
\begin{bmatrix}
1 & 0 & 0 & 0 & 1 \\
0 & 1 & 0 & 0 & 2 \\
0 & 0 & 1 & 0 & 3 \\
0 & 0 & 0 & 1 & 4
\end{bmatrix}
$$

Returning the matrix to equation form shows the solution,

$x_1 = 1$
$x_2 = 2$
$x_3 = 3$
$x_4 = 4$

A few further points about the augmented matrix may be noted here. At the start of the solution the elements of the matrix were all nonzero for this particular problem. After the first elimination the first column contained a 1 as a_{11}, its element on the diagonal of the coefficient matrix, and all other elements below a_{11} were zero. Similarly, in each succeeding elimination, an element on the diagonal, a_{ii}, was made equal to 1, and all the elements below it in that column were made equal to zero. At the end of the forward elimination all elements on the diagonal of the coefficient matrix were 1, and the elements below the diagonal were all zero. A matrix in which the diagonal elements are not all zero (not necessarily 1, however) and in which the triangle below the diagonal contains all zeros is said to be in *triangular* form. The forward elimination may be thought of as a triangularization of the coefficient matrix. The backward

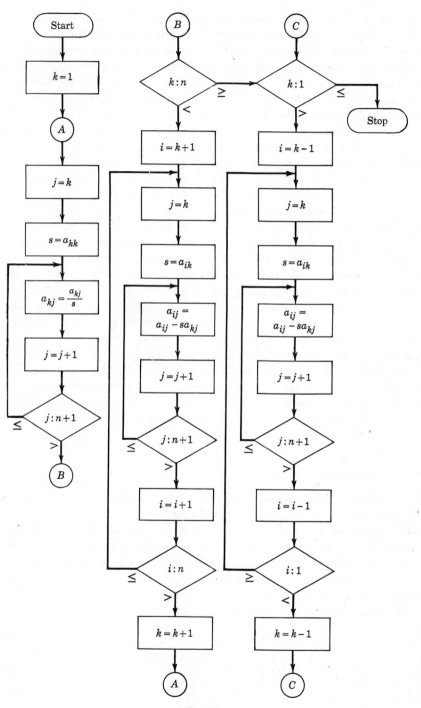

Fig. 7.1

elimination, on the other hand, did not disturb the diagonal elements; instead it eliminated all elements *above* the diagonal. At the end of the operation the matrix was in *diagonal* form, or had been diagonalized.

Many variations of the elimination method are possible. It is not necessary, for example, to divide the pivot row by its diagonal element. Instead, the divisor may be incorporated into each of the factors by which the pivot row is multiplied when it is used to modify the other rows. At the end of the triangularization or diagonalization, the diagonal elements will not, in general, be all 1's. The solution is then obtained by dividing each of the rows by the diagonal element of that row during the backward elimination. If the matrix is in diagonal form, it is necessary only to divide the elements of the constant vector by the diagonal elements.

The Gaussian elimination method, as used above to diagonalize the coefficient matrix, may easily be put into flow-chart form, as shown in Fig. 7.1. On the flow chart, k is the number of the row being used to eliminate the coefficients in the kth column in the lower rows. Before being used, however, the row is divided through by its own diagonal element a_{kk} so that the final matrix will consist of only 1's on the diagonal and the column of constant terms will give directly the values of the unknowns. This division is shown in the first column. The forward elimination is given in the three nested loops in the second column of the flow chart. The backward elimination is in the three nested loops in the third column. The indices i and j refer to the row and column numbers of an element a_{ij}. A point of interest is the step for the separate calculation of the factor s outside each loop. If this step were included within the loop, the diagonal elements would be changed before they were used to modify the other elements in the row.

7.3.2 Gauss-Jordan Method

The forward and backward eliminations may be combined into a single procedure known as the *Gauss-Jordan method*. Essentially, the kth row is used to eliminate the coefficients in the kth column, not only below it, but also above it. Thus, at the end of a single pass through the matrix, the only remaining elements are the column vector and the 1's on the diagonal.

The example of Eqs. (7.6) will be used again to demonstrate the Gauss-Jordan elimination method.

Elimination of coefficients of x_1

$$\begin{bmatrix} 1 & 1 & 1 & 1 & 10 \\ 2 & 1 & 3 & 2 & 21 \\ 1 & 3 & 2 & 1 & 17 \\ 3 & 2 & 1 & 1 & 14 \end{bmatrix} \quad \begin{matrix} R_1/1 \\ R_2 - 2R_1' \\ R_3 - R_1' \\ R_4 - 3R_1' \end{matrix} \quad \begin{bmatrix} 1 & 1 & 1 & 1 & 10 \\ 0 & -1 & 1 & 0 & 1 \\ 0 & 2 & 1 & 0 & 7 \\ 0 & -1 & -2 & -2 & -16 \end{bmatrix}$$

Elimination of coefficients of x_2

$$
\begin{bmatrix}
1 & 1 & 1 & 1 & 10 \\
0 & -1 & 1 & 0 & 1 \\
0 & 2 & 1 & 0 & 7 \\
0 & -1 & -2 & -2 & -16
\end{bmatrix}
\begin{array}{l}
R_1 - R_2' \\
R_2/(-1) \\
R_3 - 2R_2' \\
R_4 - (-1)R_2'
\end{array}
\xrightarrow{\hspace{1cm}}
\begin{bmatrix}
1 & 0 & 2 & 1 & 11 \\
0 & 1 & -1 & 0 & -1 \\
0 & 0 & 3 & 0 & 9 \\
0 & 0 & -3 & -2 & -17
\end{bmatrix}
$$

Elimination of coefficients of x_3

$$
\begin{bmatrix}
1 & 0 & 2 & 1 & 11 \\
0 & 1 & -1 & 0 & -1 \\
0 & 0 & 3 & 0 & 9 \\
0 & 0 & -3 & -2 & -17
\end{bmatrix}
\begin{array}{l}
R_1 - 2R_3' \\
R_2 - (-1)R_3' \\
R_3/3 \\
R_4 - (-1)R_3'
\end{array}
\xrightarrow{\hspace{1cm}}
\begin{bmatrix}
1 & 0 & 0 & 1 & 5 \\
0 & 1 & 0 & 0 & 2 \\
0 & 0 & 1 & 0 & 3 \\
0 & 0 & 0 & -2 & -8
\end{bmatrix}
$$

Elimination of coefficients of x_4

$$
\begin{bmatrix}
1 & 0 & 0 & 1 & 5 \\
0 & 1 & 0 & 0 & 2 \\
0 & 0 & 1 & 0 & 3 \\
0 & 0 & 0 & -2 & -8
\end{bmatrix}
\begin{array}{l}
R_1 - R_4' \\
R_2 - 0R_4' \\
R_3 - 0R_4' \\
R_4/(-2)
\end{array}
\xrightarrow{\hspace{1cm}}
\begin{bmatrix}
1 & 0 & 0 & 0 & 1 \\
0 & 1 & 0 & 0 & 2 \\
0 & 0 & 1 & 0 & 3 \\
0 & 0 & 0 & 1 & 4
\end{bmatrix}
$$

A flow chart of the Gauss-Jordan method is given in Fig. 7.2. The notation is the same as that in Fig. 7.1. The main elimination loop always starts with row 1 and proceeds to row n, bypassing row k, which is being used as the pivot row.

Some of the work indicated by the procedures of Figs. 7.1 and 7.2 is not really necessary to obtain the solution. A careful analysis points out that the storing of the zeros in the coefficient matrix is wasted effort, for these elements are never again looked at or used. In fact, the only elements of interest at the end of the operation are those of the constant vector, which has been transformed by the row operations on the coefficient matrix into the solution vector. A flow chart for this modified procedure, applied to the Gauss-Jordan elimination, is presented in Fig. 7.3. A similar modification can be made to the flow chart for the Gaussian elimination shown in Fig. 7.1.

7.3.3   Zero Pivot Elements

The diagonal element in the kth row, a_{kk}, plays a particularly important role in the elimination methods, and it is often called the *pivot element*. Since one of the steps is to divide the kth row by a_{kk}, it is mandatory that the pivot element not be zero. This may depend, however, on the order in which the equations are written. If, for example, the first equation

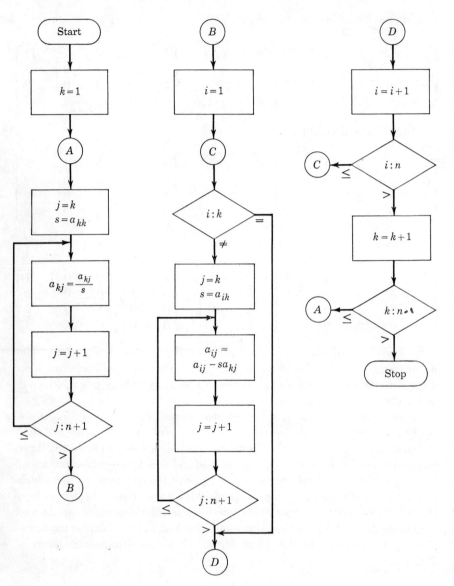

Fig. 7.2

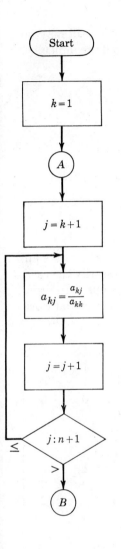

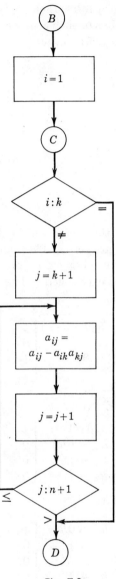

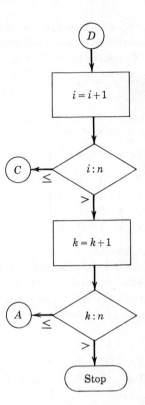

Fig. 7.3

does not happen to contain an x_1 term, then a_{11} is zero. It still may be possible to exchange the equations in such a way as to avoid the difficulty.

Consider the following system of equations:

$$x_2 + x_3 + x_4 = 5$$
$$x_1 + x_3 + x_4 = 4$$
$$x_1 + x_2 + x_4 = 6$$
$$x_1 + x_2 + x_3 = 6$$

The augmented matrix is

$$\begin{bmatrix} 0 & 1 & 1 & 1 & 5 \\ 1 & 0 & 1 & 1 & 4 \\ 1 & 1 & 0 & 1 & 6 \\ 1 & 1 & 1 & 0 & 6 \end{bmatrix}$$

At first glance it might appear that there will be difficulty at every step, since all the diagonal elements are zero. This is not necessarily the case, however, for the elements may be changed by preceding steps. In order to get started, rows 1 and 2 are exchanged. The remainder of the work proceeds smoothly.

$$\begin{bmatrix} 0 & 1 & 1 & 1 & 5 \\ 1 & 0 & 1 & 1 & 4 \\ 1 & 1 & 0 & 1 & 6 \\ 1 & 1 & 1 & 0 & 6 \end{bmatrix} \quad \xrightarrow{R_1 \leftrightarrow R_2} \quad \begin{bmatrix} 1 & 0 & 1 & 1 & 4 \\ 0 & 1 & 1 & 1 & 5 \\ 1 & 1 & 0 & 1 & 6 \\ 1 & 1 & 1 & 0 & 6 \end{bmatrix}$$

Elimination of coefficients of x_1

$$\begin{bmatrix} 1 & 0 & 1 & 1 & 4 \\ 0 & 1 & 1 & 1 & 5 \\ 1 & 1 & 0 & 1 & 6 \\ 1 & 1 & 1 & 0 & 6 \end{bmatrix} \quad \begin{matrix} R_1/1 \\ R_2 - 0R_1' \\ R_3 - R_1' \\ R_4 - R_1' \end{matrix} \xrightarrow{} \begin{bmatrix} 1 & 0 & 1 & 1 & 4 \\ 0 & 1 & 1 & 1 & 5 \\ 0 & 1 & -1 & 0 & 2 \\ 0 & 1 & 0 & -1 & 2 \end{bmatrix}$$

Elimination of coefficients of x_2

$$\begin{bmatrix} 1 & 0 & 1 & 1 & 4 \\ 0 & 1 & 1 & 1 & 5 \\ 0 & 1 & -1 & 0 & 2 \\ 0 & 1 & 0 & -1 & 2 \end{bmatrix} \quad \begin{matrix} R_1 - 0R_2' \\ R_2/1 \\ R_3 - R_2' \\ R_4 - R_2' \end{matrix} \xrightarrow{} \begin{bmatrix} 1 & 0 & 1 & 1 & 4 \\ 0 & 1 & 1 & 1 & 5 \\ 0 & 0 & -2 & -1 & -3 \\ 0 & 0 & -1 & -2 & -3 \end{bmatrix}$$

Elimination of coefficients of x_3

$$\begin{bmatrix} 1 & 0 & 1 & 1 & 4 \\ 0 & 1 & 1 & 1 & 5 \\ 0 & 0 & -2 & -1 & -3 \\ 0 & 0 & -1 & -2 & -3 \end{bmatrix} \quad \begin{matrix} R_1 - R_3' \\ R_2 - R_3' \\ R_3/(-2) \\ R_4 - (-1)R_3' \end{matrix} \xrightarrow{} \begin{bmatrix} 1 & 0 & 0 & 0.5 & 2.5 \\ 0 & 1 & 0 & 0.5 & 3.5 \\ 0 & 0 & 1 & 0.5 & 1.5 \\ 0 & 0 & 0 & -1.5 & -1.5 \end{bmatrix}$$

Elimination of coefficients of x_4

$$\begin{bmatrix} 1 & 0 & 0 & 0.5 & 2.5 \\ 0 & 1 & 0 & 0.5 & 3.5 \\ 0 & 0 & 1 & 0.5 & 1.5 \\ 0 & 0 & 0 & -1.5 & -1.5 \end{bmatrix} \begin{matrix} R_1 - 0.5R_4' \\ R_2 - 0.5R_4' \\ R_3 - 0.5R_4' \\ R_4/(-1.5) \end{matrix} \begin{bmatrix} 1 & 0 & 0 & 0 & 2 \\ 0 & 1 & 0 & 0 & 3 \\ 0 & 0 & 1 & 0 & 1 \\ 0 & 0 & 0 & 1 & 1 \end{bmatrix}$$

If a zero pivot element occurs at any step, an attempt may be made to exchange the row containing the zero with a row *below* it in which the element in that column is not zero. If all the lower elements in the column are zero, the system is singular.

7.3.4 Round-off Errors

It was mentioned earlier that the solution might depend on the order of elimination because of truncation and round-off errors brought about by having to retain a fixed number of digits in each number in the computer. If, for example, a pivot element is not zero but is small, it may contain a large relative error. This situation can be remedied, to a certain extent, by searching at each step for the element of greatest magnitude in the pivot column (in the pivot row and those below) and exchanging rows so as to use it as the pivot element. A loop to do this may readily be inserted within the main loop of the flow charts given in Figs. 7.2 and 7.3, just before the loop for division by a_{kk}. This is left as an exercise for the student.

Still better schemes can be devised for reducing the round-off error. In one method, the entire coefficient matrix is first searched for the element of maximum magnitude, and then rows and columns are exchanged so as to move that element to the pivot position a_{11}. After the elements below a_{11} have been eliminated, the new square submatrix beginning at the second column and second row is searched, the element of maximum magnitude is moved to position a_{22}, etc. Since the exchanging of columns corresponds to the exchanging of variables in the original equations, it is necessary to keep an account of the exchanges and to put the final answers back into the original order.

Another technique involves an initial scaling of the elements by dividing each row of the augmented matrix by a constant that depends on the magnitude of the elements in that row of the coefficient matrix. Several variations are possible, the simplest being to use the element of maximum magnitude in that row as the divisor. Another possibility is to use the square root of the product of the magnitudes of the two elements of smallest and largest magnitude, that is, their geometric mean. Also, after the scaling of rows has been completed, it is advantageous to scale any column in which all the elements differ by more than a limiting factor, such as 10^2, from those in another column. This scaling operation represents a change of the original variable of that column, and it must be accounted for in the final results.

7.4   ITERATION METHODS

In addition to the direct, or elimination, methods for solving a system of simultaneous linear equations, there are several indirect, or iteration, methods. Just as in the iteration method for finding a root of a polynomial, a trial value of the unknown is chosen and some scheme is used for improving it until (if all goes well) two successive values differ from each other by no more than an allowable amount. The problem is more complicated in the case of a system of equations, however, since it is necessary to assume a *set* of trial values and to go through the iteration until *all* the unknowns have converged satisfactorily.

7.4.1   Total Step Iteration

Probably the simplest method is the *total step iteration*, also called *Jacobi iteration* or the *method of simultaneous displacements*. The system of equations

$$a_{11}x_1 + a_{12}x_2 + \cdots + a_{1n}x_n = c_1$$
$$a_{21}x_1 + a_{22}x_2 + \cdots + a_{2n}x_n = c_2$$
$$\cdots \cdots \cdots \cdots \cdots \cdots \cdots \cdots \cdots \cdots$$
$$a_{n1}x_1 + a_{n2}x_2 + \cdots + a_{nn}x_n = c_n$$

is rearranged so that the first is solved for x_1, the second for x_2, etc.

$$x_1 = \frac{1}{a_{11}} (c_1 - a_{12}x_2 - a_{13}x_3 - \cdots - a_{1n}x_n)$$

$$x_2 = \frac{1}{a_{22}} (c_2 - a_{21}x_1 - a_{23}x_3 - \cdots - a_{2n}x_n) \qquad (7.7)$$

$$\cdots \cdots \cdots \cdots \cdots \cdots \cdots \cdots \cdots \cdots \cdots$$

$$x_n = \frac{1}{a_{nn}} (c_n - a_{n1}x_1 - a_{n2}x_2 - \cdots - a_{n,n-1}x_{n-1})$$

By assuming a set of values $x_1^{(1)}, x_2^{(1)}, \ldots, x_n^{(1)}$ and substituting them into the right side of the previous equation, a new set $x_1^{(2)}, x_2^{(2)}, \ldots, x_n^{(2)}$ can be calculated. Here, the superscript indicates the number of the trial or set of x values and is not to be confused with an exponent. If set 2 does not agree well enough with set 1, it can be substituted into the right side of the above equation, giving a third set $x_1^{(3)}$, $x_2^{(3)}$, . . . , $x_n^{(3)}$. The process is continued until two successive sets, say r and $r + 1$, are within the allowable deviation. This may be specified in several ways. One possible way is to calculate the sum of the absolute values of the relative deviations and to compare that with the allowable deviation. For example,

$$\left| \frac{x_1^{(r+1)} - x_1^{(r)}}{x_1^{(r+1)}} \right| + \left| \frac{x_2^{(r+1)} - x_2^{(r)}}{x_2^{(r+1)}} \right| + \cdots$$

$$+ \left| \frac{x_n^{(r+1)} - x_n^{(r)}}{x_n^{(r+1)}} \right| \leq \text{allowable deviation}$$

As a numerical example of total step iteration, consider the solution of the following system of equations:

$$3x_1 + x_2 + x_3 = 10$$
$$x_1 + 5x_2 + 2x_3 = 21$$
$$x_1 + 2x_2 + 5x_3 = 30$$

Rearranging and adding an iteration index give

$$x_1^{(r+1)} = \tfrac{1}{3}(10 - x_2^{(r)} - x_3^{(r)})$$
$$x_2^{(r+1)} = \tfrac{1}{5}(21 - x_1^{(r)} - 2x_3^{(r)})$$
$$x_3^{(r+1)} = \tfrac{1}{5}(30 - x_1^{(r)} - 2x_2^{(r)})$$

As a trial set, assume that $x_1^{(1)} = c_1/a_{11} = \tfrac{10}{3}$, $x_2^{(1)} = c_2/a_{22} = \tfrac{21}{5}$, and $x_3^{(1)} = c_3/a_{33} = \tfrac{30}{5}$. Note that the very simple assumption of all zeros for the first set would yield the above set at the end of the first iteration. Table 7.3 shows the progress of the iteration.

Table 7.3

r / Values	1	2	3	4	5	6	7	8	9
x_1	3.333	−0.067	1.745	0.560	1.609	0.838	1.146	0.920	1.052
x_2	4.200	1.111	2.752	1.623	2.268	1.791	2.124	1.924	2.048
x_3	6.000	3.654	5.569	4.550	5.219	4.771	5.116	4.921	5.046

r / Values	10	11	12	13	14	15	16	17	18
x_1	0.969	1.020	0.988	1.007	0.995	1.003	0.998	1.001	0.999+
x_2	1.971	2.018	1.989	2.007	1.996	2.003	1.998	2.001	2.000−
x_3	4.970	5.018	4.989	5.007	4.996	5.003	4.998	5.001	5.000−

7.4.2   Gauss-Seidel Iteration

Instead of using the entire set from the previous iteration in calculating the new set, it might seem advantageous to use the most recently calculated value of each variable at each step.   This is the Gauss-Seidel method.   It may be written in iteration form as

$$x_1^{(r+1)} = \frac{1}{a_{11}} (c_1 - a_{12}x_2^{(r)} - a_{13}x_3^{(r)} - \cdots - a_{1n}x_n^{(r)})$$

$$x_2^{(r+1)} = \frac{1}{a_{22}} (c_2 - a_{21}x_1^{(r+1)} - a_{23}x_3^{(r)} - \cdots - a_{2n}x_n^{(r)})$$

$$\cdots \cdots \cdots \cdots \cdots \cdots \cdots \cdots \cdots \cdots \cdots \cdots \cdots$$

$$x_n^{(r+1)} = \frac{1}{a_{nn}} (c_n - a_{n1}x_1^{(r+1)} - a_{n2}x_2^{(r+1)} - \cdots - a_{n,n-1}x_{n-1}^{(r+1)})$$

(7.8)

The previous numerical example is repeated, Table 7.4 showing the progress of the iteration.

Table 7.4

Values \ *r*	1	2	3	4	5	6	7
x_1	3.333	−0.067	0.966	1.019	1.006	1.001	1.000
x_2	4.200	1.813	1.892	1.976	1.996	2.000	2.000
x_3	6.000	5.288	5.050	5.006	5.000	5.000	5.000

A flow chart of the Gauss-Seidel method is presented in Fig. 7.4.

7.4.3   Convergence of Iteration Methods

A preliminary discussion of the convergence of the complete iteration and the Gauss-Seidel iteration is included at this point, but a more mathematically detailed discussion will be deferred to Sec. 7.5.11.   Here, the analysis will be based on geometric considerations.

As the simplest example, consider the Gauss-Seidel iteration applied to a system of only two equations,

$$f_1(x_1,x_2) = a_{11}x_1 + a_{12}x_2 = c_1$$
$$f_2(x_1,x_2) = a_{21}x_1 + a_{22}x_2 = c_2$$

(7.9)

Geometrically, these equations represent a pair of straight lines in the x_1x_2 plane, intersecting at the solution $(\bar{x}_1,\bar{x}_2)$, as shown in Fig. 7.5.   The axes have been chosen with x_2 as the abscissa, for convenience in the

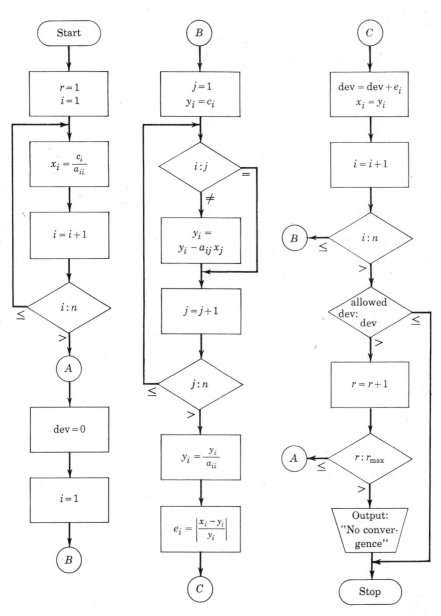

Fig. 7.4

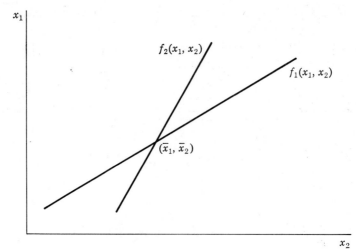

Fig. 7.5

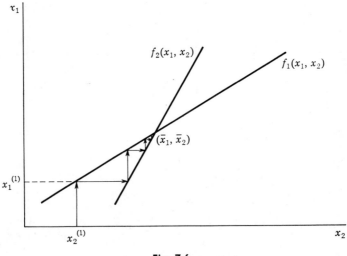

Fig. 7.6

discussion to follow, since x_2 may be regarded as the independent variable in the first equation.

In the iteration, if a value $x_2^{(1)}$ is chosen and substituted into $f_1(x_1,x_2)$, a value $x_1^{(1)}$ may be calculated, which in turn may be substituted into $f_2(x_1,x_2)$ to give a new value of x_2, etc. The process is shown geometrically in Fig. 7.6.

Under these conditions the iteration will converge to $(\bar{x}_1,\bar{x}_2)$, no matter what initial value of x_2 is chosen. A sequence starting at the right of $(\bar{x}_1, \bar{x}_2)$ is shown in Fig. 7.7.

Suppose, on the other hand, that the equations are used in the reverse order, with the starting value of x_2 substituted into $f_2(x_1, x_2)$. The result is shown in Fig. 7.8. The result is obviously diverging and will be so for any initial value of x_2 except $x_2 = \bar{x}_2$.

The criterion for convergence, in the case of two simultaneous equa-

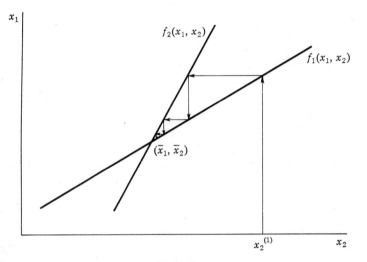

Fig. 7.7

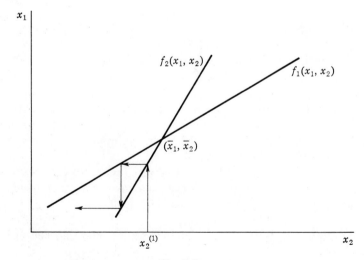

Fig. 7.8

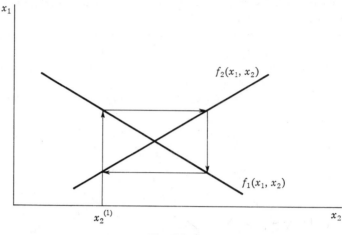

Fig. 7.9

tions, may be stated in terms of the slopes of the graphs of the functions. Simply stated, the initial substitution must be made into the function which has the lower slope $|a_{i2}/a_{i1}|$, calculated as an absolute value. If both equations are solved for x_1 in terms of x_2, as

$$x_1 = \frac{c_1}{a_{11}} - \frac{a_{12}}{a_{11}} x_2$$

$$x_1 = \frac{c_2}{a_{21}} - \frac{a_{22}}{a_{21}} x_2$$

then the first equation should be used if $|a_{12}/a_{11}| < |a_{22}/a_{21}|$ and the second should be used if $|a_{22}/a_{21}| < |a_{12}/a_{11}|$. If the slopes are equal in magnitude, the iteration will either oscillate between two points or loop in a rectangle, as shown in Fig. 7.9.

With more than two equations the geometry becomes very much more complicated. With three equations each equation represents a plane in three-dimensional space, and the solution $(\bar{x}_1, \bar{x}_2, \bar{x}_3)$ lies at the point of intersection of the three planes. With four or more equations it is not possible to construct a diagram.

In all cases, however, as will be demonstrated more fully in Sec. 7.5.11, the equations should be ordered so that the coefficient of x_1 of greatest magnitude occurs in the first equation, the coefficient of x_2 of greatest magnitude in the remaining equations occurs in the second equation, etc. This rule applies to both the complete iteration and the Gauss-Seidel iteration.

The flow chart given in Fig. 7.4 for the Gauss-Seidel iteration does not include the steps in the preliminary ordering of the equations.

7.4.4   Elimination vs. Iteration

Each of these techniques has certain advantages and disadvantages. Elimination, the direct method, has the advantage of taking a fixed number of operations for a given number of equations, regardless of the numbers involved. It has the disadvantage of accumulated round-off error, causing a possible large relative error in the solution. Although this may be minimized by reordering the equations after each step so as to keep the diagonal elements maximal, it is still a serious drawback in the handling of large systems of equations.

Iteration has the disadvantage of possibly not converging and of possibly requiring a large number of machine operations even if it does converge. Also, even though the original coefficients are used throughout the process, round-off errors may accumulate seriously. For sparse matrices (those containing a large percentage of zero elements), the number of machine operations is greatly reduced, and iteration methods gain an advantage over elimination methods.

7.5   MATRIX OPERATIONS

Although some of the terminology of matrix theory has already been used in this chapter in the section on Gaussian Elimination, no formal definitions have been given. It is the purpose of this section to fill in this gap and to present the elementary operations of matrix algebra. These will be applied to the solution of a system of simultaneous linear equations and to a further, more formal discussion of the criteria for the convergence of the iteration procedures.

7.5.1   Definitions

A matrix is a rectangular array of numbers. These numbers, which are called the *elements* of the matrix, may in general be real or complex. The position of an element in the matrix is given by specifying as subscripts of the element the row and the column in which it appears, in that order. Thus the element a_{ij} is at the intersection of the ith row and the jth column. Reference to the matrix as a whole may be made in terms of an unsubscripted name. An $m \times n$ matrix $\mathbf{A}$ is represented as

$$\mathbf{A} = \begin{bmatrix} a_{11} & a_{12} & \cdots & a_{1j} & \cdots & a_{1n} \\ a_{21} & a_{22} & \cdots & a_{2j} & \cdots & a_{2n} \\ \cdots & \cdots & \cdots & \cdots & \cdots & \cdots \\ a_{i1} & a_{i2} & \cdots & a_{ij} & \cdots & a_{in} \\ \cdots & \cdots & \cdots & \cdots & \cdots & \cdots \\ a_{m1} & a_{m2} & \cdots & a_{mj} & \cdots & a_{mn} \end{bmatrix} \tag{7.10}$$

where m and n are the *dimensions* of the columns and rows, respectively.

A matrix consisting of a single row of elements is called a *row matrix*, or a *row vector*. Only one subscript is needed to specify the position of an element. Any row of a rectangular matrix is also a row vector, and in this case the first subscript is a constant equal, of course, to the row number. A row vector **B**, consisting of n elements, may be shown as

$$\mathbf{B} = [b_1 \quad b_2 \quad \cdots \quad b_n] \tag{7.11}$$

Similarly, a matrix consisting of a single column of elements is called a *column matrix*, or a *column vector*. Again only one subscript is needed to specify the position of an element. Also, any column of a rectangular matrix is a column vector, in which case the second subscript is a constant equal to the column number. A column vector **X**, consisting of n elements, may be shown as

$$\mathbf{X} = \begin{bmatrix} x_1 \\ x_2 \\ \cdot \\ \cdot \\ \cdot \\ x_n \end{bmatrix} \tag{7.12}$$

The word *vector* is also commonly used to describe a directed line segment or a physical quantity that has direction as well as magnitude, such as force or velocity. If the vector is represented by an arrow drawn from the origin of a set of coordinates, then it may be written in matrix notation by using the coordinates of the head of the arrow as the elements of a row or column matrix (without commas). In two dimensions, for example, the vector shown in Fig. 7.10 corresponds to [3   1]. Vectors with more than three elements obviously cannot be drawn, but they may be thought of as existing in n-dimensional space.

A matrix that has the same number of rows and columns is called a *square matrix*. With such a matrix it is possible to draw a diagonal line from the upper left corner to the lower right corner, passing through elements $a_{11}, a_{22}, \ldots, a_{ii}, \ldots, a_{nn}$. These elements are called the *principal diagonal elements*. With a rectangular matrix it may still be convenient to consider a square section within the entire array and to speak of the principal diagonal elements within that square.

Several matrix forms occur so frequently that they have been given their own identifying names. A *diagonal matrix* is a square matrix containing zeros in all positions except on the principal diagonal. A diagonal matrix in which the diagonal elements are all the same number is called a *scalar matrix*. A special case of a scalar matrix is that in which the number is 1. The matrix is then called a *unit*, or *identity*, *matrix*, and

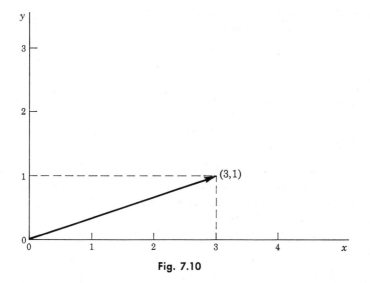

Fig. 7.10

as will be seen later, it plays the same role in many matrix operations as that of the number 1 in the algebra of real numbers.   It is symbolized by the letter **I**.   The rows or columns of the unit matrix are called *unit vectors* and are often symbolized as e_1, e_2, . . . , e_n.   An example of each of these matrix types is given here.

$$\begin{bmatrix} 5 & 0 & 0 \\ 0 & -2 & 0 \\ 0 & 0 & 3 \end{bmatrix} \qquad \begin{bmatrix} 6 & 0 & 0 \\ 0 & 6 & 0 \\ 0 & 0 & 6 \end{bmatrix} \qquad \begin{bmatrix} 1 & 0 & 0 \\ 0 & 1 & 0 \\ 0 & 0 & 1 \end{bmatrix} \qquad \begin{bmatrix} 1 \\ 0 \\ 0 \end{bmatrix} \begin{bmatrix} 0 \\ 1 \\ 0 \end{bmatrix} \begin{bmatrix} 0 \\ 0 \\ 1 \end{bmatrix}$$

Diagonal matrix   Scalar matrix   Unit matrix   Unit vectors

A *symmetric matrix* is a square matrix in which the elements of the first row are equal to those of the first column, those of the second row to those of the second column, etc.   In the subscript notation, $a_{ij} = a_{ji}$ for the entire range of i and j.   In a *skew-symmetric matrix*, on the other hand, the relation of the elements is $a_{ij} = -a_{ji}$.   For this equation to hold on the diagonal, i.e., for a_{ii} to equal $-a_{ii}$, the diagonal elements must be zero.   Examples of symmetric and skew-symmetric matrices are

$$\begin{bmatrix} 1 & 4 & -1 \\ 4 & 2 & 0 \\ -1 & 0 & 3 \end{bmatrix} \qquad \begin{bmatrix} 0 & 4 & -1 \\ -4 & 0 & 5 \\ 1 & -5 & 0 \end{bmatrix}$$

Symmetric matrix   Skew-symmetric matrix

The *transpose* of a matrix **A**, symbolized as $\mathbf{A}^T$, is obtained by rewriting the original row vectors of the matrix in column-vector form or by rewriting the original column vectors in row-vector form.   If the elements of the original matrix $\mathbf{A}$ are a_{ij} and those of the transpose $\mathbf{A}^T$ are $a_{ij}{}^T$, then $a_{ij}{}^T = a_{ji}$. This is shown in Eqs. (7.13).

$$\mathbf{A} = \begin{bmatrix} a_{11} & a_{12} & \cdots & a_{1n} \\ a_{21} & a_{22} & \cdots & a_{2n} \\ \cdots & \cdots & \cdots & \cdots \\ a_{m1} & a_{m2} & \cdots & a_{mn} \end{bmatrix} \quad \mathbf{A}^T = \begin{bmatrix} a_{11} & a_{21} & \cdots & a_{m1} \\ a_{12} & a_{22} & \cdots & a_{m2} \\ \cdots & \cdots & \cdots & \cdots \\ a_{1n} & a_{2n} & \cdots & a_{mn} \end{bmatrix} \quad (7.13)$$

7.5.2   Matrix Algebra

The usual rules by which two numbers are combined under addition, subtraction, multiplication, or division do not apply to the handling of entire matrices. Instead, these operations must be redefined for the entire matrix in terms of operations on the individual elements of the matrix. Thus, the addition of two matrices $\mathbf{A}$ and $\mathbf{B}$ to form a third matrix $\mathbf{C}$, written as

$$\mathbf{C} = \mathbf{A} + \mathbf{B}$$

does not mean anything until the rules are given under which the elements c_{ij} are to be formed from the elements a_{ij} and b_{ij}. Similarly,

$$\mathbf{C} = \mathbf{A} \times \mathbf{B}$$

is undefined without rules for combining the a_{ij}'s with the b_{ij}'s.

The operations that have been defined are matrix addition or subtraction, matrix multiplication, and multiplication of a matrix by a single, scalar quantity. Matrix division is not defined, but another operation, multiplication by an inverse, is used. The definitions have been made so as to make the operations follow the normal rules of algebra, where possible, and to make the operations useful.

7.5.3   Equality

Two matrices $\mathbf{A}$ and $\mathbf{B}$ are said to be *equal* only if they are identical. They must possess the same dimensions, and each element of the first must be equal to the corresponding element of the second, that is, $a_{ij} = b_{ij}$ for all i and j.

7.5.4   Matrix Addition

By definition, the elements c_{ij} of $\mathbf{C} = \mathbf{A} + \mathbf{B}$ are formed by the rule

$$c_{ij} = a_{ij} + b_{ij}$$

Clearly, the matrices $\mathbf{A}$, $\mathbf{B}$, and $\mathbf{C}$ must all have the same dimensions.

Also, it is apparent that matrix addition under this rule is associative and commutative, i.e.,

$$(A + B) + D = A + (B + D)$$
$$A + B = B + A$$

Subtraction is given by

$$c_{ij} = a_{ij} - b_{ij}$$

In this connection the matrix resulting from subtracting A from itself is called a *zero*, or *null, matrix*. It has the same dimensions as A, but all the elements are zero.

An example of matrix addition is shown here,

$$\begin{bmatrix} 3 & 4 & -6 & 1 \\ 2 & -5 & 2 & -1 \\ 1 & 3 & 4 & 8 \end{bmatrix} + \begin{bmatrix} 0 & 1 & 3 & 4 \\ 1 & 0 & 2 & 3 \\ 2 & 1 & 0 & 3 \end{bmatrix} = \begin{bmatrix} 3 & 5 & -3 & 5 \\ 3 & -5 & 4 & 2 \\ 3 & 4 & 4 & 11 \end{bmatrix}$$

7.5.5 Matrix Multiplication

The definition of matrix multiplication is perhaps most easily demonstrated in the example of a system of simultaneous linear equations. Consider the three equations

$$a_{11}x_1 + a_{12}x_2 + a_{13}x_3 = c_1$$
$$a_{21}x_1 + a_{22}x_2 + a_{23}x_3 = c_2$$
$$a_{31}x_1 + a_{32}x_2 + a_{33}x_3 = c_3$$

In this system we may distinguish three matrices. They are:

1. The coefficient matrix A

$$\begin{bmatrix} a_{11} & a_{12} & a_{13} \\ a_{21} & a_{22} & a_{23} \\ a_{31} & a_{32} & a_{33} \end{bmatrix}$$

2. A vector X, which we shall write in column form

$$\begin{bmatrix} x_1 \\ x_2 \\ x_3 \end{bmatrix}$$

3. A vector C, which we shall also write in column form

$$\begin{bmatrix} c_1 \\ c_2 \\ c_3 \end{bmatrix}$$

It will be most convenient to define the multiplication of A by X in such a way that the equation $AX = C$ will hold true. By looking at the original system, we can see that this will be so if, for a desired c_i, we multiply the ith row of the **A** matrix by the **X** column vector, term by term, and take the sum of the individual products. To get c_1, for example, we multiply

$$[a_{11} \quad a_{12} \quad a_{13}] \begin{bmatrix} x_1 \\ x_2 \\ x_3 \end{bmatrix} = a_{11}x_1 + a_{12}x_2 + a_{13}x_3 = c_1$$

For c_2, we take row 2 of the **A** matrix and get

$$[a_{21} \quad a_{22} \quad a_{23}] \begin{bmatrix} x_1 \\ x_2 \\ x_3 \end{bmatrix} = a_{21}x_1 + a_{22}x_2 + a_{23}x_3 = c_2$$

and, for c_3, we take row 3 of the **A** matrix,

$$[a_{31} \quad a_{32} \quad a_{33}] \begin{bmatrix} x_1 \\ x_2 \\ x_3 \end{bmatrix} = a_{31}x_1 + a_{32}x_2 + a_{33}x_3 = c_3$$

The complete multiplication yields

$$\begin{bmatrix} a_{11} & a_{12} & a_{13} \\ a_{21} & a_{22} & a_{23} \\ a_{31} & a_{32} & a_{33} \end{bmatrix} \begin{bmatrix} x_1 \\ x_2 \\ x_3 \end{bmatrix} = \begin{bmatrix} c_1 \\ c_2 \\ c_3 \end{bmatrix}$$

In many cases the second matrix will consist of more than one column. The above definition of matrix multiplication is simply extended to operate on the second matrix, one column at a time, to generate the corresponding columns of the product matrix. Such a system would arise in connection with simultaneous equations if there were several column vectors **C** to be solved with the same coefficient matrix **A** for several solution vectors **X**.

As an example of a complete matrix multiplication consider the equation **AB** = **C**, where

$$\mathbf{A} = \begin{bmatrix} a_{11} & a_{12} & a_{13} \\ a_{21} & a_{22} & a_{23} \\ a_{31} & a_{32} & a_{33} \end{bmatrix} \quad \mathbf{B} = \begin{bmatrix} b_{11} & b_{12} \\ b_{21} & b_{22} \\ b_{31} & b_{32} \end{bmatrix} \quad \mathbf{C} = \begin{bmatrix} c_{11} & c_{12} \\ c_{21} & c_{22} \\ c_{31} & c_{32} \end{bmatrix}$$

Then,

$c_{11} = a_{11}b_{11} + a_{12}b_{21} + a_{13}b_{31}$ row 1 of **A** by column 1 of **B**
$c_{21} = a_{21}b_{11} + a_{22}b_{21} + a_{23}b_{31}$ row 2 of **A** by column 1 of **B**
$c_{31} = a_{31}b_{11} + a_{32}b_{21} + a_{33}b_{31}$ row 3 of **A** by column 1 of **B**
$c_{12} = a_{11}b_{12} + a_{12}b_{22} + a_{13}b_{32}$ row 1 of **A** by column 2 of **B**
$c_{22} = a_{21}b_{12} + a_{22}b_{22} + a_{23}b_{32}$ row 2 of **A** by column 2 of **B**
$c_{32} = a_{31}b_{12} + a_{32}b_{22} + a_{33}b_{32}$ row 3 of **A** by column 2 of **B**

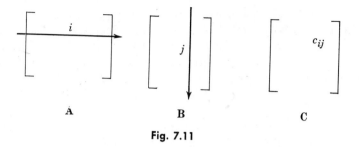

Fig. 7.11

Note that the subscripts on c give the row of the **A** matrix and the column of the **B** matrix that are being used, as shown in Fig. 7.11.

In order that two matrices may be multiplied, it is apparent that the row dimension of the first must equal the column dimension of the second. Such matrices are said to be *conformable*.

By definition, then, the elements c_{ij} of the matrix multiplication $\mathbf{C} = \mathbf{A} \times \mathbf{B}$ are formed by the rule

$$c_{ij} = \sum_{k=1}^{n} a_{ik}b_{kj} \qquad i = 1, 2, \ldots, m; j = 1, 2, \ldots, l \qquad (7.14)$$

where the dimensions of **A** are $m \times n$, of **B** are $n \times l$, and of **C** are $m \times l$.

The summation sign used above is the mathematical notation for $a_{i1}b_{1j} + a_{i2}b_{2j} + \cdots + a_{in}b_{nj}$. As indicated at the right, the summation is to be carried out for all i from 1 to n and for all j from 1 to l, thus yielding ml sums which are the terms of the **C** matrix.

As a numerical example, consider the multiplication of **A** by **B**, where

$$\mathbf{A} = \begin{bmatrix} 3 & 4 & -6 \\ 2 & -5 & 2 \\ 1 & 3 & 4 \\ 2 & 1 & 3 \end{bmatrix} \quad \text{and} \quad \mathbf{B} = \begin{bmatrix} 0 & 1 \\ 1 & 0 \\ 2 & 1 \end{bmatrix}$$

The result is

$$\begin{bmatrix} -8 & -3 \\ -1 & 4 \\ 11 & 5 \\ 7 & 5 \end{bmatrix}$$

Suppose, instead, that the multiplication were attempted in the reverse order, that is, $\mathbf{B} \times \mathbf{A}$. With these particular matrices it would not be possible, since they are not conformable in that order. In general, even if the multiplication is possible, the result of $\mathbf{A} \times \mathbf{B}$ does not equal that of $\mathbf{B} \times \mathbf{A}$. Hence, matrix multiplication is not commutative, and it is necessary to specify the order of the matrices being multiplied. *Pre-multiplication* of **B** by **A** means forming the product **AB**; *postmultiplication* of **B** by **A** means forming the product **BA**.

For matrices that are properly conformable, however, matrix multiplication is associative, and

$$\mathbf{A(BC)} = \mathbf{(AB)C}$$

This may be proved by writing out the indicated multiplications and may be seen in the following numerical example:

Let

$$\mathbf{A} = \begin{bmatrix} 3 & 4 & -6 \\ 2 & -5 & 2 \\ 1 & 3 & 4 \end{bmatrix} \qquad \mathbf{B} = \begin{bmatrix} 0 & 1 & 2 \\ 1 & 0 & 1 \\ 2 & 1 & 0 \end{bmatrix} \qquad \mathbf{C} = \begin{bmatrix} 1 & 0 & 2 \\ 1 & 0 & 2 \\ 2 & 1 & 0 \end{bmatrix}$$

Then,

$$\mathbf{BC} = \begin{bmatrix} 5 & 2 & 2 \\ 3 & 1 & 2 \\ 3 & 0 & 6 \end{bmatrix} \quad \text{and} \quad \mathbf{A(BC)} = \begin{bmatrix} 9 & 10 & -22 \\ 1 & -1 & 6 \\ 26 & 5 & 32 \end{bmatrix}$$

$$\mathbf{AB} = \begin{bmatrix} -8 & -3 & 10 \\ -1 & 4 & -1 \\ 11 & 5 & 5 \end{bmatrix} \quad \text{and} \quad \mathbf{(AB)C} = \begin{bmatrix} 9 & 10 & -22 \\ 1 & -1 & 6 \\ 26 & 5 & 32 \end{bmatrix}$$

Multiplication of a square matrix **A** by the identity matrix **I** yields the same matrix **A**. In this case the multiplication is commutative and

$$\mathbf{IA} = \mathbf{AI} = \mathbf{A}$$

For example, let **A** and **I** be given by

$$\mathbf{A} = \begin{bmatrix} 3 & 4 & -6 \\ 2 & -5 & 2 \\ 1 & 3 & 4 \end{bmatrix} \qquad \mathbf{I} = \begin{bmatrix} 1 & 0 & 0 \\ 0 & 1 & 0 \\ 0 & 0 & 1 \end{bmatrix}$$

Then,

$$\mathbf{IA} = \mathbf{AI} = \mathbf{A} = \begin{bmatrix} 3 & 4 & -6 \\ 2 & -5 & 2 \\ 1 & 3 & 4 \end{bmatrix}$$

7.5.6   Scalar Multiplication

The product of a scalar c and a matrix **A** is defined as a matrix whose elements are the products $c(a_{ij})$. Thus, if

$$\mathbf{A} = \begin{bmatrix} 1 & 0 \\ 3 & 4 \end{bmatrix} \qquad \text{then } 2\mathbf{A} = \begin{bmatrix} 2 & 0 \\ 6 & 8 \end{bmatrix}$$

Scalar multiplication is readily seen to possess the following distributive, associative, and commutative properties:

1. $(c + d)\mathbf{A} = c\mathbf{A} + d\mathbf{A}$
2. $c(\mathbf{A} + \mathbf{B}) = c\mathbf{A} + c\mathbf{B}$
3. $(cd)\mathbf{A} = c(d\mathbf{A})$
4. $\mathbf{A}(c\mathbf{B}) = (c\mathbf{A})\mathbf{B} = c(\mathbf{AB})$

A square matrix may easily be designed so that a scalar multiplication may be replaced by a matrix multiplication. The desired form is just the scalar matrix with all the diagonal elements equal to the scalar c. Using this idea, the above example becomes

$$2\mathbf{A} = \begin{bmatrix} 2 & 0 \\ 0 & 2 \end{bmatrix} \begin{bmatrix} 1 & 0 \\ 3 & 4 \end{bmatrix} = \begin{bmatrix} 2 & 0 \\ 6 & 8 \end{bmatrix}$$

The converse is also true; i.e., a scalar matrix may be replaced by a scalar constant in multiplications.

7.5.7 Equivalence of Operations

Premultiplication of a matrix $\mathbf{B}$ by a matrix $\mathbf{A}$ is equivalent to performing a series of *row* operations on $\mathbf{B}$. To alter a matrix $\mathbf{B}$ by multiplying the jth row of $\mathbf{B}$ by a constant c and adding the result to the ith row of $\mathbf{B}$, it is merely necessary to set a_{ij} equal to c and all diagonal elements of $\mathbf{A}$ equal to 1. Suppose, for example, that it is desired to eliminate the elements in the second and third rows below the diagonal in the first column of the augmented matrix shown below, as the first step in a Gaussian elimination process. It could be accomplished by a *matrix multiplication*,

$$\begin{bmatrix} 1 & 0 & 0 \\ -4 & 1 & 0 \\ -7 & 0 & 1 \end{bmatrix} \begin{bmatrix} 1 & 2 & 3 & 0 \\ 4 & 5 & 6 & 2 \\ 7 & 8 & 9 & 3 \end{bmatrix} = \begin{bmatrix} 1 & 2 & 3 & 0 \\ 0 & -3 & -6 & 2 \\ 0 & -6 & -12 & 3 \end{bmatrix}$$

The values in $\mathbf{A}$ indicate that row 1 of $\mathbf{B}$ is being used as the pivot row. It is subtracted four times from row 2 and seven times from row 3.

If matrix $\mathbf{A}$ is used as a postmultiplier of $\mathbf{B}$, the operations are performed on the *columns* of $\mathbf{B}$. The following example shows elimination of the elements to the right of the diagonal in the first row of $\mathbf{B}$:

$$\begin{bmatrix} 1 & 2 & 3 \\ 4 & 5 & 6 \\ 7 & 8 & 9 \end{bmatrix} \begin{bmatrix} 1 & -2 & -3 \\ 0 & 1 & 0 \\ 0 & 0 & 1 \end{bmatrix} = \begin{bmatrix} 1 & 0 & 0 \\ 4 & -3 & -6 \\ 7 & -6 & -12 \end{bmatrix}$$

Here, column 1 of $\mathbf{B}$ is being subtracted twice from column 2 and three times from column 3.

If the elements of the ith row of **B** are to be multiplied by a constant c, **A** should have all 1's on the diagonal except for $a_{ii} = c$ and should be used as a postmultiplier of **B**.

It would clearly be possible to program either Gaussian elimination or the Gauss-Jordan modification of that process in terms of a series of matrix multiplications instead of the row operations previously described in Sec. 7.3. Such a program would require more computer storage, however, since it would be necessary to reserve space for three matrices rather than one.

7.5.8   Orthogonality and Orthonormality

Two vectors are said to be *orthogonal* if their product equals zero. Any two of the unit vectors are orthogonal, such as e_1 and e_2, since

$$[1 \quad 0 \quad 0] \begin{bmatrix} 0 \\ 1 \\ 0 \end{bmatrix} = 0$$

A system of vectors all of which are orthogonal to each other is called an *orthogonal system*. For vectors in two- or three-dimensional space, orthogonality means simply that the vectors are mutually perpendicular. As an example, the vectors $[1 \quad 1 \quad 1]$, $[1 \quad 2 \quad -3]$, and $[5 \quad -4 \quad -1]$, which may be considered as representing three line segments drawn from the origin to the points with coordinates $(1,1,1)$, $(1,2,-3)$, and $(5,-4,-1)$, form an orthogonal system.

The lengths of these three vectors, still regarded as line segments, may be calculated by the formula

$$l = \sqrt{x^2 + y^2 + z^2}$$

giving $l_1 = \sqrt{3}$, $l_2 = \sqrt{14}$, and $l_3 = \sqrt{42}$. The vectors may be reduced to unit length, or *normalized*, by dividing by the respective values of l. The system is then said to be *orthonormal*. It is convenient, as will be shown in Sec. 7.5.11, to define an *orthogonal matrix* as a matrix consisting of a set of orthonormal vectors, which may be placed either column-wise or row-wise. Following this definition, the matrix from the above example

$$\begin{bmatrix} \dfrac{1}{\sqrt{3}} & \dfrac{1}{\sqrt{14}} & \dfrac{5}{\sqrt{42}} \\ \dfrac{1}{\sqrt{3}} & \dfrac{2}{\sqrt{14}} & \dfrac{-4}{\sqrt{42}} \\ \dfrac{1}{\sqrt{3}} & \dfrac{-3}{\sqrt{14}} & \dfrac{-1}{\sqrt{42}} \end{bmatrix}$$

is orthogonal.

7.5.9   The Inverse of a Matrix

In the algebra of real numbers the inverse of any nonzero number a is defined, such that $a^{-1} \cdot a = a \cdot a^{-1} = 1$. The question then arises as to whether there is a similar behavior with arrays or whether it is possible to find an inverse of $\mathbf{A}$ such that

$$\mathbf{A}^{-1}\mathbf{A} = \mathbf{A}\mathbf{A}^{-1} = \mathbf{I} \tag{7.15}$$

where $\mathbf{I}$ is the identity, or unit, matrix. On the assumption for the moment that an inverse may exist, then a system of simultaneous equations might be solved as follows:

$$\mathbf{A}\mathbf{X} = \mathbf{C}$$

Premultiplying both sides by the inverse of the coefficient matrix, $\mathbf{A}^{-1}$, gives

$$\mathbf{A}^{-1}\mathbf{A}\mathbf{X} = \mathbf{A}^{-1}\mathbf{C}$$

Since $\mathbf{A}^{-1}\mathbf{A} = \mathbf{I}$, and since the scalar matrix $\mathbf{I}$ can be replaced by the scalar constant 1,

$$\mathbf{X} = \mathbf{A}^{-1}\mathbf{C} \tag{7.16}$$

Thus, the result of the multiplication of the inverse of $\mathbf{A}$ and the column vector $\mathbf{C}$ is the column vector of the x values.

Let us now consider the problem of finding the inverse of a matrix. We may immediately limit ourselves to square matrices, since, for both $\mathbf{A}^{-1}\mathbf{A}$ and $\mathbf{A}\mathbf{A}^{-1}$ multiplications to be possible, the column and row dimensions must be equal. Denoting the inverse of $\mathbf{A}$ by $\mathbf{B}$, we may attempt to solve for the elements of $\mathbf{B}$ by writing the equation $\mathbf{A}\mathbf{B} = \mathbf{I}$ in array form. As an example, assume that $\mathbf{A}$ is a 3×3 matrix. Then, we have

$$\begin{bmatrix} a_{11} & a_{12} & a_{13} \\ a_{21} & a_{22} & a_{23} \\ a_{31} & a_{32} & a_{33} \end{bmatrix} \begin{bmatrix} b_{11} & b_{12} & b_{13} \\ b_{21} & b_{22} & b_{23} \\ b_{31} & b_{32} & b_{33} \end{bmatrix} = \begin{bmatrix} 1 & 0 & 0 \\ 0 & 1 & 0 \\ 0 & 0 & 1 \end{bmatrix}$$

From the properties of matrix multiplication this may be regarded as three distinct multiplications of $\mathbf{A}$ by the individual columns of $\mathbf{B}$ to yield the corresponding columns of $\mathbf{I}$.

$$\mathbf{A}\begin{bmatrix} b_{11} \\ b_{21} \\ b_{31} \end{bmatrix} = \begin{bmatrix} 1 \\ 0 \\ 0 \end{bmatrix} \qquad \mathbf{A}\begin{bmatrix} b_{12} \\ b_{22} \\ b_{32} \end{bmatrix} = \begin{bmatrix} 0 \\ 1 \\ 0 \end{bmatrix} \qquad \mathbf{A}\begin{bmatrix} b_{13} \\ b_{23} \\ b_{33} \end{bmatrix} = \begin{bmatrix} 0 \\ 0 \\ 1 \end{bmatrix}$$

The problem is now one of solving three systems of three simultaneous equations for the unknowns b_{ij}. Many methods may be used, including

Gauss-Jordan elimination.   As a specific example, let us find the inverse of

$$A = \begin{bmatrix} 1 & 1 & 2 \\ 1 & 2 & 3 \\ 2 & 3 & 1 \end{bmatrix}$$

Proceeding as above, let B represent the inverse of A, giving

$$\begin{bmatrix} 1 & 1 & 2 \\ 1 & 2 & 3 \\ 2 & 3 & 1 \end{bmatrix} \begin{bmatrix} b_{11} & b_{12} & b_{13} \\ b_{21} & b_{22} & b_{23} \\ b_{31} & b_{32} & b_{33} \end{bmatrix} = \begin{bmatrix} 1 & 0 & 0 \\ 0 & 1 & 0 \\ 0 & 0 & 1 \end{bmatrix}$$

The system to be solved for b_{11}, b_{21}, and b_{31} is

$$\begin{bmatrix} 1 & 1 & 2 \\ 1 & 2 & 3 \\ 2 & 3 & 1 \end{bmatrix} \begin{bmatrix} b_{11} \\ b_{21} \\ b_{31} \end{bmatrix} = \begin{bmatrix} 1 \\ 0 \\ 0 \end{bmatrix}$$

Since this is the matrix representation of a system of three simultaneous linear equations in b_{11}, b_{21}, and b_{31}, we may proceed to solve it by Gauss-Jordan elimination.   The augmented matrix is

$$\begin{bmatrix} 1 & 1 & 2 & \vdots & 1 \\ 1 & 2 & 3 & \vdots & 0 \\ 2 & 3 & 1 & \vdots & 0 \end{bmatrix}$$

The steps in the solution are

$$\begin{bmatrix} 1 & 1 & 2 & 1 \\ 1 & 2 & 3 & 0 \\ 2 & 3 & 1 & 0 \end{bmatrix} \xrightarrow[\begin{array}{c} R_1/1 \\ R_2 - R_1' \\ R_3 - 2R_1' \end{array}]{} \begin{bmatrix} 1 & 1 & 2 & 1 \\ 0 & 1 & 1 & -1 \\ 0 & 1 & -3 & -2 \end{bmatrix} \xrightarrow[\begin{array}{c} R_1 - R_2' \\ R_2/1 \\ R_3 - R_2' \end{array}]{}$$

$$\begin{bmatrix} 1 & 0 & 1 & 2 \\ 0 & 1 & 1 & -1 \\ 0 & 0 & -4 & -1 \end{bmatrix}$$

$$\begin{bmatrix} 1 & 0 & 1 & 2 \\ 0 & 1 & 1 & -1 \\ 0 & 0 & -4 & -1 \end{bmatrix} \xrightarrow[\begin{array}{c} R_1 - R_3' \\ R_2 - R_3' \\ R_3/(-4) \end{array}]{} \begin{bmatrix} 1 & 0 & 0 & \frac{7}{4} \\ 0 & 1 & 0 & -\frac{5}{4} \\ 0 & 0 & 1 & \frac{1}{4} \end{bmatrix} \begin{array}{l} b_{11} = \frac{7}{4} \\ b_{21} = -\frac{5}{4} \\ b_{31} = \frac{1}{4} \end{array}$$

It is not necessary to repeat the entire above operation in solving for the remainder of the b_{ij}'s.   Since the matrix A is the same in all cases, and since we have already found the row operations necessary to diagonalize it, we merely need to apply those same operations to the column vectors of the identity matrix that would have been used in forming the augmented matrix. Thus, the operations on column 2 yield

$$
\begin{bmatrix} 0 \\ 1 \\ 0 \end{bmatrix} \xrightarrow{R_1/1} \begin{bmatrix} 0 \\ 1 \\ 0 \end{bmatrix} \xrightarrow{R_1 - R_2'} \begin{bmatrix} -1 \\ 1 \\ -1 \end{bmatrix} \xrightarrow{R_1 - R_3'} \begin{bmatrix} -\frac{5}{4} \\ \frac{3}{4} \\ \frac{1}{4} \end{bmatrix} \quad \begin{array}{l} b_{12} = -\frac{5}{4} \\ b_{22} = \frac{3}{4} \\ b_{32} = \frac{1}{4} \end{array}
$$

with the row operations $R_2 - R_1'$, $R_3 - 2R_1'$ on the first; $R_2/1$, $R_3 - R_2'$ on the second; $R_2 - R_3'$, $R_3/(-4)$ on the third.

and on column 3

$$
\begin{bmatrix} 0 \\ 0 \\ 1 \end{bmatrix} \xrightarrow{R_1/1} \begin{bmatrix} 0 \\ 0 \\ 1 \end{bmatrix} \xrightarrow{R_1 - R_2'} \begin{bmatrix} 0 \\ 0 \\ 1 \end{bmatrix} \xrightarrow{R_1 - R_3'} \begin{bmatrix} \frac{1}{4} \\ \frac{1}{4} \\ -\frac{1}{4} \end{bmatrix} \quad \begin{array}{l} b_{13} = \frac{1}{4} \\ b_{23} = \frac{1}{4} \\ b_{33} = -\frac{1}{4} \end{array}
$$

Collecting the results gives

$$
\mathbf{B} = \mathbf{A}^{-1} = \begin{bmatrix} \frac{7}{4} & -\frac{5}{4} & \frac{1}{4} \\ -\frac{5}{4} & \frac{3}{4} & \frac{1}{4} \\ \frac{1}{4} & \frac{1}{4} & -\frac{1}{4} \end{bmatrix}
$$

The reader may satisfy himself that $\mathbf{B}$ is the inverse of $\mathbf{A}$ by checking that $\mathbf{AB} = \mathbf{I}$.

Assume now that $\mathbf{A}$ is the coefficient matrix of the equations

$$
\begin{aligned}
x_1 + x_2 + 2x_3 &= 8 \\
x_1 + 2x_2 + 3x_3 &= 11 \\
2x_1 + 3x_2 + x_3 &= 15
\end{aligned}
$$

Then,

$$
\mathbf{X} = \mathbf{A}^{-1}\mathbf{C} = \begin{bmatrix} \frac{7}{4} & -\frac{5}{4} & \frac{1}{4} \\ -\frac{5}{4} & \frac{3}{4} & \frac{1}{4} \\ \frac{1}{4} & \frac{1}{4} & -\frac{1}{4} \end{bmatrix} \begin{bmatrix} 8 \\ 11 \\ 15 \end{bmatrix} = \begin{bmatrix} 4 \\ 2 \\ 1 \end{bmatrix} \quad \begin{array}{l} x_1 = 4 \\ x_2 = 2 \\ x_3 = 1 \end{array}
$$

If the system of simultaneous equations is singular, the inverse of the coefficient matrix does not exist. This is clear if one considers that the same row operations on the coefficient matrix are used in the Gauss-Jordan elimination method applied to the direct solution of the equations as in the calculation of the inverse of the coefficient matrix. If these operations lead to a zero term on the diagonal which cannot be removed by exchanging rows, the system is singular and the inverse matrix does not exist. The matrix is also said to be singular.

The solution of a system of simultaneous equations by matrix inversion and multiplication is most valuable when several systems are to be solved, all of which have the same coefficient matrix but different right-hand-side column vectors. It is necessary to calculate the inverse matrix only once in this case and then to use it as a premultiplier of each of the right-hand-side column vectors.

The flow charts given in Sec. 7.3 for Gaussian elimination and Gauss-

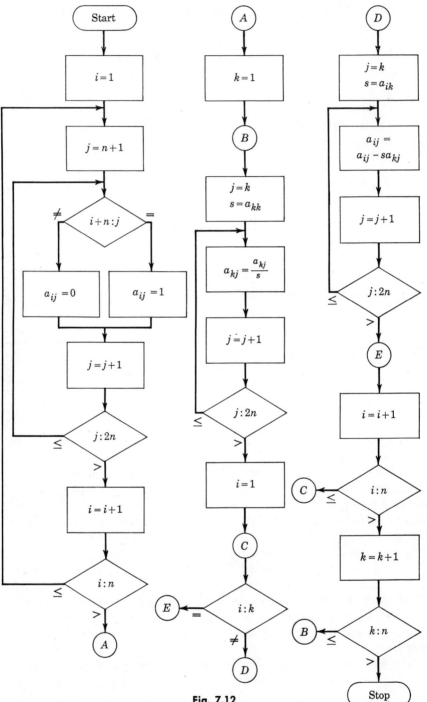

Fig. 7.12

Jordan elimination are easily modified to calculate an inverse matrix. Instead of the coefficient matrix being augmented by a single column vector, it is augmented by the n columns of the identity matrix. Each row is then considered to consist of $2n$ elements. At the end of the elimination, when the coefficient matrix has been converted to the identity matrix, the inverted coefficient matrix will occupy the positions held by the original identity matrix. Thus,

$$\begin{bmatrix} a_{11} & a_{12} & a_{13} & 1 & 0 & 0 \\ a_{21} & a_{22} & a_{23} & 0 & 1 & 0 \\ a_{31} & a_{32} & a_{33} & 0 & 0 & 1 \end{bmatrix}$$

is transformed into

$$\begin{bmatrix} 1 & 0 & 0 & b_{11} & b_{12} & b_{13} \\ 0 & 1 & 0 & b_{21} & b_{22} & b_{23} \\ 0 & 0 & 1 & b_{31} & b_{32} & b_{33} \end{bmatrix}$$

A flow chart for the inversion of an $n \times n$ matrix by Gauss-Jordan elimination is shown in Fig. 7.12. The first loop generates the augmenting identity matrix; the remainder of the procedure is similar to that given in Fig. 7.2, except for the necessary changes in indexing. An additional loop could be added, if desired, to move the elements of the inverse matrix to locations b_{ij}.

7.5.10   Inverse of an Orthogonal Matrix

For an orthogonal matrix, as defined in Sec. 7.5.8, the inverse can be shown to be equal to the transpose. Hence, if $\mathbf{A}$ is orthogonal,

$$\mathbf{A}^{-1} = \mathbf{A}^T \qquad \text{and} \qquad \mathbf{A}^T\mathbf{A} = \mathbf{I}$$

This relationship holds because of the orthonormal character of the vectors comprising the matrix $\mathbf{A}$. If we let $\mathbf{C} = \mathbf{A}^T\mathbf{A}$, the elements c_{ij} are given by

$$c_{ij} = \sum_{k=1}^{n} (a_{ik})^T(a_{kj}) = \sum_{k=1}^{n} a_{ki}a_{kj} \tag{7.17}$$

The diagonal elements c_{ii} are thus equal to the lengths of the vectors of $\mathbf{A}$, all of which are unity; the nondiagonal elements come from products of different vectors of $\mathbf{A}$ and are therefore equal to zero. Hence, $\mathbf{C} = \mathbf{I}$.

7.5.11   Convergence of Iteration Methods

Among the methods that have been discussed for solving a system of simultaneous linear equations were those of complete iteration and Gauss-

Seidel iteration. The conditions for convergence of these iterations are examined most easily in terms of matrix operations. The following discussion is by no means complete but is intended to give some insight into the problem.

Consider first the method of complete iteration used on the system $\mathbf{AX} = \mathbf{C}$. The system must be put into a form suitable for iteration, namely,

$$\mathbf{X}^{(r+1)} = f(\mathbf{X}^{(r)})$$

where r is the iteration index, or counter. Many rearrangements are possible, such as the one previously used in Eq. (7.7). In expanded form

$$x_1{}^{(r+1)} = \frac{1}{a_{11}} \left(c_1 - a_{12}x_2{}^{(r)} - a_{13}x_3{}^{(r)} - \cdots - a_{1n}x_n{}^{(r)} \right)$$

$$x_2{}^{(r+1)} = \frac{1}{a_{22}} \left(c_2 - a_{21}x_1{}^{(r)} - a_{23}x_3{}^{(r)} - \cdots - a_{2n}x_n{}^{(r)} \right) \qquad (7.7)$$

$$\cdots \cdots \cdots \cdots \cdots \cdots \cdots \cdots \cdots \cdots \cdots$$

$$x_n{}^{(r+1)} = \frac{1}{a_{nn}} \left(c_n - a_{n1}x_1{}^{(r)} - a_{n2}x_2{}^{(r)} - \cdots - a_{n,n-1}x_{n-1}{}^{(r)} \right)$$

The system may be written in matrix form as

$$
\begin{bmatrix} x_1{}^{(r+1)} \\ x_2{}^{(r+1)} \\ \cdots \\ x_n{}^{(r+1)} \end{bmatrix} =
\begin{bmatrix} c_1/a_{11} \\ c_2/a_{22} \\ \cdots \\ c_n/a_{nn} \end{bmatrix}
$$

$$
+ \begin{bmatrix} 0 & -a_{12}/a_{11} & -a_{13}/a_{11} & \cdots & -a_{1n}/a_{11} \\ -a_{21}/a_{22} & 0 & -a_{23}/a_{22} & \cdots & -a_{2n}/a_{22} \\ \cdots & \cdots & \cdots & \cdots & \cdots \\ -a_{n1}/a_{nn} & -a_{n2}/a_{nn} & -a_{n3}/a_{nn} & \cdots & 0 \end{bmatrix}
\begin{bmatrix} x_1{}^{(r)} \\ x_2{}^{(r)} \\ \cdots \\ x_n{}^{(r)} \end{bmatrix}
$$

or

$$\mathbf{X}^{(r+1)} = \mathbf{D} + \mathbf{B}\mathbf{X}^{(r)}$$

where $\mathbf{D}$ and $\mathbf{B}$ are the altered $\mathbf{C}$ and $\mathbf{A}$ matrices shown above.

Let the assumed starting values of x_i be denoted by the vector $\mathbf{X}^{(1)}$. Then by the iteration process the improved values are given by $\mathbf{X}^{(2)}$, where

$$\mathbf{X}^{(2)} = \mathbf{D} + \mathbf{B}\mathbf{X}^{(1)}$$

A second iteration yields

$$\mathbf{X}^{(3)} = \mathbf{D} + \mathbf{B}\mathbf{X}^{(2)} = \mathbf{D} + \mathbf{B}(\mathbf{D} + \mathbf{B}\mathbf{X}^{(1)})$$

and the rth iteration gives

$$\mathbf{X}^{(r+1)} = \mathbf{D} + \mathbf{B}\mathbf{X}^{(r)} = \mathbf{D} + \mathbf{B}\mathbf{D} + \mathbf{B}^2\mathbf{D} + \cdots + \mathbf{B}^{r-1}\mathbf{D} + \mathbf{B}^r\mathbf{X}^{(1)}$$

$$(7.18)$$

Each term in the above equation represents a column matrix, and all these matrices have the same dimension, with the result that the indicated addition is possible. Also, note that the superscripts on $\mathbf{B}$ are true exponents; that is, $\mathbf{B}^3$ means $\mathbf{B} \cdot \mathbf{B} \cdot \mathbf{B}$, whereas the superscript on $\mathbf{X}$ is merely the iteration index. It is possible to raise the matrix $\mathbf{B}$ to successive powers because it is square and therefore conformable with itself. Furthermore, it can be shown that powers of the same matrix are commutative in multiplication; hence $\mathbf{B}^r$ may be calculated either as $\mathbf{B} \cdot \mathbf{B}^{r-1}$ or as $\mathbf{B}^{r-1} \cdot \mathbf{B}$.

By removing the factor $\mathbf{D}$ as a postmultiplier, Eq. (7.18) becomes

$$\mathbf{X}^{(r+1)} = (\mathbf{I} + \mathbf{B} + \mathbf{B}^2 + \cdots + \mathbf{B}^{r-1})\mathbf{D} + \mathbf{B}^r\mathbf{X}^{(1)}$$

If the sequence of iterates $\mathbf{X}^{(1)}$, $\mathbf{X}^{(2)}$, . . . , $\mathbf{X}^{(r)}$, $\mathbf{X}^{(r+1)}$ is to converge to the solution $\mathbf{X}$, the series $\mathbf{I} + \mathbf{B} + \mathbf{B}^2 + \cdots + \mathbf{B}^{r-1}$ must approach a limit. The matrix $\mathbf{B}$ must be such, therefore, that $\mathbf{B}^r \rightarrow 0$ as $r \rightarrow \infty$. A discussion of the necessary *and* sufficient conditions for this to hold true is beyond the scope of this book, but several *sufficient* conditions can be stated.[1] Essentially, these have to do with the matrix $\mathbf{B}$ consisting of elements that are "sufficiently small." Any one of the following conditions is sufficient:

1. The sum of the absolute values of the elements in each row is less than 1.

$$\sum_{j=1}^{n} |b_{ij}| < 1 \qquad i = 1, 2, \ldots, n \tag{7.19}$$

2. The sum of the absolute values of the elements in each column is less than 1.

$$\sum_{i=1}^{n} |b_{ij}| < 1 \qquad j = 1, 2, \ldots, n \tag{7.20}$$

3. The sum of the squares of the elements is less than 1.

$$\sum_{\substack{i=1 \\ j=1}}^{n} (b_{ij})^2 < 1 \tag{7.21}$$

The summation in this case is for all i and j; that is, for each value of j, i goes through the range from 1 to n.

These conditions may also be stated in terms of the original coefficient matrix $\mathbf{A}$, in which case they become

1. $$\sum_{\substack{j \neq i \\ j=1}}^{n} \left| \frac{a_{ij}}{a_{ii}} \right| < 1 \qquad i = 1, 2, \ldots, n \tag{7.22}$$

---

[1] The conditions for convergence are thoroughly discussed in V. N. Faddeeva, "Computational Methods of Linear Algebra," Dover Publications, New York, 1959.

2. $\displaystyle\sum_{\substack{i \neq j \\ i=1}}^{n} \left| \frac{a_{ij}}{a_{ii}} \right| < 1 \qquad j = 1, 2, \ldots, n$ $\qquad\qquad$ (7.23)

3. $\displaystyle\sum_{\substack{i \neq j \\ i=1 \\ j=1}}^{n} \left(\frac{a_{ij}}{a_{ii}} \right)^2 < 1$ $\qquad\qquad$ (7.24)

In the above expressions note that the summations exclude the diagonal elements.

In more general terms it may be stated that the iteration will converge if the coefficient matrix **A** is "strongly diagonal," that is to say, that the diagonal elements are much larger than the nondiagonal elements. This statement is consistent with the convergence criteria given in Eqs. (7.19) to (7.24).

If the iteration process is convergent, the solution vector **X** may be obtained as the result of the rth iteration, i.e.,

$$\mathbf{X}^{(r+1)} = \mathbf{D} + \mathbf{BX}^{(r)}$$

or alternatively in terms of the limit of

$$\mathbf{X}^{(r+1)} = (\mathbf{I} + \mathbf{B} + \mathbf{B}^2 + \cdots + \mathbf{B}^{r-1})\mathbf{D} + \mathbf{B}^r\mathbf{X}^{(1)}$$

On the assumption that $\mathbf{X}^{(1)}$ is finite, $\mathbf{B}^r\mathbf{X}^{(1)}$ tends to zero, and the solution is given by

$$\mathbf{X} = (\mathbf{I} + \mathbf{B} + \mathbf{B}^2 + \cdots + \mathbf{B}^{r-1})\mathbf{D}$$

As a numerical example, let us solve the following equations by iteration.

$$x_1 + 2x_2 = 5$$
$$2x_1 + x_2 = 4$$

As arranged, the coefficient matrix is *not* strongly diagonal. The equations should first be rearranged as

$$2x_1 + x_2 = 4$$
$$x_1 + 2x_2 = 5$$

In iterative form

$$x_1 = \frac{1}{2}(4 - x_2) = 2 - \frac{x_2}{2}$$

$$x_2 = \frac{1}{2}(5 - x_1) = 2.5 - \frac{x_1}{2}$$

or

$$\begin{bmatrix} x_1 \\ x_2 \end{bmatrix} = \begin{bmatrix} 2 \\ 2.5 \end{bmatrix} + \begin{bmatrix} 0 & -\frac{1}{2} \\ -\frac{1}{2} & 0 \end{bmatrix} \begin{bmatrix} x_1 \\ x_2 \end{bmatrix}$$
$$\mathbf{X}^{(r+1)} = \quad\; \mathbf{D} \;\; + \qquad\quad \mathbf{B} \qquad\quad \mathbf{X}^{(r)}$$

As a check on the possible convergence, we may apply any one of the sufficient conditions to matrix **B**. The sums of the row elements are obviously less than 1.

Table 7.5

	b_{11}	b_{12}	b_{21}	b_{22}
B^1	0	-0.5	-0.5	0
B^2	0.25	0	0	0.25
B^3	0	-0.125	-0.125	0
B^4	0.0625	0	0	0.0625
B^5	0	-0.0312	-0.0312	0
B^6	0.0156	0	0	0.0156
B^7	0	-0.0078	-0.0078	0
B^8	0.0039	0	0	0.0039
B^9	0	-0.0020	-0.0020	0
B^{10}	0.0010	0	0	0.0010
B^{11}	0	-0.0005	-0.0005	0
B^{12}	0.0002	0	0	0.0002

The elements of the powers of **B** are listed in Table 7.5. The sum of **B**^{1} through **B**^{12} is

$$\begin{bmatrix} 0.3332 & -0.6665 \\ -0.6665 & 0.3332 \end{bmatrix}$$

and adding **I** gives

$$\begin{bmatrix} 1.3332 & -0.6665 \\ -0.6665 & 1.3332 \end{bmatrix}$$

Using this matrix as a premultiplier of **D** gives the solution

$$\begin{bmatrix} 1.3332 & -0.6665 \\ -0.6665 & 1.3332 \end{bmatrix} \begin{bmatrix} 2 \\ 2.5 \end{bmatrix} = \begin{bmatrix} 1.000 \\ 2.000 \end{bmatrix}$$

Because of the accumulation of round-off errors in the summing of powers of the **B** matrix, it is preferable to obtain the solution by iteration, i.e.,

$$X^{(r+1)} = D + BX^{(r)}$$

The convergence of the Gauss-Seidel iteration scheme may be studied in a manner similar to that applied to the complete iteration, but the analysis is more complex. A result of the analysis, however, is that conditions 1 and 2 listed above are also sufficient for the convergence of the Gauss-Seidel iteration.

7.6   DETERMINANTS

Although determinants find little practical use in the computer programming of the solution of large systems of simultaneous linear equations, they are very important in the theoretical analysis of these equations and are useful for solving small systems by hand calculation. This section includes a summary of the definitions and basic operations that are needed in the evaluation of a determinant.

By definition, a determinant is the *value* of a square matrix, calculated according to a certain rule for combining the elements of the matrix. The determinant of a matrix A is denoted as det $\mathbf{A}$ or more often by enclosing the matrix and matrix name within vertical bars. Thus,

$$|\mathbf{A}| = \begin{vmatrix} a_{11} & a_{12} & \cdots & a_{1n} \\ a_{21} & a_{22} & \cdots & a_{2n} \\ \cdots & \cdots & \cdots & \cdots \\ a_{n1} & a_{n2} & \cdots & a_{nn} \end{vmatrix} \tag{7.25}$$

The vertical bars are used in place of the brackets ordinarily placed around a matrix, in order to indicate that the determinant value of the matrix, which is a scalar number, is to be used rather than the matrix itself.

The rule by which the elements are combined in the evaluation is as follows:

The determinant value of a square matrix of dimension $n \times n$ is obtained by forming the sum of the products of all possible combinations of the elements taken n at a time, one element being taken from each row and each column. This implies the restriction that two or more elements must not be taken from the same row or column. If the elements are taken from the rows in order, that is, $a_{1j_1}, a_{2j_2}, a_{3j_3}, \ldots, a_{nj_n}$, then the sign of each product is given by $(-1)^p$, where p is the total number of inversions in the sequence of the column subscripts $j_1, j_2, j_3, \ldots, j_n$, that is, the total number of times that any subscript j_i precedes a smaller subscript.

Consider the evaluation of a third-order determinant,

$$\begin{vmatrix} a_{11} & a_{12} & a_{13} \\ a_{21} & a_{22} & a_{23} \\ a_{31} & a_{32} & a_{33} \end{vmatrix} = (-1)^0 a_{11}a_{22}a_{33} + (-1)^1 a_{11}a_{23}a_{32} + (-1)^1 a_{12}a_{21}a_{33} \\ + (-1)^2 a_{12}a_{23}a_{31} + (-1)^2 a_{13}a_{21}a_{32} + (-1)^3 a_{13}a_{22}a_{31}$$

or

$$\det \mathbf{A} = a_{11}a_{22}a_{33} - a_{11}a_{23}a_{32} - a_{12}a_{21}a_{33} + a_{12}a_{23}a_{31} + a_{13}a_{21}a_{32} \\ - a_{13}a_{22}a_{31}$$

To take the sixth term as an example, the j subscript values are 3, 2, 1. Since 3 is greater than both 2 and 1 and 2 is greater than 1, there are three inversions; hence, the sign is minus.

The total combinations possible in forming the products is

$$n(n-1)(n-2)(\cdot \cdot \cdot)(2)(1) = n!$$

inasmuch as there are n columns from which to select the first element, $n-1$ from which to select the second element, etc. Obviously, the labor involved in evaluating even a relatively small determinant by this method would be prohibitive.

Before illustrating more efficient and practical ways of evaluating a determinant, it will be helpful to list some of the properties of a determinant and the operations that may be performed on it. These follow from the expansion rule given above.

1. If all the elements in any row or column are zero, the value of the determinant is zero.

2. If the elements of two rows or columns are equal, the value of the determinant is zero.

3. If any row or column is proportional to another row or column, the value of the determinant is zero.

4. If all the elements of any row or column are multiplied by a scalar c, the value of the determinant is multiplied by c.

5. Adding any row or column multiplied by a scalar to another row or column does not change the value of the determinant.

6. Rows may be interchanged with columns without changing the value of the determinant, if they are kept in the original order.

7. If any two rows or columns are interchanged, the sign of the determinant is changed.

Two further definitions are also needed.

1. The *minor* of any element a_{ij} of a determinant $|A|$ of order n is the determinant of order $n-1$ that remains when row i and column j are removed from A.

2. The *cofactor* of any element a_{ij} of a determinant $|A|$ is the product of $(-1)^{i+j}$ and the minor of a_{ij}.

Based on these definitions, a determinant may be expanded in terms of its cofactors. With any row or column as a base, the determinant is equal to the sum of the products of the elements of that row or column and their cofactors. Thus, a fourth-order determinant may be expanded into four third-order determinants as shown in the following example.   Column 1 is taken as the base.

$$\begin{vmatrix} 1 & -1 & 0 & 1 \\ 0 & 0 & 0 & 1 \\ 2 & 5 & 6 & 0 \\ 3 & 4 & 0 & -2 \end{vmatrix} = 1 \begin{vmatrix} 0 & 0 & 1 \\ 5 & 6 & 0 \\ 4 & 0 & -2 \end{vmatrix} - 0 \begin{vmatrix} -1 & 0 & 1 \\ 5 & 6 & 0 \\ 4 & 0 & -2 \end{vmatrix}$$

$$+ 2 \begin{vmatrix} -1 & 0 & 1 \\ 0 & 0 & 1 \\ 4 & 0 & -2 \end{vmatrix} - 3 \begin{vmatrix} -1 & 0 & 1 \\ 0 & 0 & 1 \\ 5 & 6 & 0 \end{vmatrix}$$

A better choice for the base would have been column 3 or row 2.   Using the latter and omitting the zero products give

$$\begin{vmatrix} 1 & -1 & 0 & 1 \\ 0 & 0 & 0 & 1 \\ 2 & 5 & 6 & 0 \\ 3 & 4 & 0 & -2 \end{vmatrix} = \begin{vmatrix} 1 & -1 & 0 \\ 2 & 5 & 6 \\ 3 & 4 & 0 \end{vmatrix}$$

This might be further expanded in terms of column 3 as

$$\begin{vmatrix} 1 & -1 & 0 \\ 2 & 5 & 6 \\ 3 & 4 & 0 \end{vmatrix} = -6 \begin{vmatrix} 1 & -1 \\ 3 & 4 \end{vmatrix} = -6[4 - (-3)] = -42$$

The value of a determinant in triangular form is easily obtained by expanding in terms of cofactors, with column 1 as a base.   To illustrate, consider the expansion of a 4×4 triangular determinant.

$$\begin{vmatrix} a_{11} & a_{12} & a_{13} & a_{14} \\ 0 & a_{22} & a_{23} & a_{24} \\ 0 & 0 & a_{33} & a_{34} \\ 0 & 0 & 0 & a_{44} \end{vmatrix} = a_{11} \begin{vmatrix} a_{22} & a_{23} & a_{24} \\ 0 & a_{33} & a_{34} \\ 0 & 0 & a_{44} \end{vmatrix} = a_{11}a_{22} \begin{vmatrix} a_{33} & a_{34} \\ 0 & a_{44} \end{vmatrix}$$

$$= a_{11}a_{22}a_{33}a_{44}$$

Generalizing, the value of an $n \times n$ triangular determinant is

$$a_{11}a_{22}a_{33} \cdot \cdot \cdot a_{nn}$$

Determinants may also have other triangular forms, as illustrated below:

$$\begin{vmatrix} a_{11} & 0 & 0 & 0 \\ a_{21} & a_{22} & 0 & 0 \\ a_{31} & a_{32} & a_{33} & 0 \\ a_{41} & a_{42} & a_{43} & a_{44} \end{vmatrix} \quad \begin{vmatrix} 0 & 0 & 0 & a_{14} \\ 0 & 0 & a_{23} & a_{24} \\ 0 & a_{32} & a_{33} & a_{34} \\ a_{41} & a_{42} & a_{43} & a_{44} \end{vmatrix} \quad \begin{vmatrix} a_{11} & a_{12} & a_{13} & a_{14} \\ a_{21} & a_{22} & a_{23} & 0 \\ a_{31} & a_{32} & 0 & 0 \\ a_{41} & 0 & 0 & 0 \end{vmatrix}$$

In each case the value of the determinant is just the product of the diagonal coefficients with the proper sign. For example,

$$\begin{vmatrix} 0 & 0 & 2 \\ 0 & -4 & 3 \\ 6 & 1 & -3 \end{vmatrix} = -(2)(-4)(6) = 48$$

7.6.1 Gaussian Elimination

It has just been shown that the value of a triangular determinant is easily calculated. Thus, the Gaussian elimination method used to solve simultaneous equations may also be used to evaluate determinants. However, by rule 4, the value of the determinant will be changed if the rows are divided by their diagonal elements. Consequently, it is necessary either not to take out these factors or else to save them as a factor to be used in the final evaluation. The two methods are illustrated in the following example: If the diagonal elements are not factored out,

$$\begin{vmatrix} 2 & 2 & 2 & 4 \\ 1 & 2 & 4 & 1 \\ 4 & 0 & 1 & 2 \\ 2 & 0 & 3 & 1 \end{vmatrix} \quad \begin{array}{c} R_1 \\ R_2 - R_1/2 \\ R_3 - 2R_1 \\ R_4 - R_1 \end{array} \quad \begin{vmatrix} 2 & 2 & 2 & 4 \\ 0 & 1 & 3 & -1 \\ 0 & -4 & -3 & -6 \\ 0 & -2 & 1 & -3 \end{vmatrix}$$

$$\begin{vmatrix} 2 & 2 & 2 & 4 \\ 0 & 1 & 3 & -1 \\ 0 & -4 & -3 & -6 \\ 0 & -2 & 1 & -3 \end{vmatrix} \quad \begin{array}{c} R_1 \\ R_2 \\ R_3 - (-4R_2) \\ R_4 - (-2R_2) \end{array} \quad \begin{vmatrix} 2 & 2 & 2 & 4 \\ 0 & 1 & 3 & -1 \\ 0 & 0 & 9 & -10 \\ 0 & 0 & 7 & -5 \end{vmatrix}$$

$$\begin{vmatrix} 2 & 2 & 2 & 4 \\ 0 & 1 & 3 & -1 \\ 0 & 0 & 9 & -10 \\ 0 & 0 & 7 & -5 \end{vmatrix} \quad \begin{array}{c} R_1 \\ R_2 \\ R_3 \\ R_4 - \frac{7}{9}R_3 \end{array} \quad \begin{vmatrix} 2 & 2 & 2 & 4 \\ 0 & 1 & 3 & -1 \\ 0 & 0 & 9 & -10 \\ 0 & 0 & 0 & \frac{25}{9} \end{vmatrix}$$

The value of the determinant is, then,

$(2)(1)(9)(\frac{25}{9}) = 50$

If the diagonal elements are factored out,

$$
\begin{vmatrix}
2 & 2 & 2 & 4 \\
1 & 2 & 4 & 1 \\
4 & 0 & 1 & 2 \\
2 & 0 & 3 & 1
\end{vmatrix}
\begin{array}{l}
R_1/2 \\
R_2 - R_1' \\
R_3 - 4R_1' \\
R_4 - 2R_1'
\end{array}
\quad 2
\begin{vmatrix}
1 & 1 & 1 & 2 \\
0 & 1 & 3 & -1 \\
0 & -4 & -3 & -6 \\
0 & -2 & 1 & -3
\end{vmatrix}
$$

$$
2
\begin{vmatrix}
1 & 1 & 1 & 2 \\
0 & 1 & 3 & -1 \\
0 & -4 & -3 & -6 \\
0 & -2 & 1 & -3
\end{vmatrix}
\begin{array}{l}
R_1 \\
R_2/1 \\
R_3 - (-4R_2') \\
R_4 - (-2R_2')
\end{array}
\quad (2)(1)
\begin{vmatrix}
1 & 1 & 1 & 2 \\
0 & 1 & 3 & -1 \\
0 & 0 & 9 & -10 \\
0 & 0 & 7 & -5
\end{vmatrix}
$$

$$
(2)(1)
\begin{vmatrix}
1 & 1 & 1 & 2 \\
0 & 1 & 3 & -1 \\
0 & 0 & 9 & -10 \\
0 & 0 & 7 & -5
\end{vmatrix}
\begin{array}{l}
R_1 \\
R_2 \\
R_3/9 \\
R_4 - 7R_3'
\end{array}
\quad (2)(1)(9)
\begin{vmatrix}
1 & 1 & 1 & 2 \\
0 & 1 & 3 & -1 \\
0 & 0 & 1 & -\frac{10}{9} \\
0 & 0 & 0 & \frac{25}{9}
\end{vmatrix}
$$

$$
(2)(1)(9)
\begin{vmatrix}
1 & 1 & 1 & 2 \\
0 & 1 & 3 & -1 \\
0 & 0 & 1 & -\frac{10}{9} \\
0 & 0 & 0 & \frac{25}{9}
\end{vmatrix}
\begin{array}{l}
R_1 \\
R_2 \\
R_3 \\
R_4/\frac{25}{9}
\end{array}
\quad (2)(1)(9)(\tfrac{25}{9})
\begin{vmatrix}
1 & 1 & 1 & 2 \\
0 & 1 & 3 & -1 \\
0 & 0 & 1 & -\frac{10}{9} \\
0 & 0 & 0 & 1
\end{vmatrix}
$$

In this case the value of the remaining determinant is simply 1, and it follows that the value of the original determinant is the product of the factors which were saved.

$$(2)(1)(9)(\tfrac{25}{9}) = 50$$

7.6.2   Simultaneous Equations

Determinants may be used to solve a system of simultaneous linear equations. According to Cramer's rule, the x_i's are given by

$$x_i = \frac{|\mathbf{D}_i|}{|\mathbf{D}|} \tag{7.26}$$

where $|\mathbf{D}|$ is the determinant of the coefficient matrix of the system and each $|\mathbf{D}_i|$ is a determinant obtained by replacing the ith column of the coefficient matrix with the column vector of the system. Consider the equations

$$
\begin{aligned}
x_1 + x_2 + 2x_3 &= 14 \\
2x_1 + x_2 + x_3 &= 10 \\
x_1 + 2x_2 + x_3 &= 12
\end{aligned}
$$

Then,

$$|\mathbf{D}| = \begin{vmatrix} 1 & 1 & 2 \\ 2 & 1 & 1 \\ 1 & 2 & 1 \end{vmatrix} = 4 \qquad |\mathbf{D}_1| = \begin{vmatrix} 14 & 1 & 2 \\ 10 & 1 & 1 \\ 12 & 2 & 1 \end{vmatrix} = 4$$

$$|\mathbf{D}_2| = \begin{vmatrix} 1 & 14 & 2 \\ 2 & 10 & 1 \\ 1 & 12 & 1 \end{vmatrix} = 12 \qquad |\mathbf{D}_3| = \begin{vmatrix} 1 & 1 & 14 \\ 2 & 1 & 10 \\ 1 & 2 & 12 \end{vmatrix} = 20$$

$$x_1 = \tfrac{4}{4} = 1$$
$$x_2 = \tfrac{12}{4} = 3$$
$$x_3 = \tfrac{20}{4} = 5$$

The work required to evaluate any one of the above determinants by such a method as Gaussian elimination is nearly equal to that required to solve the entire system of equations by the same elimination method. Essentially, the use of Cramer's rule involves the handling of the columns of the coefficient matrix n times in solving for the n unknowns one at a time, whereas straightforward elimination avoids this duplication of effort by solving for all the unknowns together. Cramer's rule is therefore inefficient as a method for solving a large system of equations.

Determinants may also be used to advantage in discussing the theory of simultaneous equations. In considering two equations, as was done in Sec. 7.2, the solutions, Eq. (7.3), may be written in determinant form,

$$x_1 = \frac{a_{22}c_1 - a_{12}c_2}{a_{11}a_{22} - a_{21}a_{12}} = \frac{\begin{vmatrix} c_1 & a_{12} \\ c_2 & a_{22} \end{vmatrix}}{\begin{vmatrix} a_{11} & a_{12} \\ a_{21} & a_{22} \end{vmatrix}} = \frac{|\mathbf{D}_1|}{|\mathbf{D}|}$$

$$x_2 = \frac{a_{11}c_2 - a_{21}c_1}{a_{11}a_{22} - a_{21}a_{12}} = \frac{\begin{vmatrix} a_{11} & c_1 \\ a_{21} & c_2 \end{vmatrix}}{\begin{vmatrix} a_{11} & a_{12} \\ a_{21} & a_{22} \end{vmatrix}} = \frac{|\mathbf{D}_2|}{|\mathbf{D}|}$$

provided that $|\mathbf{D}|$ does not equal zero. In general a system of n equations possesses a unique, nontrivial solution if the determinant of the coefficient matrix (often called *the determinant of the system*) does not vanish. If the system contains two or more equations that are dependent, the determinant will contain two or more rows that are proportional to each other and by rule 3 (page 257) its value is zero. There is then no unique solution of the equations.

For a system of homogeneous equations the right-hand-side column vector is 0, and therefore all the determinants in the numerators of the solution are zero, by rule 1 (page 257).   If the determinant of the system is different from 0, the only possible solution is for all $x_i = 0$.   An important case is that in which the determinant of the system is also 0, as discussed in the following section.

7.7   CHARACTERISTIC ROOTS OF A DETERMINANT

Many vibration problems in engineering involve a system of equations of the form

$$a_{11}x_1 + a_{12}x_2 + \cdots + a_{1n}x_n = \lambda x_1$$
$$a_{21}x_1 + a_{22}x_2 + \cdots + a_{2n}x_n = \lambda x_2$$
$$\cdots\cdots\cdots\cdots\cdots\cdots\cdots\cdots$$
$$a_{n1}x_1 + a_{n2}x_2 + \cdots + a_{nn}x_n = \lambda x_n$$

An example of this type is presented in Sec. 7.8.3.

The equations may be written in homogeneous form as

$$(a_{11} - \lambda)x_1 + \qquad a_{12}x_2 + \cdots + \qquad a_{1n}x_n = 0$$
$$a_{21}x_1 + (a_{22} - \lambda)x_2 + \cdots + \qquad a_{2n}x_n = 0 \qquad\qquad (7.27)$$
$$\cdots\cdots\cdots\cdots\cdots\cdots\cdots\cdots\cdots\cdots$$
$$a_{n1}x_1 + \qquad a_{n2}x_2 + \cdots + (a_{nn} - \lambda)x_n = 0$$

One possible solution is that all $x_i = 0$, as pointed out previously. For a nontrivial solution to exist, the determinant of the system of the homogeneous equations must be zero.   That determinant is

$$\begin{vmatrix} a_{11} - \lambda & a_{12} & \cdots & a_{1n} \\ a_{21} & a_{22} - \lambda & \cdots & a_{2n} \\ \cdots & \cdots & \cdots & \cdots \\ a_{n1} & a_{n2} & \cdots & a_{nn} - \lambda \end{vmatrix}$$

Expansion of the determinant gives an nth-order polynomial in λ called the *characteristic equation*, the roots of which are $\lambda_1, \lambda_2, \ldots, \lambda_n$.   They are called the *characteristic roots* of the determinant.   In vibration analysis they are related to the frequencies of vibration.   For each of the n values of λ, and only for these values, a nontrivial set of x_i's may be calculated.   These solutions are called the *characteristic vectors*.

As an example, consider the solution of the equations

$$x_1 + x_2 \qquad\quad = \lambda x_1$$
$$2x_1 + x_2 + \quad x_3 = \lambda x_2$$
$$x_2 + 2x_3 = \lambda x_3$$

The determinant to be solved for λ is

$$\begin{vmatrix} 1 - \lambda & 1 & 0 \\ 2 & 1 - \lambda & 1 \\ 0 & 1 & 2 - \lambda \end{vmatrix} = 0$$

Expansion yields

$$(1 - \lambda)(1 - \lambda)(2 - \lambda) - 2(2 - \lambda) - 1(1 - \lambda) = 0$$

or

$$\lambda^3 - 4\lambda^2 + 2\lambda + 3 = 0$$

The roots are

$$\lambda_1 = \frac{1}{2} - \frac{\sqrt{5}}{2} \approx -0.61803$$

$$\lambda_2 = \frac{1}{2} + \frac{\sqrt{5}}{2} \approx 1.61803$$

$$\lambda_3 = 3.00000$$

Substitution of the values of λ_i back into the equations produces systems containing dependent equations. At least one of the values of x is arbitrary in each system, say x_i. The characteristic vector $\mathbf{X}_i$ may be expressed in terms of it.

For $\lambda_1 = -0.61803$, the equations may be solved to yield $x_1 = x_1$, $x_2 = -1.61803x_1$, and $x_3 = 0.61803x_1$. In vector notation

$$\mathbf{X}_1 = x_1 \begin{bmatrix} 1 \\ -1.61803 \\ 0.61803 \end{bmatrix} \quad \text{or simply} \quad \begin{bmatrix} 1 \\ -1.61803 \\ 0.61803 \end{bmatrix}$$

Similarly, for $\lambda_2 = 1.61803$ and $\lambda_3 = 3.00000$, the vectors are

$$\mathbf{X}_2 = \begin{bmatrix} 1.61803 \\ 1 \\ -2.61803 \end{bmatrix} \quad \mathbf{X}_3 = \begin{bmatrix} 0.50000 \\ 1.00000 \\ 1 \end{bmatrix}$$

7.7.1 Trace of a Matrix

When the determinant containing λ shown in Eq. (7.27) is expanded, a polynomial is obtained, which may be written as

$$(-1)^n\lambda^n + (-1)^{n-1}b_1\lambda^{n-1} + b_2\lambda^{n-2} + \cdots + b_n = 0$$

where, in particular, the coefficients of λ^n and λ^{n-1} come only from the diagonal product $(a_{11} - \lambda)(a_{22} - \lambda)(\cdot\ \cdot\ \cdot)(a_{nn} - \lambda)$. Carrying out this expansion shows that

$$b_1 = \sum_{i=1}^{n} a_{ii}$$

According to Newton's relations (Sec. 6.2.2), however, b_1 is also equal to the algebraic sum of the roots of the polynomial. Hence, for a square matrix, the sum of the diagonal elements equals the sum of the characteristic roots. This sum is often referred to as the *trace*, or *spur*, of the matrix.

7.7.2   Jacobi's Method for Getting Characteristic Roots

For a large matrix, the straightforward expansion of the determinant containing λ, followed by the solution of the resulting polynomial, is theoretically possible but is practically very time-consuming. Instead, many methods have been devised in which the original matrix is altered to a form either that is very easily expanded or that yields the characteristic roots directly. One of the most widely used methods for handling symmetric matrices is that originally due to Jacobi, which has since been used in several modified forms.

Consider first a diagonal matrix and its characteristic determinant, such as

$$\begin{bmatrix} 3 & 0 & 0 \\ 0 & 4 & 0 \\ 0 & 0 & 5 \end{bmatrix} \quad \begin{vmatrix} 3-\lambda & 0 & 0 \\ 0 & 4-\lambda & 0 \\ 0 & 0 & 5-\lambda \end{vmatrix}$$

The characteristic polynomial is given by $(3 - \lambda)(4 - \lambda)(5 - \lambda) = 0$, and the characteristic roots are obviously $\lambda_1 = 3$, $\lambda_2 = 4$, and $\lambda_3 = 5$. It is also clear that the characteristic roots of any diagonal or triangular matrix are just the values on the diagonal. If, then, a matrix can be transformed into a diagonal or triangular form, in such a way that the characteristic roots are not changed, the problem of finding the roots may conceivably be reduced.

The transformation to be used is known as a *similarity transformation*. Two matrices, **A** and **B**, are called similar if they are related by

$$\mathbf{B} = \mathbf{U}^{-1}\mathbf{A}\mathbf{U}$$

where **U** is a nonsingular matrix. To show that **A** and **B** have the same characteristic roots, it is necessary only to substitute the above equation into the determinant equation

$$|\mathbf{B} - \lambda\mathbf{I}| = 0$$

giving

$$|\mathbf{U}^{-1}\mathbf{A}\mathbf{U} - \lambda\mathbf{I}| = 0$$

Since $\lambda\mathbf{I}$ may be replaced by $\lambda\mathbf{U}^{-1}\mathbf{I}\mathbf{U} = \mathbf{U}^{-1}\lambda\mathbf{I}\mathbf{U}$, we may factor out $\mathbf{U}^{-1}$ and $\mathbf{U}$, as

$$|\mathbf{U}^{-1}|\,|\mathbf{A} - \lambda\mathbf{I}|\,|\mathbf{U}| = 0$$

Hence,

$$|\mathbf{A} - \lambda\mathbf{I}| = 0$$

and the same values of λ satisfy the polynomials from both $\mathbf{A}$ and $\mathbf{B}$.

Assume, now, that a matrix $\mathbf{U}$ has been found that transforms $\mathbf{A}$ into the diagonal matrix $\mathbf{D} = [\lambda_1 \quad \lambda_2 \quad \cdots \quad \lambda_n]$. Then,

$$\mathbf{U}^{-1}\mathbf{A}\mathbf{U} = \mathbf{D}$$

and premultiplication by $\mathbf{U}$ yields

$$\mathbf{A}\mathbf{U} = \mathbf{U}\mathbf{D}$$

The meaning of this equation may be made clearer by writing it in expanded form for a third-order matrix,

$$\begin{bmatrix} a_{11} & a_{12} & a_{13} \\ a_{21} & a_{22} & a_{23} \\ a_{31} & a_{32} & a_{33} \end{bmatrix}\begin{bmatrix} u_{11} & u_{12} & u_{13} \\ u_{21} & u_{22} & u_{23} \\ u_{31} & u_{32} & u_{33} \end{bmatrix} = \begin{bmatrix} u_{11} & u_{12} & u_{13} \\ u_{21} & u_{22} & u_{23} \\ u_{31} & u_{32} & u_{33} \end{bmatrix}\begin{bmatrix} \lambda_1 & 0 & 0 \\ 0 & \lambda_2 & 0 \\ 0 & 0 & \lambda_3 \end{bmatrix}$$

$$= \begin{bmatrix} \lambda_1 u_{11} & \lambda_2 u_{12} & \lambda_3 u_{13} \\ \lambda_1 u_{21} & \lambda_2 u_{22} & \lambda_3 u_{23} \\ \lambda_1 u_{31} & \lambda_2 u_{32} & \lambda_3 u_{33} \end{bmatrix}$$

Comparing this equation with $\mathbf{A}\mathbf{X}_i = \lambda_i\mathbf{X}_i$

$$\begin{bmatrix} a_{11} & a_{12} & a_{13} \\ a_{21} & a_{22} & a_{23} \\ a_{31} & a_{32} & a_{33} \end{bmatrix}\begin{bmatrix} x_1 \\ x_2 \\ x_3 \end{bmatrix} = \lambda_i \begin{bmatrix} x_1 \\ x_2 \\ x_3 \end{bmatrix} = \begin{bmatrix} \lambda_i x_1 \\ \lambda_i x_2 \\ \lambda_i x_3 \end{bmatrix}$$

shows that the ith column of the $\mathbf{U}$ matrix equals the characteristic vector corresponding to the ith characteristic root λ_i. Therefore, the method yields both the characteristic roots and characteristic vectors at the same time.

As a preliminary application of Jacobi's method, let us restrict ourselves to diagonalizing a second-order matrix. A convenient choice for $\mathbf{U}$ is the orthogonal matrix

$$\mathbf{U} = \begin{bmatrix} \cos\theta & \sin\theta \\ -\sin\theta & \cos\theta \end{bmatrix}$$

Because of the orthogonality of $\mathbf{U}$, as discussed in Sec. 7.5.10,

$$\mathbf{U}^{-1} = \mathbf{U}^T = \begin{bmatrix} \cos\theta & -\sin\theta \\ \sin\theta & \cos\theta \end{bmatrix}$$

Carrying out the transformation yields

$$\mathbf{U^{-1}AU} = \begin{bmatrix} \cos\theta & -\sin\theta \\ \sin\theta & \cos\theta \end{bmatrix} \begin{bmatrix} a_{11} & a_{12} \\ a_{21} & a_{22} \end{bmatrix} \begin{bmatrix} \cos\theta & \sin\theta \\ -\sin\theta & \cos\theta \end{bmatrix}$$

$$= \begin{bmatrix} a_{11}\cos^2\theta - 2a_{12}\sin\theta\cos\theta & (a_{11} - a_{22})\sin\theta\cos\theta \\ \qquad + a_{22}\sin^2\theta & \qquad + a_{12}(\cos^2\theta - \sin^2\theta) \\ (a_{11} - a_{22})\sin\theta\cos\theta & a_{11}\sin^2\theta + 2a_{12}\sin\theta\cos\theta \\ \qquad + a_{12}(\cos^2\theta - \sin^2\theta) & \qquad + a_{22}\cos^2\theta \end{bmatrix}$$

Note that symmetry of $\mathbf{A}$ has been assumed, that is, $a_{12} = a_{21}$, and that the transformed matrix is also symmetrical.

The angle θ is arbitrary and may be chosen so as to annihilate the off-diagonal elements. Setting either of those terms equal to zero and using double-angle trigonometric identities give

$$\frac{a_{11} - a_{22}}{2}\sin 2\theta + a_{12}\cos 2\theta = 0 \tag{7.28}$$

or

$$\tan 2\theta = \frac{-2a_{12}}{a_{11} - a_{22}} \qquad \text{where } |\theta| \le \frac{\pi}{4}$$

We may also make the substitution $\sin 2\theta = \sqrt{1 - \cos^2 2\theta}$ into Eq. (7.28) and solve for $\cos 2\theta$, getting

$$\cos 2\theta = \frac{|a_{11} - a_{22}|}{\sqrt{(a_{11} - a_{22})^2 + 4a_{12}{}^2}}$$

The functions of θ are obtained from trigonometric identities as

$$\sin\theta = (\text{sign of } \tan 2\theta)\sqrt{\frac{1 - \cos 2\theta}{2}}$$

$$\cos\theta = \sqrt{\frac{1 + \cos 2\theta}{2}}$$

where

$$\begin{aligned} \textit{Sign of } \tan 2\theta &= \text{sign of } (-2a_{12}) & \text{if } a_{11} - a_{22} \ge 0 \\ &= \text{sign of } (+2a_{12}) & \text{if } a_{11} - a_{22} < 0 \end{aligned}$$

As an example, let

$$\mathbf{A} = \begin{bmatrix} 1 & 2 \\ 2 & 1 \end{bmatrix}$$

Then,

$$\cos 2\theta = 0$$

Sign of $\tan 2\theta = (-)$

$$\sin \theta = -\frac{\sqrt{2}}{2} \qquad \cos \theta = \frac{\sqrt{2}}{2}$$

$$\mathbf{U} = \begin{bmatrix} \sqrt{2}/2 & -\sqrt{2}/2 \\ \sqrt{2}/2 & \sqrt{2}/2 \end{bmatrix}$$

$$\mathbf{U}^{-1}\mathbf{A}\mathbf{U} = \begin{bmatrix} \frac{1}{2} + 4(\frac{1}{2}) + \frac{1}{2} & 0 \\ 0 & \frac{1}{2} - 4(\frac{1}{2}) + \frac{1}{2} \end{bmatrix} = \begin{bmatrix} 3 & 0 \\ 0 & -1 \end{bmatrix}$$

The roots are $\lambda_1 = 3$, $\lambda_2 = -1$.   The characteristic vectors are given by the first and second columns of the $\mathbf{U}$ matrix as

$$\mathbf{X}_1 = \begin{bmatrix} \sqrt{2}/2 \\ \sqrt{2}/2 \end{bmatrix} \qquad \mathbf{X}_2 = \begin{bmatrix} -\sqrt{2}/2 \\ \sqrt{2}/2 \end{bmatrix}$$

Since one of the values in each vector $\mathbf{X}_i$ is arbitrary, say x_i, we may divide by it and get

$$\mathbf{X}_1 = \begin{bmatrix} 1 \\ 1 \end{bmatrix} \qquad \mathbf{X}_2 = \begin{bmatrix} -1 \\ 1 \end{bmatrix}$$

Carrying out the multiplications $(\mathbf{A} - \lambda_i)\mathbf{X}_i$ verifies that the product does come out equal to zero for the above vectors.

For a matrix that is larger than second-order, it is necessary to annihilate one pair of off-diagonal elements at a time, a_{ij} and a_{ji} (where i is assumed to be less than j).   The $\mathbf{U}$ matrix, which is again orthogonal, is formed from the unit matrix by setting $u_{ii} = \cos \theta$, $u_{ij} = \sin \theta$, $u_{ji} = -\sin \theta$, $u_{jj} = \cos \theta$.

To annihilate elements a_{13} and a_{31} in a fourth-order matrix, for example,

$$\mathbf{U} = \begin{bmatrix} \cos \theta & 0 & \sin \theta & 0 \\ 0 & 1 & 0 & 0 \\ -\sin \theta & 0 & \cos \theta & 0 \\ 0 & 0 & 0 & 1 \end{bmatrix}$$

If $\mathbf{B} = \mathbf{U}^{-1}\mathbf{A}\mathbf{U}$ in the annihilation of elements a_{ij} and a_{ji}, only the ith and jth rows and columns of $\mathbf{B}$ differ from those of $\mathbf{A}$ and need to be calculated.   Thus,

$$b_{ii} = a_{ii} \cos^2 \theta - 2a_{ij} \sin \theta \cos \theta + a_{jj} \sin^2 \theta$$
$$b_{ij} = b_{ji} = 0$$
$$b_{jj} = a_{ii} \sin^2 \theta + 2a_{ij} \sin \theta \cos \theta + a_{jj} \cos^2 \theta$$

ith row: $\quad b_{ik} = a_{ik} \cos \theta - a_{jk} \sin \theta$

jth row: $\quad b_{jk} = a_{jk} \cos \theta + a_{ik} \sin \theta$ $\qquad$ for $k = 1$ to n, except for

ith column: $\quad b_{ki} = a_{ki} \cos \theta - a_{kj} \sin \theta$ $\qquad\qquad\qquad k = i, k = j$

jth column: $\quad b_{kj} = a_{kj} \cos \theta + a_{ki} \sin \theta$

Inasmuch as only two elements of **A** are annihilated in a single transformation, an iterative process is needed to reduce the matrix to its final diagonal form. In each successive step the remaining off-diagonal pair of largest magnitude is annihilated. Previously annihilated elements will not necessarily remain zero during the iteration, but it can be shown that the iteration does converge. Scanning the matrix to find the largest element at each step is, however, relatively time-consuming. It is simpler to annihilate the elements according to a predefined pattern, and this method has also been shown to be convergent. The iteration is stopped when all off-diagonal elements have been reduced to a desired level.

The sequence of transformations may be written as

$$\mathbf{U}_k{}^{-1}\mathbf{U}_{k-1}^{-1} \cdot \cdot \cdot \mathbf{U}_2{}^{-1}\mathbf{U}_1{}^{-1}\mathbf{A}\mathbf{U}_1\mathbf{U}_2 \cdot \cdot \cdot \mathbf{U}_{k-1}\mathbf{U}_k$$

Since $\mathbf{U}_k{}^{-1} = \mathbf{U}_k{}^T$, and $\mathbf{U}_j{}^T\mathbf{U}_i{}^T = (\mathbf{U}_i\mathbf{U}_j)^T$, the overall transformation is equivalent to

$$(\mathbf{U}_1\mathbf{U}_2 \cdot \cdot \cdot \mathbf{U}_{k-1}\mathbf{U}_k)^{-1}\mathbf{A}(\mathbf{U}_1\mathbf{U}_2 \cdot \cdot \cdot \mathbf{U}_{k-1}\mathbf{U}_k)$$

If we designate the sequence $\mathbf{U}_1\mathbf{U}_2 \cdot \cdot \cdot \mathbf{U}_{k-1}\mathbf{U}_k$ as **U**, then the characteristic vectors are given, as before, by the columns of **U**.

As a second example, let us apply Jacobi's method to finding the characteristic roots of the symmetric matrix

$$\mathbf{A} = \begin{bmatrix} 1 & 2 & 1 \\ 2 & 1 & 1 \\ 1 & 1 & 1 \end{bmatrix}$$

The steps may be summarized as

1. Annihilation of a_{12} and a_{21}

$$\cos 2\theta = \frac{|a_{11} - a_{22}|}{\sqrt{(a_{11} - a_{22})^2 + 4a_{12}{}^2}} = 0$$

Sign of $\tan 2\theta = (-)$

$$\sin \theta = -\frac{\sqrt{2}}{2} \qquad \cos \theta = \frac{\sqrt{2}}{2}$$

$$\mathbf{U}_1 = \begin{bmatrix} \sqrt{2}/2 & -\sqrt{2}/2 & 0 \\ \sqrt{2}/2 & \sqrt{2}/2 & 0 \\ 0 & 0 & 1 \end{bmatrix} \qquad \mathbf{U}_1{}^{-1}\mathbf{A}\mathbf{U}_1 = \begin{bmatrix} 3 & 0 & \sqrt{2} \\ 0 & -1 & 0 \\ \sqrt{2} & 0 & 1 \end{bmatrix}$$

2. Annihilation of a_{13} and a_{31}

$$\cos 2\theta = \frac{|a_{11} - a_{33}|}{\sqrt{(a_{11} - a_{33})^2 + 4a_{13}{}^2}} = \frac{\sqrt{3}}{3}$$

$$\sin \theta = -\sqrt{\frac{1 - \sqrt{3}/3}{2}} \qquad \cos \theta = \sqrt{\frac{1 + \sqrt{3}/3}{2}}$$

$$\mathbf{U}_2 = \begin{bmatrix} \sqrt{(1 + \sqrt{3}/3)/2} & 0 & -\sqrt{(1 - \sqrt{3}/3)/2} \\ 0 & 1 & 0 \\ \sqrt{(1 - \sqrt{3}/3)/2} & 0 & \sqrt{(1 + \sqrt{3}/3)/2} \end{bmatrix}$$

$$\mathbf{U}_2{}^{-1}\mathbf{U}_1{}^{-1}\mathbf{A}\mathbf{U}_1\mathbf{U}_2 = \begin{bmatrix} 2 + \sqrt{3} & 0 & 0 \\ 0 & -1 & 0 \\ 0 & 0 & 2 - \sqrt{3} \end{bmatrix}$$

3. Since the matrix is now in diagonal form, the roots are $\lambda_1 = 2 + \sqrt{3}$, $\lambda_2 = -1$, and $\lambda_3 = 2 - \sqrt{3}$.

4. The characteristic vectors are the columns of the matrix formed from the product $\mathbf{U}_1\mathbf{U}_2$ and are

$$\mathbf{X}_1 = \begin{bmatrix} \sqrt{(1 + \sqrt{3}/3)/4} \\ \sqrt{(1 + \sqrt{3}/3)/4} \\ \sqrt{(1 - \sqrt{3}/3)/2} \end{bmatrix} \qquad \mathbf{X}_2 = \begin{bmatrix} -\sqrt{2}/2 \\ \sqrt{2}/2 \\ 0 \end{bmatrix}$$

$$\mathbf{X}_3 = \begin{bmatrix} -\sqrt{(1 - \sqrt{3}/3)/4} \\ -\sqrt{(1 - \sqrt{3}/3)/4} \\ \sqrt{(1 + \sqrt{3}/3)/2} \end{bmatrix}$$

or, more simply,

$$\mathbf{X}_1 = \begin{bmatrix} 1 \\ 1 \\ \sqrt{2(1 - \sqrt{3}/3)/(1 + \sqrt{3}/3)} \end{bmatrix} \qquad \mathbf{X}_2 = \begin{bmatrix} -1 \\ 1 \\ 0 \end{bmatrix}$$

$$\mathbf{X}_3 = \begin{bmatrix} -\sqrt{(1 - \sqrt{3}/3)/(2 + 2\sqrt{3}/3)} \\ -\sqrt{(1 - \sqrt{3}/3)/(2 + 2\sqrt{3}/3)} \\ 1 \end{bmatrix}$$

7.7.3   Rotation of Vectors

The equations for the change of coordinates of a vector $\mathbf{V}$ being rotated through an angle θ may be obtained from Fig. 7.13 as

$$\begin{aligned} x' &= x \cos \theta - y \sin \theta \\ y' &= x \sin \theta + y \cos \theta \end{aligned} \qquad (7.29)$$

Equations (7.29) may be written in matrix form in two ways,

$$\begin{bmatrix} x' \\ y' \end{bmatrix} = \begin{bmatrix} \cos \theta & -\sin \theta \\ \sin \theta & \cos \theta \end{bmatrix} \begin{bmatrix} x \\ y \end{bmatrix} \qquad (7.30)$$

$$[x' \quad y'] = [x \quad y] \begin{bmatrix} \cos \theta & \sin \theta \\ -\sin \theta & \cos \theta \end{bmatrix} \qquad (7.31)$$

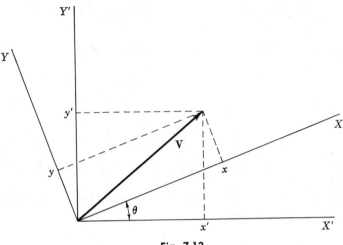

Fig. 7.13

Jacobi's method is thus seen to be equivalent to planar rotations of the column vectors and the row vectors of a matrix. In this process the lengths of the vectors are not changed. Hence, the sum of the squares of the elements of the matrix does not change during the annihilation of the off-diagonal elements. What is lost by them is gained by the diagonal elements. Also, since the trace, or sum of the diagonal elements, equals the sum of the characteristic roots, it is invariant in the transformation. These two facts may be used as checks in a computer program for carrying out Jacobi's method.[1]

In the final example of Sec. 7.7.2, the sum of the squares is 15; the trace is 3.

7.8　APPLICATIONS

Many engineering problems in the areas of electrical circuitry, structural analysis, and vibration analysis require the solving of simultaneous equations. Some examples will now be presented.

[1] A complete discussion of Jacobi's method, along with flow charts for programming the method, is given in J. Greenstadt, "Mathematical Methods for Digital Computers," A. Ralston and H. S. Wilf, eds., chap. 7, John Wiley & Sons, Inc., New York, 1960. A fast and accurate algorithm for using Jacobi's method is presented in F. J. Corbato, On the Coding of Jacobi's Method for Computing Eigenvalues and Eigenvectors of Real Symmetric Matrices, *J. Assoc. Comp. Mach.*, **10**(2): 123–125 (1963).

7.8.1   Electrical Circuit

An example of an electrical circuit requiring the solution of a system of simultaneous equations is shown in Fig. 7.14. In this circuit are five resistances $R_1, \ldots, R_5$ and two generators producing voltages E_1 and E_2. The currents $I_1, \ldots, I_5$ are indicated in the figure and represent the unknowns. Three equations are obtained by using Kirchhoff's law, which states that the sum of the voltages around a closed loop must be zero.

$$
\begin{aligned}
R_1 I_1 + R_2 I_2 \qquad\qquad\qquad -E_1 &= 0 \\
-R_2 I_2 + R_3 I_3 + R_4 I_4 \qquad\qquad &= 0 \\
-R_4 I_4 + R_5 I_5 + E_2 &= 0
\end{aligned}
\tag{7.32a}
$$

In addition, the fact that the sums of the current flowing into junctions A and B must be equal to zero produces the two additional equations

$$
\begin{aligned}
I_1 - I_2 - I_3 \qquad\quad &= 0 \\
I_3 - I_4 - I_5 &= 0
\end{aligned}
\tag{7.32b}
$$

Thus, a system of five equations for the five unknown currents $I_1, \ldots, I_5$ is formed.

7.8.2   Structural Analysis

In the design of buildings, bridges, ships, aircraft, etc., the engineer must often solve simultaneous equations in order to analyze the structure for maximum stresses and strains. As an illustration, Fig. 7.15 represents a building frame acted upon by uniformly distributed roof and floor loads w (units of force/length) and lateral wind loads P_1 and P_2 (units of force). The symbols L and I represent length and moment of inertia of each structural member, respectively, and in this example these are assumed to be the same for all members. The moment of inertia is a measure of the ability of a beam or column to resist bending deformations.

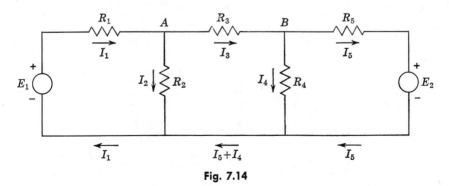

Fig. 7.14

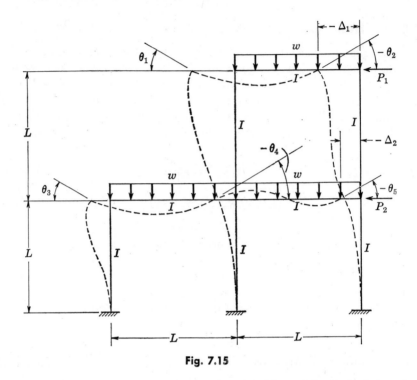

Fig. 7.15

The first step in the engineering analysis of this problem is to determine the angular rotations $\theta_1, \ldots, \theta_5$ of the joints of the structure and horizontal displacements Δ_1 and Δ_2, as indicated on the deflected structure drawn with dashed lines in Fig. 7.15. There are several methods for accomplishing this, and one of them, the *slope-deflection method*, produces the following system of simultaneous equations:

$$
\begin{aligned}
4\theta_1 + \theta_2 \quad\quad + \theta_4 \quad\quad\quad - \frac{3}{L}\Delta_1 + \frac{3}{L}\Delta_2 &= \frac{wL^3}{24EI} \\
\theta_1 + 4\theta_2 \quad\quad\quad + \theta_5 - \frac{3}{L}\Delta_1 + \frac{3}{L}\Delta_2 &= -\frac{wL^3}{24EI} \\
4\theta_3 + \theta_4 \quad\quad\quad\quad - \frac{3}{L}\Delta_2 &= \frac{wL^3}{24EI} \\
\theta_1 \quad\quad + \theta_3 + 8\theta_4 + \theta_5 - \frac{3}{L}\Delta_1 \quad\quad &= 0 \quad\quad\quad (7.33) \\
\theta_2 \quad\quad + \theta_4 + 6\theta_5 - \frac{3}{L}\Delta_1 \quad\quad &= -\frac{wL^3}{24EI} \\
\theta_1 + \theta_2 \quad\quad + \theta_4 + \theta_5 - \frac{4}{L}\Delta_1 + \frac{4}{L}\Delta_2 &= \frac{P_1 L^2}{6EI} \\
\theta_3 + \theta_4 + \theta_5 \quad\quad - \frac{6}{L}\Delta_2 &= \frac{(P_1 + P_2)L^2}{6EI}
\end{aligned}
$$

The symbol E represents the modulus of elasticity of the structural material, which, on the assumption that the frame is made of steel, has a value of 30,000,000 psi.   Once these rotations and deflections are known, the maximum stresses in the members can be determined by using the theory of strength of materials.

7.8.3   Vibration Analysis

The solutions of many vibration problems in engineering are determined by solving a system of simultaneous equations.   As an illustration, Fig. 7.16a represents an airplane wing which is to be analyzed for vibrations.   Since the cross section of the wing varies throughout its length, an exact analysis is exceedingly difficult.   A very accurate analysis can be made, however, if the actual wing is replaced by a beam of negligible weight supporting several point masses $m_1, \ldots, m_n$, as shown in Fig. 7.16b.   The free undamped motion of each mass m_i is expressed by the equation

$$y_i = A_i \sin \omega t \tag{7.34}$$

where t represents time, ω the frequency in radians per unit of time, and y_i the deflection of mass m_i, as shown in Fig. 7.16c.   Using Eq. (7.34) in equations derived from the theory of strength of materials and Newton's

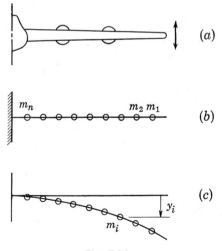

Fig. 7.16

laws of motion results in a system of equations of the form

$$
\begin{aligned}
(a_{11}m_1 - p)A_1 + \quad & a_{12}m_2A_2 + \cdots + & a_{1n}m_nA_n = 0 \\
a_{21}m_1A_1 + (a_{22}m_2 - p)A_2 + \cdots + & a_{2n}m_nA_n = 0 \\
\cdots\cdots\cdots\cdots\cdots\cdots\cdots\cdots\cdots\cdots\cdots\cdots\cdots\cdots\cdots & \\
a_{n1}m_1A_1 + \quad & a_{n2}m_2A_2 + \cdots + (a_{nn}m_n - p)A_n = 0
\end{aligned} \tag{7.35}
$$

where the a_{ij}'s are called *influence coefficients* and

$$
p = \frac{1}{\omega^2} \tag{7.36}
$$

Since the equations are homogeneous, the only way a solution can exist, other than having all the A_i's identically equal to zero, is to require that the determinant of the coefficients be zero. Setting the determinant equal to zero produces a polynomial in p of the form

$$
b_1p^n + b_2p^{n-1} + \cdots + b_np + b_{n+1} = 0
$$

Such an equation is called a *frequency* equation. The roots p_i of this polynomial can be determined by using the methods of Chap. 6, and the corresponding natural frequencies ω_i can then be calculated from Eq. (7.36). Actually, a continuous body such as an airplane wing has an infinite number of natural frequencies, and the above procedure gives approximate values to the n lowest frequencies. The accuracy depends upon the number of masses m_i used.

Now suppose that an airplane wing is acted upon by a given periodic wind force $P \sin \beta t$ distributed as shown in Fig. 7.17. Here β represents the frequency of the applied load. The governing equations in this case are

$$
\begin{aligned}
(a_{11}m_1 - q)A_1 + a_{12}m_2A_2 + \cdots + a_{1n}m_nA_n \\
= -q(a_{11}P_1 + a_{12}P_2 + \cdots + a_{1n}P_n) \\
a_{21}m_1A_1 + (a_{22}m_2 - q)A_2 + \cdots + a_{2n}m_nA_n \\
= -q(a_{21}P_1 + a_{22}P_2 + \cdots + a_{2n}P_n) \quad (7.37) \\
\cdots\cdots\cdots\cdots\cdots\cdots\cdots\cdots\cdots\cdots\cdots\cdots\cdots\cdots\cdots \\
a_{n1}m_1A_1 + a_{n2}m_2A_2 + \cdots + (a_{nn}m_n - q)A_n \\
= -q(a_{n1}P_1 + a_{n2}P_2 + \cdots + a_{nn}P_n)
\end{aligned}
$$

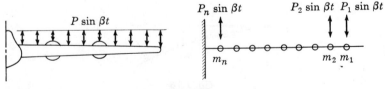

Fig. 7.17

whereas in Eq. (7.34)

$$y_i = A_i \sin \beta t$$

and

$$q = \frac{1}{\beta^2}$$

The system of Eqs. (7.37) can be solved for the amplitudes A_i.   By comparing Eqs. (7.37) with Eqs. (7.35), it is readily ascertained that if β happens to equal one of the natural frequencies ω_i the amplitudes will be infinitely large.   Such a phenomenon is known as *resonance* and is something to be avoided in an airplane wing.

Vibrations occurring in other systems such as hulls of ships, bridges, electrical machinery, propeller shafts, and many others can often be analyzed in a similar manner.

PROBLEMS

Solve Probs. 7.1 to 7.14, using any of the methods described in the text.   In several of the problems the system of equations is singular.

7.1 $3x_1 - 7x_2 = 10$
$4x_1 + x_2 = 3$

♦7.2 $x_1 + 3x_2 = 5$
$4x_1 - x_2 = 12$

7.3 $x_1 + 2x_2 + 3x_3 = 6$
$3x_1 + 10x_2 - 4x_3 = -29$
$-2x_1 - 4x_2 + x_3 = 9$

♦7.4 $3x_1 + 2x_2 + x_3 = 5$
$2x_1 + 4x_2 + 6x_3 = -2$
$-9x_1 + 6x_2 - 3x_3 = 4$

7.5 $x_1 \quad\quad + x_3 = 0$
$x_1 + x_2 \quad\quad = 4$
$x_2 + x_3 = -4$

♦7.6 $x_1 + 2x_2 + 3x_3 + 4x_4 = 16$
$x_1 - 2x_2 - 3x_3 + 4x_4 = 2$
$x_1 + x_2 + x_3 + x_4 = 6$
$x_1 - x_2 - x_3 - x_4 = -4$

7.7 $x_1 + x_2 + x_3 + x_4 = 4$
$x_1 - x_2 + x_3 - x_4 = 0$
$x_1 \quad\quad + x_3 \quad\quad = 2$
$x_2 \quad\quad - x_4 = 0$

◆7.8 $10x_1 + x_2 - 2x_3 = 9$
        $2x_1 + 20x_2 - x_3 = 21$
        $x_1 - 2x_2 + 10x_3 = 9$

7.9 $10x_1 - 2x_2 + x_3 = 9$
        $x_1 - 10x_2 + x_3 = -16$
        $2x_1 - x_2 + 10x_3 = 30$

◆7.10 $2x_1 + 2x_2 + 4x_3 - 2x_4 = 10$
        $x_1 + 3x_2 + 2x_3 + x_4 = 17$
        $3x_1 + x_2 + 3x_3 + x_4 = 18$
        $x_1 + 3x_2 + 4x_3 + 2x_4 = 27$

7.11 $x_1 + 2x_2 + 3x_3 + 4x_4 = 10$
        $3x_1 + x_2 + 7x_3 + 5x_4 + 3x_5 = 19$
        $2x_1 + 4x_2 + x_3 + 4x_4 + x_5 = 12$
        $5x_1 + x_4 + 4x_5 = 10$
        $x_1 + x_2 + x_3 + 2x_4 + 3x_5 = 8$

◆7.12 $x_1 + 2x_2 + 3x_3 + 5x_4 = 6$
        $2x_1 + 3x_2 + 4x_3 + 8x_4 = 2$
        $3x_1 + 5x_2 + 7x_3 + 2x_4 = 4$
        $x_1 + 2x_2 + 4x_3 + 6x_4 = 10$

7.13 $x_1 + 2x_2 + 3x_3 + 4x_4 = 1$
        $2x_1 + x_2 + x_4 = 2$
        $6x_1 + x_3 = 6$
        $3x_1 - 3x_2 - 2x_3 - 5x_4 = 3$

◆7.14 $x_1 + 3x_2 + x_3 + x_5 + 2x_6 + 4x_7 = 1$
        $2x_2 + x_4 + 2x_6 = 3$
        $3x_3 + x_4 + x_5 + x_7 = 5$
        $2x_2 + 2x_4 + 4x_6 + 3x_7 = 1$
        $2x_1 + 6x_2 + 2x_3 + 6x_5 + 5x_6 + 9x_7 = 3$
        $x_1 + 5x_2 + x_3 + x_4 + x_5 + 5x_6 + 6x_7 = 7$
        $x_1 + 4x_3 + x_4 + 2x_5 + 2x_6 + 8x_7 = 8$

In Probs. 7.15 to 7.19, perform the matrix addition **A + B**.

7.15 $\mathbf{A} = \begin{bmatrix} 12 & -8 \\ 42 & 31 \end{bmatrix}$ $\mathbf{B} = \begin{bmatrix} -3 & 0 \\ 17 & -17 \end{bmatrix}$

◆7.16 $\mathbf{A} = \begin{bmatrix} -7 & 8 & -9 \\ 4 & 5 & -6 \\ 1 & 2 & 3 \end{bmatrix}$ $\mathbf{B} = \begin{bmatrix} 6 & -4 & -6 \\ 2 & 0 & 4 \\ 8 & 8 & 2 \end{bmatrix}$

7.17 $\mathbf{A} = \begin{bmatrix} 3 & 5 & -7 \\ -6 & 8 & 0 \\ -9 & 1 & 9 \\ 2 & 4 & 8 \end{bmatrix}$ $\mathbf{B} = \begin{bmatrix} 7 & 5 & 0 \\ -8 & -6 & 8 \\ 5 & 3 & -7 \\ 9 & 1 & -2 \end{bmatrix}$

7.18 $\mathbf{A} = \begin{bmatrix} 5 & 1 & -7 & 8 & 9 \\ 9 & 0 & 3 & -2 & 2 \end{bmatrix}$ $\mathbf{B} = \begin{bmatrix} 5 & -5 & -1 & 1 & 2 \\ -9 & 8 & 1 & 4 & -2 \end{bmatrix}$

7.19 $\quad \mathbf{A} = \begin{bmatrix} -9 \\ 2 \\ -8 \\ 1 \end{bmatrix} \quad \mathbf{B} = \begin{bmatrix} 7 \\ 2 \\ 1 \\ -4 \end{bmatrix}$

In Probs. 7.20 to 7.24, perform the matrix multiplication $\mathbf{AB}$.

7.20 $\quad \mathbf{A} = \begin{bmatrix} 9 & -6 & 5 \end{bmatrix} \quad \mathbf{B} = \begin{bmatrix} 7 \\ 8 \\ 3 \end{bmatrix}$

7.21 $\quad \mathbf{A} = \begin{bmatrix} 0 & 0 & -4 \\ 2 & -7 & 4 \\ 7 & 1 & -9 \end{bmatrix} \quad \mathbf{B} = \begin{bmatrix} 5 \\ 5 \\ 0 \end{bmatrix}$

7.22 $\quad \mathbf{A} = \begin{bmatrix} -8 & 9 & 6 \\ 6 & 0 & 0 \\ -5 & -4 & -9 \end{bmatrix} \quad \mathbf{B} = \begin{bmatrix} 6 & 0 & 1 \\ 5 & 1 & 7 \\ 6 & 4 & 2 \end{bmatrix}$

♦7.23 $\quad \mathbf{A} = \begin{bmatrix} -1 & 2 \\ 7 & 4 \\ 2 & 2 \\ 2 & 5 \end{bmatrix} \quad \mathbf{B} = \begin{bmatrix} 9 & -2 & 8 & 1 \\ 1 & 4 & 9 & -4 \end{bmatrix}$

7.24 $\quad \mathbf{A} = \begin{bmatrix} 8 & 6 & 4 & -4 \\ -2 & 4 & 9 & 4 \\ -5 & 8 & 2 & 1 \\ 3 & 4 & 4 & 4 \end{bmatrix} \quad \mathbf{B} = \begin{bmatrix} 6 \\ -4 \\ 4 \\ -4 \end{bmatrix}$

7.25 Perform the matrix multiplications $\mathbf{AB}$ and $\mathbf{BA}$.

$$\mathbf{A} = \begin{bmatrix} 6 & 1 & 3 \\ 4 & 8 & 9 \\ 3 & -6 & -6 \end{bmatrix} \quad \mathbf{B} = \begin{bmatrix} -2 & 0 & 5 \\ 8 & 5 & 9 \\ 4 & 0 & -4 \end{bmatrix}$$

7.26 Perform the matrix multiplications $\mathbf{A(BC)}$ and $\mathbf{(AB)C}$.

$$\mathbf{A} = \begin{bmatrix} 8 & 8 & 7 \\ 1 & -1 & 9 \\ 5 & -4 & 1 \end{bmatrix} \quad \mathbf{B} = \begin{bmatrix} 3 & -7 & 9 \\ 4 & 8 & 9 \\ 6 & 8 & -9 \end{bmatrix} \quad \mathbf{C} = \begin{bmatrix} 9 & -6 & 2 \\ 8 & 5 & 0 \\ -7 & 4 & 8 \end{bmatrix}$$

7.27 Perform the matrix multiplication $(3\mathbf{A})(2\mathbf{B})$.

$$\mathbf{A} = \begin{bmatrix} 8 & 2 & 4 \\ 6 & 0 & -1 \\ 4 & 7 & 7 \end{bmatrix} \quad \mathbf{B} = \begin{bmatrix} -4 & 1 & -6 \\ 0 & 6 & 0 \\ 5 & 1 & 4 \end{bmatrix}$$

7.28 Demonstrate that the vectors $\begin{bmatrix} 1 & 1 & 1 \end{bmatrix}$, $\begin{bmatrix} 1 & 2 & -3 \end{bmatrix}$, and $\begin{bmatrix} 5 & -4 & -1 \end{bmatrix}$ are mutually perpendicular (orthogonal).

◆**7.29**   The length of the given vector **V** is 10. Construct a vector orthogonal to **V**, and form the corresponding orthogonal matrix.

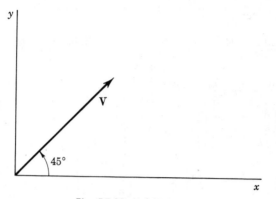

Fig. P7.29 and P7.30

7.30   The length of the given vector **V** is 10. Construct two vectors orthogonal to **V** and to each other, and form the corresponding orthogonal matrix.

7.31   Construct a vector orthogonal to the given vector **V**, and form the corresponding orthogonal matrix.

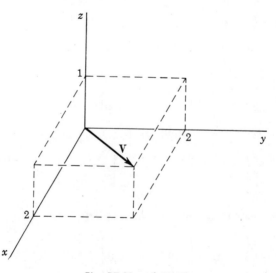

Fig. P7.31 and P7.32

7.32   Construct two vectors orthogonal to the given vector **V** and to each other, and form the corresponding orthogonal matrix.

7.33, 7.34, ◆7.35, 7.36–7.42, ◆7.43, 7.44–7.46 Invert the coefficient matrix of the simultaneous equations given in Probs. 7.1 to 7.14.   Calculate the solution of the equations by using the inverted matrix.

7.47   Solve the equations

$$x_1 + 2x_2 + 3x_3 = c_1$$
$$3x_1 + 10x_2 - 4x_3 = c_2$$
$$-2x_1 - 4x_2 + x_3 = c_3$$

for the following sets of c_i's:

	1	2	3	4	5
c_1	6	6	14	-6	4
c_2	-29	9	11	-9	-1
c_3	9	-5	-7	5	-1

7.48   Solve the equations

$$3x_1 + 2x_2 + x_3 = c_1$$
$$2x_1 + 4x_2 + 6x_3 = c_2$$
$$-9x_1 + 6x_2 - 3x_3 = c_3$$

for the following sets of c_i's:

	1	2	3	4	5
c_1	5	14	6	-6	2
c_2	-2	20	20	-12	4
c_3	4	-42	-18	18	-6

7.49–7.53   Demonstrate the validity of the relations

$$\mathbf{A}^{-1} = \mathbf{A}^T \qquad \mathbf{A}^T\mathbf{A} = \mathbf{I}$$

by performing the indicated operations with the orthogonal matrices formed in Probs. 7.28 to 7.32.

7.54   Apply the iteration-convergence criterion given by Eqs. (7.22) to (7.24) to the equations in Prob. 7.8.

7.55   Apply the iteration-convergence criterion given by Eqs. (7.22) to (7.24) to the equations in Prob. 7.9.

7.56   Apply the iteration-convergence criterion given by Eqs. (7.22) to (7.24) to the equations in Prob. 7.10.   How should these equations be rearranged?

In Probs. 7.57 to 7.62 evaluate the given determinants.

♦**7.57** $\begin{vmatrix} 1 & 0 \\ 6 & -8 \end{vmatrix}$

7.58 $\begin{vmatrix} 3 & 7 \\ -4 & 2 \end{vmatrix}$

♦**7.59** $\begin{vmatrix} 1 & 4 & 8 \\ 0 & -2 & 10 \\ 0 & 0 & 5 \end{vmatrix}$

7.60 $\begin{vmatrix} 4 & -3 & 2 \\ 0 & 0 & 0 \\ 6 & 1 & 4 \end{vmatrix}$

♦**7.61** $\begin{vmatrix} 1 & 2 & 3 & -1 \\ 4 & 1 & 0 & 0 \\ 0 & 2 & 2 & 0 \\ 1 & 1 & 1 & 2 \end{vmatrix}$

7.62 $\begin{vmatrix} 1 & 4 & 8 & -2 \\ 0 & 0 & 1 & 0 \\ 5 & 1 & 0 & -10 \\ 1 & 0 & -2 & -2 \end{vmatrix}$

7.63, ♦**7.64, 7.65–7.68,** ♦**7.69, 7.70–7.73,** ♦**7.74, 7.75, 7.76**   Evaluate the determinants of the coefficient matrix of the simultaneous equations given in Probs. 7.1 to 7.14.

7.77–7.82   Solve Probs. 7.1 to 7.6 by applying Cramer's rule.

In Probs. 7.83 to 7.91, determine the characteristic roots and vectors of the given matrix.

7.83 $\begin{bmatrix} 1 & 1 \\ 2 & 0 \end{bmatrix}$

♦**7.84** $\begin{bmatrix} 3 & 7 \\ 4 & 1 \end{bmatrix}$

7.85 $\begin{bmatrix} 1 & 2 \\ 4 & -1 \end{bmatrix}$

7.86 $\begin{bmatrix} 2 & 0 & 1 \\ 0 & 2 & 0 \\ 1 & 0 & 2 \end{bmatrix}$

♦**7.87** $\begin{bmatrix} 1 & 1 & 0 \\ 1 & 2 & 1 \\ 0 & 1 & 3 \end{bmatrix}$

7.88 $\begin{bmatrix} 4 & 2 & 1 \\ 2 & 1 & 2 \\ 1 & 2 & 8 \end{bmatrix}$

7.89 $\begin{bmatrix} 3 & 0 & 0 & 1 \\ 0 & 3 & 0 & 0 \\ 0 & 0 & 1 & 0 \\ 1 & 0 & 0 & 3 \end{bmatrix}$

♦7.90 $\begin{bmatrix} 2 & 1 & 1 & 0 \\ 1 & 2 & 0 & 1 \\ 1 & 0 & 2 & 0 \\ 0 & 1 & 0 & 3 \end{bmatrix}$

7.91 $\begin{bmatrix} 5 & 1 & 2 & 1 \\ 1 & 4 & 1 & 1 \\ 2 & 1 & 3 & 4 \\ 1 & 1 & 4 & 4 \end{bmatrix}$

7.92 The circuit shown is often used in electrical measurements and is known as a *Wheatstone bridge*. The governing equations are obtained from Kirchhoff's laws and are:

Closed loop through battery and along ABD

$$I_1 R_1 + I_4 R_4 - E = 0 \tag{1}$$

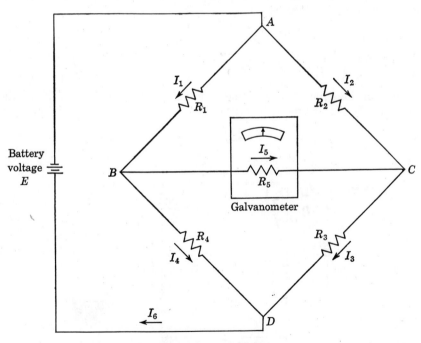

Fig. P7.92 and P7.93

Closed loop ABC

$$I_1R_1 + I_5R_5 - I_2R_2 = 0 \tag{2}$$

Closed loop BCD

$$I_5R_5 + I_3R_3 - I_4R_4 = 0 \tag{3}$$

Junction A

$$I_6 = I_1 + I_2 \tag{4}$$

Junction B

$$I_1 = I_5 + I_4 \tag{5}$$

Junction C

$$I_2 + I_5 = I_3 \tag{6}$$

where the R_i's represent resistances, the I_i's currents, and E the applied voltage. If E and all the R_i's are known, the six equations can be solved for the six unknown currents. Reduce the system to three equations for the three currents I_1, I_2, and I_5 by solving Eqs. (5) and (6) for I_4 and I_3 and substituting into Eqs. (1), (2), and (3).

♦**7.93** Determine the currents in the Wheatstone bridge described in Prob. 7.92, using the following data:

$E = 20$ volts
$R_1 = 10$ ohms
$R_2 = R_3 = R_4 = R_5 = 100$ ohms

7.94 In the structure shown, the following system of equations defines the values of the angular rotations of the joints θ_1 and θ_2 and the horizontal dis-

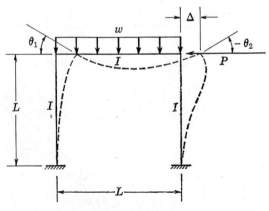

Fig. P7.94 and P7.95

placement Δ:

$$4\theta_1 + \theta_2 - \frac{3}{L}\Delta = \frac{wL^3}{24EI}$$

$$\theta_1 + 4\theta_2 - \frac{3}{L}\Delta = \frac{-wL^3}{24EI}$$

$$\theta_1 + \theta_2 - \frac{4}{L}\Delta = \frac{PL^2}{6EI}$$

Determine the values of θ_1, θ_2, and Δ in terms of w, P, L, E, and I. The assumed positive directions of the unknown quantities are indicated by dashed lines. Sketch the deflected shape of the structure.

7.95 Use the following sets of data to determine the angular rotations θ_1 and θ_2 and displacement Δ given by the equations in Prob. 7.94. The units of w are kips per inch (1 kip = 1000 lb); of P, kips; of L, inches; of E, kips per square inch; of I, inches⁴; of θ_1, θ_2, radians; and of Δ, inches.

	1	2	3	4
w	0.1	0	0.1	0.1
P	0	20	20	20
L	200	200	200	200
E	30,000	30,000	30,000	30,000
I	150	150	150	100

7.96 Use the following data to solve the system of equations given in Sec. 7.8.2:

w	P_1	P_2	L	E	I
0.1	20	20	200	30,000	200

The units are the same as given in Prob. 7.95.

◆**7.97** Let an airplane wing be represented by the two masses, as shown. With the given data determine the approximate values of the first two natural fre-

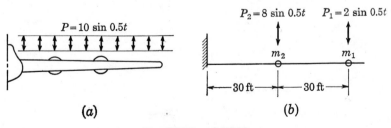

(a)　　　　　(b)

Fig. P7.97 and P7.98

quencies. Also, determine the amplitude ratio A_2/A_1 for each frequency. Sketch the deflection of the wing for each frequency at an instant when $\sin \omega t = 1$, assuming $A_2 = 2$ ft.

$$m_1 = 0.155 \qquad a_{11} = 4.15 \qquad a_{12} = 1.30$$
$$m_2 = 0.621 \qquad a_{21} = 1.30 \qquad a_{22} = 0.518$$

7.98 Let an airplane wing subjected to a periodic force be represented by two masses, as shown. Determine the vibration amplitudes A_1 and A_2, using the data of the previous problem. The units are such that the forces are expressed in kips (1 kip = 1000 lb) and the amplitudes in inches. Sketch the deflection of the wing at an instant when $\sin 0.5t = 1$.

8   INTERPOLATION

8.1   INTRODUCTION

Mathematical functions are often described in tabular form. That is, for prescribed values x_1, x_2, . . . , x_n of the independent variable x, corresponding functional values $f(x_1)$, $f(x_2)$, . . . , $f(x_n)$ are given. The logarithmic and trigonometric functions are examples of functions which are presented in tabular form. The process of passing a curve through the given points in order to determine functional values $f(x)$ for values of x not explicitly shown in the table is called *interpolation*.

8.1.1   Graphical Interpolation

One common method of interpolation is accomplished by plotting the function from the tabular data and, by using a drafting instrument such as a French curve, constructing a smooth curve through the plotted points. To illustrate, assume that the data given in Table 8.1 were taken from a rocket fired vertically from the surface of the earth. The speed vs. time graph plotted from these tabulated values is given in Fig. 8.1. Interpolation can be accomplished by reading the graph directly. For example, when the time is 150 sec, the speed can be interpolated as 0.45 mps.

Table 8.1

Time, sec	Speed, mps
0	0.0000
60	0.0824
120	0.2747
180	0.6502
240	1.3851
300	3.2229

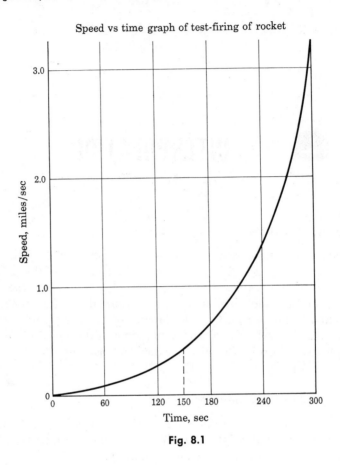

Speed vs time graph of test-firing of rocket

Fig. 8.1

In Fig. 8.1, the curve is passed through all the given points. This tacitly means that the given functional values are exact. Any data obtained experimentally, however, always contain some measurement errors which may be greater at certain measured points than others. Interpolation does not account for such errors and thus assumes that the given data represent exact functional values. Methods of accounting for measurement errors are considered in Chap. 12.

8.1.2   Linear Interpolation

Most engineering students are introduced to linear interpolation when studying logarithms and the trigonometric functions. Table 8.2, an abbreviated sine table, will be used to illustrate this method. The sine of

22° is obtained by using linear interpolation, as follows:

$$\frac{\sin 22° - 0.34202}{0.50000 - 0.34202} = \frac{22 - 20}{30 - 20}$$

$$\sin 22° = 0.37362$$

Table 8.2

x, deg	$f(x) = \sin x$
0	0.00000
10	0.17365
20	0.34202
30	0.50000
40	0.64279

The actual value of sin 22° to five decimal places is 0.37461, showing an error in the third significant digit when linear interpolation is used.

A general linear-interpolation formula can be derived geometrically by using similar triangles, as illustrated in Fig. 8.2. Triangle ABD is similar to triangle ACE, and it follows that

$$\frac{DB}{EC} = \frac{BA}{CA}$$

$$\frac{f(x) - f(x_1)}{f(x_2) - f(x_1)} = \frac{x - x_1}{x_2 - x_1}$$

Solving for $f(x)$ gives

$$f(x) = f(x_1) + \frac{f(x_2) - f(x_1)}{x_2 - x_1} (x - x_1) \tag{8.1}$$

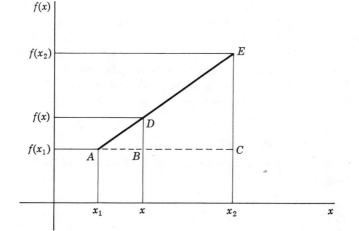

Fig. 8.2

An alternative derivation starts with the equation of the straight line in the form

$$f(x) = a_1 x + a_2$$

The coefficients a_1 and a_2 are determined by the substitution of the functional values $f(x_1)$ and $f(x_2)$ into this expression.   Thus,

$$f(x_1) = a_1 x_1 + a_2$$
$$f(x_2) = a_1 x_2 + a_2$$

Solving these two simultaneous equations for a_1 and a_2 gives

$$a_1 = \frac{f(x_2) - f(x_1)}{x_2 - x_1}$$

$$a_2 = f(x_1) - \frac{f(x_2) - f(x_1)}{x_2 - x_1} x_1$$

The interpolating function is then

$$f(x) = \frac{f(x_2) - f(x_1)}{x_2 - x_1} x + f(x_1) - \frac{f(x_2) - f(x_1)}{x_2 - x_1} x_1$$

which can easily be rearranged to agree with Eq. (8.1).

8.1.3   Polynomial Interpolation

It seems reasonable that a more accurate interpolation than linear interpolation can be obtained by replacing the straight line by a higher-order polynomial-interpolation function.   Polynomial interpolation can be stated as follows:

> Given $n + 1$ points at which functional values are known, find an nth-order polynomial passing through the $n + 1$ points.

The nth-order polynomial which passes through $n + 1$ points given in tabular form is called an *interpolating polynomial*.   A discussion of whether or not the interpolating polynomial is an accurate representation of the true function at points other than those given is presented in Sec. 8.11.   In order to differentiate between the true function and the interpolating polynomial, the following notation will be adopted:

$$f(x) = \text{true function}$$
$$p(x) = \text{interpolating polynomial}$$

The following relations are then true, from the definition of the interpolating polynomial:

$$p(x_1) = f(x_1)$$
$$p(x_2) = f(x_2)$$
$$\cdots \cdots \cdots$$
$$p(x_{n+1}) = f(x_{n+1}) \,.$$

(8.2)

The obvious way of determining the interpolating polynomial is to assume it in the form

$$p(x) = a_1x^n + a_2x^{n-1} + \cdots + a_nx + a_{n+1} \qquad (8.3)$$

and to substitute the $n + 1$ functional values into this polynomial, producing a system of $n + 1$ algebraic, simultaneous linear equations which may be solved for the coefficients $a_1, a_2, \ldots, a_{n+1}$. Solution of this system of equations leads to the interpolating polynomial, and the general problem of interpolation appears to be solved. While this is certainly true, there are methods available which obtain the interpolating polynomial in quicker and more efficient ways. Before the general case is considered, an interpolating parabola ($n = 2$) will be examined, since it gives much insight into the problem.

8.2   INTERPOLATING PARABOLAS

Suppose that the data given in Table 8.3 are used to obtain the value of

Table 8.3

x, deg	$f(x) = \sin x$
20	0.34202
25	0.42262
30	0.50000

$\sin 22°$. Assuming an interpolating parabola of the form

$$p(x) = a_1x^2 + a_2x + a_3$$

and substituting functional values directly from the table give the system of equations

$$400a_1 + 20a_2 + a_3 = 0.34202$$
$$625a_1 + 25a_2 + a_3 = 0.42262$$
$$900a_1 + 30a_2 + a_3 = 0.50000$$

Solving the system produces

$$a_1 = -0.0000644 \qquad a_2 = 0.01902 \qquad a_3 = -0.01258$$

and the interpolating parabola is

$$p(x) = -0.0000644x^2 + 0.01902x - 0.01258 \qquad (8.4)$$

For $x = 22°$, $p(22) = 0.37465$, which compares favorably with the true value 0.37461.

Although the above procedure is straightforward, it requires the solution of three simultaneous equations.   A system of three equations which is easier to solve is obtained by assuming the interpolating parabola in the form

$$p(x) = a_1 + a_2(x - 20) + a_3(x - 20)(x - 25)$$

Substitution of the tabulated values leads to

$$0.34202 = a_1$$
$$0.42262 = a_1 + 5a_2$$
$$0.50000 = a_1 + 10a_2 + (10)(5)a_3$$

These equations are in triangular form (Chap. 7) and can be solved easily by successive substitution, producing

$$a_1 = 0.34202 \qquad a_2 = 0.01612 \qquad a_3 = -0.0000644$$

The interpolating parabola is then

$$p(x) = 0.34202 + 0.01612(x - 20) - 0.0000644(x - 20)(x - 25) \quad (8.5)$$

and, at $x = 22$, $p(22) = 0.37465$.

As another example consider again the rocket problem given on page 285. An interpolating parabola passing through the following three points,

Time t	Speed
60	0.0824
120	0.2747
180	0.6502

is assumed in the form

$$p(t) = a_1 + a_2(t - 60) + a_3(t - 60)(t - 120)$$

and according to the procedure given in the previous example,

$$0.0824 = a_1$$
$$0.2747 = a_1 + 60a_2$$
$$0.6502 = a_1 + 120a_2 + (120)(60)a_3$$

Solving these equations gives

$$a_1 = 0.0824 \qquad a_2 = 0.003205 \qquad a_3 = 0.00002544$$

and the interpolating parabola is

$$p(t) = 0.0824 + 0.003205(t - 60) + 0.00002544(t - 60)(t - 120)$$

The interpolated value of the speed when the time is 150 sec is obtained by direct substitution as follows:

$$p(150) = 0.0824 + 0.003205(150 - 60)$$
$$+ 0.00002544(150 - 60)(150 - 120)$$
$$p(150) = 0.4395 \text{ mps}$$

8.3 GREGORY-NEWTON INTERPOLATION FORMULA

In the previous section it was shown that, if an interpolating parabola was expressed in a special form, the coefficients of the parabola were obtained quite easily. The procedure used there is a special case of the more general Gregory-Newton method, which will now be presented.

As stated in the introduction, the general method consists in passing an nth-order polynomial $p(x)$ through $n + 1$ points. These $n + 1$ points are given in a tabular form,

x	$f(x)$
x_1	$f(x_1)$
x_2	$f(x_2)$
.	.
.	.
.	.
x_{n+1}	$f(x_{n+1})$

where the values in the table are assumed to be ordered, that is, $x_{i+1} > x_i$. Let $p(x)$ take the form

$$p(x) = a_1 + a_2(x - x_1) + a_3(x - x_1)(x - x_2)$$
$$+ a_4(x - x_1)(x - x_2)(x - x_3) + \cdots$$
$$+ a_{n+1}(x - x_1)(x - x_2)(\cdots)(x - x_n) \tag{8.6}$$

Following the procedure given in Sec. 8.2, the functional values $f(x_1)$, $f(x_2)$, ..., $f(x_{n+1})$ are substituted into the above equation.

$$f(x_1) = a_1$$
$$f(x_2) = a_1 + a_2(x_2 - x_1)$$
$$f(x_3) = a_1 + a_2(x_3 - x_1) + a_3(x_3 - x_1)(x_3 - x_2)$$
$$f(x_4) = a_1 + a_2(x_4 - x_1) + a_3(x_4 - x_1)(x_4 - x_2)$$
$$+ a_4(x_4 - x_1)(x_4 - x_2)(x_4 - x_3)$$
$$\cdots\cdots\cdots\cdots\cdots\cdots\cdots\cdots\cdots\cdots\cdots$$
$$f(x_{n+1}) = a_1 + a_2(x_{n+1} - x_1) + a_3(x_{n+1} - x_1)(x_{n+1} - x_2)$$
$$+ a_4(x_{n+1} - x_1)(x_{n+1} - x_2)(x_{n+1} - x_3) + \cdots$$
$$+ a_{n+1}(x_{n+1} - x_1)(x_{n+1} - x_2)(\cdots)(x_{n+1} - x_n)$$

The coefficients $a_1, a_2, \ldots, a_{n+1}$ can now be found by successive substitution, and the interpolating polynomial is determined.

8.3.1   Equal Intervals

Often tabular data are presented at equal intervals of the independent variable x. The above equations may then be simplified as follows: Taking the interval between any two x's as h, then

$$h = x_2 - x_1 = x_3 - x_2 = \cdots = x_{n+1} - x_n$$

Substituting h into the previous system of equations gives

$$f(x_1) = a_1$$
$$f(x_1 + h) = a_1 + a_2(h)$$
$$f(x_1 + 2h) = a_1 + a_2(2h) + a_3(2h)(h)$$
$$f(x_1 + 3h) = a_1 + a_2(3h) + a_3(3h)(2h) + a_4(3h)(2h)(h)$$
$$\cdots \cdots \cdots \cdots \cdots \cdots \cdots \cdots \cdots \cdots \cdots \cdots$$
$$f(x_1 + nh) = a_1 + a_2(nh) + a_3(nh)[(n-1)h] + a_4(nh)[(n-1)h][(n-2)h]$$
$$+ \cdots + a_{n+1}(nh)[(n-1)h](\cdots)(h)$$

Solving for the coefficients $a_1, a_2, \ldots, a_{n+1}$ yields

$$a_1 = f(x_1) \qquad \leftarrow DIFFERENCE$$

$$a_2 = \frac{f(x_1 + h) - f(x_1)}{h}$$

$$a_3 = \frac{f(x_1 + 2h) - f(x_1) - \{[f(x_1 + h) - f(x_1)]/h\}2h}{2h^2}$$

$$= \frac{f(x_1 + 2h) - 2f(x_1 + h) + f(x_1)}{2h^2}$$

$$a_4 = \frac{f(x_1 + 3h) - f(x_1) - \{[f(x_1 + h) - f(x_1)]/h\}3h - \{[f(x_1 + 2h) - 2f(x_1 + h) + f(x_1)]/2h^2\}6h^2}{6h^3}$$

$$= \frac{f(x_1 + 3h) - 3f(x_1 + 2h) + 3f(x_1 + h) - f(x_1)}{6h^3}$$

$$\cdots \cdots \cdots \cdots \cdots \cdots \cdots \cdots \cdots \cdots \cdots \cdots$$

$$a_{n+1} = \frac{f(x_1 + nh) - f(x_1) - \{[f(x_1 + h) - f(x_1)]/h\}nh + \cdots}{(nh)[(n-1)h](\cdots)(h)}$$

It is possible to simplify these expressions by noticing that the denominator of each coefficient a_i is $h^{i-1}(i-1)!$. Further simplification in notation is obtained by defining the following terms.

$$\Delta f(x_1) = f(x_1 + h) - f(x_1)$$
$$\Delta^2 f(x_1) = f(x_1 + 2h) - 2f(x_1 + h) + f(x_1)$$
$$\Delta^3 f(x_1) = f(x_1 + 3h) - 3f(x_1 + 2h) + 3f(x_1 + h) - f(x_1)$$
$$\cdots \cdots \cdots \cdots \cdots \cdots \cdots \cdots \cdots \cdots \cdots \cdots$$

$$(8.7)$$

The coefficients can then be written as

$a_1 = f(x_1)$

$a_2 = \dfrac{\Delta f(x_1)}{1! \, h}$

$a_3 = \dfrac{\Delta^2 f(x_1)}{2! \, h^2}$

$a_4 = \dfrac{\Delta^3 f(x_1)}{3! \, h^3}$

. . . .

$a_{n+1} = \dfrac{\Delta^n f(x_1)}{n! \, h^n}$

or, for any coefficient a_i,

$$a_i = \frac{\Delta^{i-1} f(x_1)}{(i-1)! \, h^{i-1}}$$

With these expressions for the coefficients, the interpolating polynomial can now be written as

$$p(x) = f(x_1) + \frac{\Delta f(x_1)}{1! \, h}(x - x_1) + \frac{\Delta^2 f(x_1)}{2! \, h^2}(x - x_1)(x - x_2)$$

$$+ \frac{\Delta^3 f(x_1)}{3! \, h^3}(x - x_1)(x - x_2)(x - x_3) + \cdots$$

$$+ \frac{\Delta^n f(x_1)}{n! \, h^n}(x - x_1)(x - x_2)(\cdots)(x - x_n)$$

Added convenience is obtained by making the substitution

$x = x_1 + uh$

It then follows that

$x - x_1 = uh$
$x - x_2 = x_1 + uh - (x_1 + h) = h(u - 1)$
$x - x_3 = x_1 + uh - (x_1 + 2h) = h(u - 2)$
. . . .
$x - x_n = x_1 + uh - [x_1 + (n - 1)h] = h[u - (n - 1)]$

When these relations are substituted into the equation for $p(x)$, the h's cancel and the polynomial becomes

$$p(u) = f(x_1) + \frac{\Delta f(x_1)}{1!} u + \frac{\Delta^2 f(x_1)}{2!} u(u - 1) + \frac{\Delta^3 f(x_1)}{3!} u(u - 1)(u - 2)$$

$$+ \cdots + \frac{\Delta^n f(x_1)}{n!} u(u - 1)(\cdots)[u - (n - 1)] \quad (8.8)$$

This expression for the polynomial $p(x)$ is the *Gregory-Newton interpolation formula*.

The general term for $\Delta^k f(x_1)$ may be obtained through deductive reasoning. By examining the expressions for $\Delta f(x_1)$, $\Delta^2 f(x_1)$, and $\Delta^3 f(x_1)$ given in Eqs. (8.7), it may be seen that all the functional values between $f(x_1 + kh)$ and $f(x_1)$ appear in $\Delta^k f(x_1)$. Further, the coefficients of these functional values are the same as those in the expansion of $(a - b)^n$. Therefore, they can be expressed in terms of the binomial coefficients. For example,

$$\Delta^4 f(x_1) = \binom{4}{0} f(x_1 + 4h) - \binom{4}{1} f(x_1 + 3h) + \binom{4}{2} f(x_1 + 2h)$$
$$- \binom{4}{3} f(x_1 + h) + \binom{4}{4} f(x_1)$$
$$= f(x_1 + 4h) - 4f(x_1 + 3h) + 6f(x_1 + 2h) - 4f(x_1 + h) + f(x_1)$$

Now, with a procedure secured for determining $\Delta^k f(x_1)$, the coefficients of the Gregory-Newton interpolation formula are completely determined.

Again consider the data on the speed of a rocket which were given in Table 8.1 and are repeated here. A fifth-order polynomial will now be

Time, sec	Speed, mps
0	0.0000
60	0.0824
120	0.2747
180	0.6502
240	1.3851
300	3.2229

constructed to pass through the six given points. The differences $\Delta^i f(0)$ are determined as follows:

$\Delta f(0) = 0.0824 - 0.0000 = 0.0824$
$\Delta^2 f(0) = 0.2747 - 2(0.0824) + 0.0000 = 0.1099$
$\Delta^3 f(0) = 0.6502 - 3(0.2747) + 3(0.0824) - 0.0000 = 0.0733$
$\Delta^4 f(0) = 1.3851 - 4(0.6502) + 6(0.2747) - 4(0.0824) + 0.0000 = 0.1029$
$\Delta^5 f(0) = 3.2229 - 5(1.3851) + 10(0.6502) - 10(0.2747)$
$$+ 5(0.0824) - 0.0000 = 0.4644$$

Substitution into the Gregory-Newton formula gives

$$p(u) = 0.0824u + 0.05495u(u - 1) + 0.01222u(u - 1)(u - 2)$$
$$+ 0.004288u(u - 1)(u - 2)(u - 3)$$
$$+ 0.003870u(u - 1)(u - 2)(u - 3)(u - 4)$$

To apply the formula, the value of u is substituted directly into the equation. For example, if it is desired to determine the interpolated speed when the time is 150 sec,

$$u = \frac{x - x_1}{h} = \frac{150 - 0}{60} = 2.5$$

and direct substitution gives $p(2.5) = 0.4365$ mph. This is in reasonable agreement with the value 0.45 obtained graphically in Sec. 8.1.

A linear interpolation between the speeds given at 120 sec and 180 sec produces a value of 0.4625, while the parabolic interpolation done in Sec. 8.2 gives 0.4395.

8.4   FINITE DIFFERENCES

A finite difference (or, simply, difference) of the function $f(x)$ is the value of the function at one point x_1 minus the value at a second point x_2. Algebraically, this is represented as

$$f(x_1) - f(x_2)$$

Geometrically, differences are represented by intervals along the vertical axis, as illustrated in Fig. 8.3. The quantities $\Delta f(x_i)$, $\nabla f(x_i)$, and $\delta f(x_i)$ are all differences. In particular, $\Delta f(x_i)$ is called a *forward difference*,

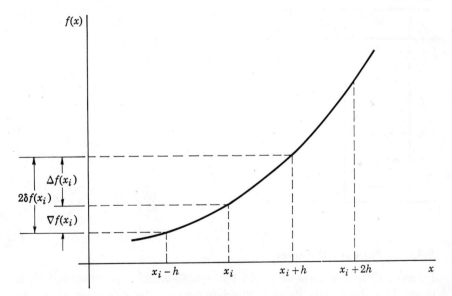

Fig. 8.3

$\nabla f(x_i)$ a *backward difference*, and $\delta f(x_i)$ a *central difference*. The central difference contains a factor of 2 since it spans two intervals.

General methods of obtaining forward, backward, and central differences will now be discussed. It will also be shown that finite differences provide a convenient means of determining interpolating polynomials.

8.4.1 Difference Tables

Consider a function $f(x)$ which is given in tabular form for equally spaced values of x, that is, $x_1, x_2, x_3, \ldots , x_m$. *First differences* are formed by subtracting successive functional values. That is, the subtractions

$$f(x_{i+1}) - f(x_i)$$

are carried out for all integer values of i from 1 to $m - 1$. The process is illustrated in Table 8.4. *Second differences* are differences of the first differences, *third differences* are differences of the second differences, etc. If we let D^k represent the kth difference, D^k is obtained by subtracting two successive values of D^{k-1}. To illustrate, Table 8.5 represents a complete difference table formed from Table 8.4.

Table 8.4

x	$f(x)$	Differences
0	0	
		10
1	10	
		10
2	20	
		30
3	50	
		10
4	60	
		40
5	100	

8.4.2 Forward Differences

Forward differences are obtained by first choosing a point x_i and then proceeding diagonally downward from $f(x_i)$ in the difference table, as illustrated in Fig. 8.4. By letting $\Delta^k f(x_i)$ symbolize a forward difference,

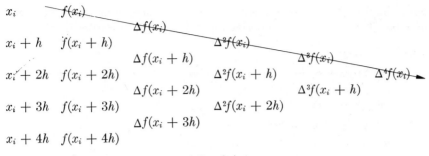

Fig. 8.4

the forward differences in Table 8.5, at the point $x = 0$, are

$\Delta f(0) = 10$
$\Delta^2 f(0) = 0$
$\Delta^3 f(0) = 20$
$\Delta^4 f(0) = -60$
$\Delta^5 f(0) = 150$

The forward differences at the point $x = 3$ are

$\Delta f(3) = 10$
$\Delta^2 f(3) = 30$

Thus, in designating a forward difference $\Delta^k f(x_i)$, both the *order* k and the point x_i must be specified.

General algebraic expressions for forward differences can be easily obtained.   Upon letting x_i be the point at which the forward differences

Table 8.5

x	$f(x)$	D^1	D^2	D^3	D^4	D^5
0	0					
		10				
1	10		0			
		10		20		
2	20		20		-60	
		30		-40		150
3	50		-20		90	
		10		50		
4	60		30			
		40				
5	100					

are to be found and h be the spacing between successive points, the first four forward differences are derived as follows: Referring to Fig. 8.4,

$$\Delta f(x_i) = f(x_i + h) - f(x_i)$$

$$
\begin{aligned}
\Delta^2 f(x_i) &= \Delta f(x_i + h) - \Delta f(x_i) \\
&= [f(x_i + 2h) - f(x_i + h)] - [f(x_i + h) - f(x_i)] \\
&= f(x_i + 2h) - 2f(x_i + h) + f(x_i)
\end{aligned}
$$

$$
\begin{aligned}
\Delta^3 f(x_i) &= \Delta^2 f(x_i + h) - \Delta^2 f(x_i) \\
&= [f(x_i + 3h) - 2f(x_i + 2h) + f(x_i + h)] \\
&\qquad - [f(x_i + 2h) - 2f(x_i + h) + f(x_i)] \quad (8.9) \\
&= f(x_i + 3h) - 3f(x_i + 2h) + 3f(x_i + h) - f(x_i)
\end{aligned}
$$

$$
\begin{aligned}
\Delta^4 f(x_i) &= \Delta^3 f(x_i + h) - \Delta^3 f(x_i) \\
&= [f(x_i + 4h) - 3f(x_i + 3h) + 3f(x_i + 2h) - f(x_i + h)] \\
&\qquad - [f(x_i + 3h) - 3f(x_i + 2h) + f(x_i + h) - f(x_i)] \\
&= f(x_i + 4h) - 4f(x_i + 3h) + 6f(x_i + 2h) \\
&\qquad\qquad - 4f(x_i + h) + f(x_i)
\end{aligned}
$$

The coefficients of the functional values in $\Delta^k f(x_i)$ are the same as those in $(a - b)^k$. Thus, the algebraic expression for the kth forward difference at point x_i is

$$
\Delta^k f(x_i) = f(x_i + kh) - \binom{k}{1} f[x_i + (k - 1)h]
$$

$$
+ \binom{k}{2} f[x_i + (k - 2)h] + \cdots + (-1)^k f(x_i) \quad (8.10)
$$

8.4.3   Application to Interpolation

If i is set equal to 1 in Eq. (8.10), the forward differences at the point x_1 correspond to the expressions appearing in the coefficients of the Gregory-Newton interpolation formula, Eq. (8.8). The use of the forward-difference symbol in Sec. 8.3 is therefore justified, and a difference table represents a convenient way of obtaining the coefficients of the Gregory-Newton formula. By using Table 8.5, for example, the interpolating polynomial can be immediately written down as follows:

$$
\begin{aligned}
p(u) = 0 &+ \frac{10}{1!} u + \frac{0}{2!} u(u - 1) + \frac{20}{3!} u(u - 1)(u - 2) \\
&+ \frac{-60}{4!} u(u - 1)(u - 2)(u - 3) \\
&\qquad\qquad + \frac{150}{5!} u(u - 1)(u - 2)(u - 3)(u - 4) \\
= 10u &+ 3.3333u(u - 1)(u - 2) - 2.5u(u - 1)(u - 2)(u - 3) \\
&\qquad\qquad + 1.25u(u - 1)(u - 2)(u - 3)(u - 4)
\end{aligned}
$$

$x_i - 4h \quad f(x_i - 4h)$

$\qquad\qquad\qquad \nabla f(x_i - 3h)$

$x_i - 3h \quad f(x_i - 3h) \qquad\qquad\qquad \nabla^2 f(x_i - 2h)$

$\qquad\qquad\qquad \nabla f(x_i - 2h) \qquad\qquad\qquad \nabla^3 f(x_i - h)$

$x_i - 2h \quad f(x_i - 2h) \qquad\qquad\qquad \nabla^2 f(x_i - h) \qquad\qquad\qquad \nabla^4 f(x_i)$

$\qquad\qquad\qquad \nabla f(x_i - h) \qquad\qquad\qquad \nabla^3 f(x_i)$

$x_i - h \quad f(x_i - h) \qquad\qquad\qquad \nabla^2 f(x_i)$

$\qquad\qquad\qquad \nabla f(x_i)$

$x_i \qquad f(x_i)$

Fig. 8.5

8.4.4   Backward Differences

As the name implies, backward differences are obtained by choosing a point x_i and proceeding diagonally upward from $f(x_i)$ in the difference table, as illustrated in Fig. 8.5.   Letting $\nabla^k f(x_i)$ symbolize a backward difference, the backward differences at point $x = 5$ in Table 8.5 are

$\nabla f(5) = 40$
$\nabla^2 f(5) = 30$
$\nabla^3 f(5) = 50$
$\nabla^4 f(5) = 90$
$\nabla^5 f(5) = 150$

The backward differences at the point $x = 3$ are

$\nabla f(3) = 30$
$\nabla^2 f(3) = 20$
$\nabla^3 f(3) = 20$

By proceeding in a manner similar to that used for forward differences, general algebraic expressions for backward differences can be developed.

$$\nabla f(x_i) = f(x_i) - f(x_i - h)$$
$$\nabla^2 f(x_i) = f(x_i) - 2f(x_i - h) + f(x_i - 2h)$$
$$\nabla^3 f(x_i) = f(x_i) - 3f(x_i - h) + 3f(x_i - 2h) - f(x_i - 3h)$$
$$\nabla^4 f(x_i) = f(x_i) - 4f(x_i - h) + 6f(x_i - 2h) - 4f(x_i - 3h)$$
$$+ f(x_i - 4h) \quad (8.11)$$

$\cdots \cdots \cdots \cdots \cdots \cdots \cdots$

$$\nabla^k f(x_i) = f(x_i) - \binom{k}{1} f(x_i - h) + \binom{k}{2} f(x_i - 2h)$$
$$- \cdots + (-1)^k f(x_i - kh)$$

It should be noted that all forward and backward differences appear in a complete difference table and that a forward difference at one point may be a backward difference at another point.   In Table 8.5

$$\Delta^3 f(0) = \nabla^3 f(3) = 20$$

In applying finite differences, it is quite clear that forward differences are convenient if the point x_i is near the beginning of the table and backward differences are convenient for a point x_i near the end of a table.

8.4.5   Central Differences

Forward and backward differences have the common disadvantage that they are one-sided.   The forward differences of a point x_i are all in terms of functional values evaluated at points $x_j > x_i$.   Similarly, the backward differences are all in terms of functional values evaluated at points $x_j < x_i$. Intuitively, for a point near the center of a table, it seems desirable to express the differences in terms of functional values on both sides of a given point.   Thus, some sort of averaging of forward and backward differences is needed, and central differences are a means to this end.

Central differences, symbolized by $\delta f(x_i)$, are obtained by choosing a point x_i and proceeding horizontally from $f(x_i)$ in the difference table, as illustrated in Fig. 8.6.   Immediately, a problem arises in that there are no central differences of odd order.   Two methods of solving this problem are illustrated in Fig. 8.7a and b.   In Fig. 8.7a, a central difference of odd order is obtained by taking the forward difference of the preceding central difference of even order.   In Fig. 8.7b, backward differences are used in a similar manner to obtain the central differences of odd order. Upon letting $\delta_f f(x_i)$ and $\delta_b f(x_i)$ be the central differences represented in Fig. 8.7a and b, respectively, the central differences at point $x = 3$ in

$$
\begin{array}{lll}
x_i - 2h & f(x_i - 2h) & \\
& & \cdots \\
x_i - h & f(x_i - h) & \delta^2 f(x_i - h) \\
& & \cdots \qquad\qquad \cdots \\
x_i & f(x_i) \text{———} \delta^2 f(x_i) \text{————} \delta^4 f(x_i) \rightarrow \\
& & \cdots \qquad\qquad \cdots \\
x_i + h & f(x_i + h) & \delta^2 f(x_i + h) \\
& & \cdots \\
x_i + 2h & f(x_i + 2h) &
\end{array}
$$

Fig. 8.6

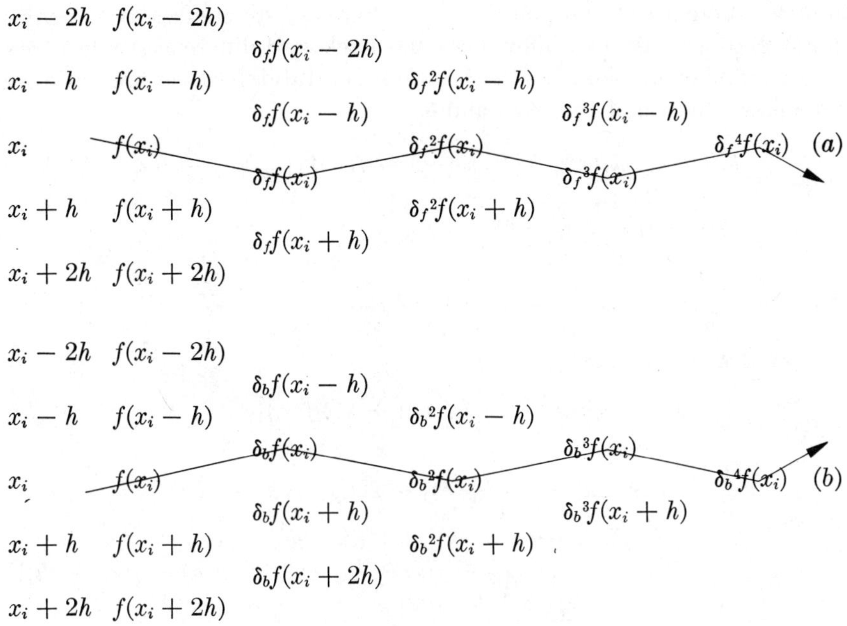

$x_i - 2h \quad f(x_i - 2h)$

$\qquad\qquad\qquad \delta_f f(x_i - 2h)$

$x_i - h \quad f(x_i - h) \qquad\qquad\qquad \delta_f^2 f(x_i - h)$

$\qquad\qquad\qquad \delta_f f(x_i - h) \qquad\qquad\qquad \delta_f^3 f(x_i - h)$

$x_i \quad\quad f(x_i) \qquad\qquad\qquad \delta_f^2 f(x_i) \qquad\qquad\qquad\qquad \delta_f^4 f(x_i) \quad (a)$

$\qquad\qquad\quad \delta_f f(x_i) \qquad\qquad\qquad \delta_f^3 f(x_i)$

$x_i + h \quad f(x_i + h) \qquad\qquad\qquad \delta_f^2 f(x_i + h)$

$\qquad\qquad\qquad \delta_f f(x_i + h)$

$x_i + 2h \quad f(x_i + 2h)$

$x_i - 2h \quad f(x_i - 2h)$

$\qquad\qquad\qquad \delta_b f(x_i - h)$

$x_i - h \quad f(x_i - h) \qquad\qquad\qquad \delta_b^2 f(x_i - h)$

$\qquad\qquad\quad \delta_b f(x_i) \qquad\qquad\qquad \delta_b^3 f(x_i)$

$x_i \quad\quad f(x_i) \qquad\qquad\qquad \delta_b^2 f(x_i) \qquad\qquad\qquad\qquad \delta_b^4 f(x_i) \quad (b)$

$\qquad\qquad\quad \delta_b f(x_i + h) \qquad\qquad\qquad \delta_b^3 f(x_i + h)$

$x_i + h \quad f(x_i + h) \qquad\qquad\qquad \delta_b^2 f(x_i + h)$

$\qquad\qquad\qquad \delta_b f(x_i + 2h)$

$x_i + 2h \quad f(x_i + 2h)$

Fig. 8.7

Table 8.5 are

$$\delta_f f(3) = 10 \qquad\qquad \delta_b f(3) = 30$$
$$\delta_f^2 f(3) = -20 \qquad\quad \delta_b^2 f(3) = -20$$
$$\delta_f^3 f(3) = 50 \qquad\qquad \delta_b^3 f(3) = -40$$
$$\delta_f^4 f(3) = 90 \qquad\qquad \delta_b^4 f(3) = 90$$
$$\qquad\qquad\qquad\qquad\quad \delta_b^5 f(3) = 150$$

The next step should be obvious. To obtain true central differences of odd order, the forward and backward central differences of odd order as represented in Fig. 8.7a and b are averaged. The central differences in the previous example become

$$\delta f(3) = \frac{10 + 30}{2} = 20$$
$$\delta^2 f(3) = -20$$
$$\delta^3 f(3) = \frac{50 - 40}{2} = 5$$
$$\delta^4 f(3) = 90$$

Note that the use of central differences allows the determination of a fourth difference at the point $x = 3$. Forward differences at the same point produce only two differences and backward differences only three.

Algebraic expressions for the first four central differences are obtained as follows, by use of Fig. 8.7a and b:

$$\delta f(x_i) = \frac{\delta_f f(x_i) + \delta_b f(x_i)}{2} = \frac{[f(x_i + h) - f(x_i)] + [f(x_i) - f(x_i - h)]}{2}$$

$$= \frac{f(x_i + h) - f(x_i - h)}{2}$$

$$\delta^2 f(x_i) = \delta_f f(x_i) - \delta_b f(x_i) = \Delta f(x_i) - \nabla f(x_i)$$
$$= [f(x_i + h) - f(x_i)] - [f(x_i) - f(x_i - h)]$$
$$= f(x_i + h) - 2f(x_i) + f(x_i - h)$$

$$\delta^3 f(x_i) = \frac{\delta_f{}^3 f(x_i) + \delta_b{}^3 f(x_i)}{2} = \frac{\Delta[\delta^2 f(x_i)] + \nabla[\delta^2 f(x_i)]}{2}$$

$$= \frac{1}{2}\{[f(x_i + 2h) - f(x_i + h)] - 2[f(x_i + h) - f(x_i)]$$
$$+ [f(x_i) - f(x_i - h)] + [f(x_i + h) - f(x_i)]$$
$$- 2[f(x_i) - f(x_i - h)] + [f(x_i - h) - f(x_i - 2h)]\}$$
$$= \frac{f(x_i + 2h) - 2f(x_i + h) + 2f(x_i - h) - f(x_i - 2h)}{2}$$

$$\delta^4 f(x_i) = \delta_f{}^3 f(x_i) - \delta_b{}^3 f(x_i) = \Delta[\delta^2 f(x_i)] - \nabla[\delta^2 f(x_i)]$$
$$= \{[f(x_i + 2h) - f(x_i + h)] - 2[f(x_i + h) - f(x_i)]$$
$$+ [f(x_i) - f(x_i - h)]\} - \{[f(x_i + h) - f(x_i)]$$
$$- 2[f(x_i) - f(x_i - h)] + [f(x_i - h) - f(x_i - 2h)]\}$$
$$= f(x_i + 2h) - 4f(x_i + h) + 6f(x_i) - 4f(x_i - h) + f(x_i - 2h)$$

It may be noted that the coefficients of the central differences of even order follow the binomial pattern. Therefore, a general expression for $\delta^k f(x_i)$, where k is even, is

$$\delta^k f(x_i) = f\left(x_i + \frac{kh}{2}\right) - \binom{k}{1} f\left[x_i + \frac{(k - 2)h}{2}\right] + \cdots + f\left(x_i - \frac{kh}{2}\right)$$

$$(8.12a)$$

The odd-order central differences can then be obtained by

$$\delta^k f(x_i) = \frac{\Delta[\delta^{k-1} f(x_i)] + \nabla[\delta^{k-1} f(x_i)]}{2}$$

$$(8.12b)$$

Two methods of calculating particular finite differences will now be demonstrated. In Table 8.6 suppose that it is desired to obtain $\Delta^2 f(10)$, $\nabla^3 f(40)$, $\delta^2 f(30)$, and $\delta^3 f(30)$. The differences will be obtained first by direct application of the formulas and then by construction of the difference table. The function represents sin x, where x is given in degrees.

Table 8.6

x	$f(x)$
0	0.00000
10	0.17365
20	0.34202
30	0.50000
40	0.64279
50	0.76604

By formula: The formulas for forward differences are given in Eqs. (8.9). In this example $h = 10$.

$$\Delta^2 f(10) = f(30) - 2f(20) + f(10)$$
$$= 0.50000 - 2(0.34202) + 0.17365$$
$$= -0.01039$$

The formulas for backward differences are given in Eqs. (8.11).

$$\nabla^3 f(40) = f(40) - 3f(30) + 3f(20) - f(10)$$
$$= 0.64279 - 3(0.50000) + 3(0.34202) - 0.17365$$
$$= -0.00480$$

The formulas for central differences are given in Eqs. (8.12).

$$\delta^2 f(30) = f(40) - 2f(30) + f(20)$$
$$= 0.64279 - 2(0.50000) + 0.34202$$
$$= -0.01519$$

$$\delta^3 f(30) = \frac{\Delta[\delta^2 f(30)] + \nabla[\delta^2 f(30)]}{2}$$

$$= \frac{\Delta[f(40) - 2f(30) + f(20)] + \nabla[f(40) - 2f(30) + f(20)]}{2}$$

$$= \frac{1}{2} \{[f(50) - f(40)] - 2[f(40) - f(30)] + [f(30) - f(20)]$$
$$+ [f(40) - f(30)] - 2[f(30) - f(20)] + [f(20) - f(10)]\}$$

$$= \frac{f(50) - 2f(40) + 2f(20) - f(10)}{2}$$

$$= \frac{0.76604 - 2(0.64279) + 2(0.34202) - 0.17365}{2}$$

$$= -0.00458$$

By use of difference table:

x	$f(x)$	D^1	D^2	D^3	D^4	D^5
0	0.00000					
		0.17365				
10	0.17365		−0.00528			
		0.16837		−0.00511		
20	0.34202		−0.01039		0.00031	
		0.15798		−0.00480		0.00014
30	0.50000		−0.01519		0.00045	
		0.14279		−0.00435		
40	0.64279		−0.01954			
		0.12325				
50	0.76604					

The first three answers are read directly from the table (underlined),

$\Delta^2 f(10) = -0.01039$
$\nabla^3 f(40) = -0.00480$
$\delta^2 f(30) = -0.01519$

The third central difference at $x = 30$ is obtained by averaging the two entries in the D^3 column directly above and below the horizontal row defined by $x = 30$.

$$\delta^3 f(30) = \frac{-0.00480 - 0.00435}{2} = -0.00458$$

Two advantages of using the difference table are that no formula need be memorized and that no multiplications are necessary.

8.4.6 Special Property of Polynomials

Consider an nth-order polynomial in the form of Eq. (8.3),

$$f(x) = a_1 x^n + a_2 x^{n-1} + \cdots + a_n x + a_{n+1} \qquad (8.3)$$

The value of the function at the point $x + h$ is given by

$$f(x + h) = a_1(x + h)^n + a_2(x + h)^{n-1} + \cdots + a_n(x + h) + a_{n+1}$$

For any given values of x and h, an algebraic expression for the first forward difference at the point x can be constructed by using the above two relations.

$$\Delta f(x) = f(x + h) - f(x)$$
$$= a_1[(x + h)^n - x^n] + a_2[(x + h)^{n-1} - x^{n-1}]$$
$$+ \cdots + a_n(x + h - x) + (a_{n+1} - a_{n+1})$$
$$= a_1 h \binom{n}{1} x^{n-1} + \left[a_1 h^2 \binom{n}{2} + a_2 h \binom{n-1}{1} \right] x^{n-2}$$
$$+ \cdots + (a_1 h^n + a_2 h^{n-1} + \cdots + a_n h)$$

One important property can be immediately deduced from this equation. The first differences of an nth-order polynomial can be expressed as a polynomial of order $n - 1$.

As an illustration, consider the function $f(x) = x^2 - 3x + 1$, which is tabulated along with its first differences in Table 8.7. The equation for the first differences is obtained as follows:

$$\Delta f(x) = [(x + h)^2 - 3(x + h) + 1] - [x^2 - 3x + 1] = 2hx + (h^2 - 3h)$$

Letting $h = 1$ gives

$$\Delta f(x) = 2x - 2$$

Substituting $x = 0, 1, 2, 3,$ and 4 into this equation gives all the entries in the first difference column of the table.

Table 8.7

x	$f(x)$	D^1
0	1	
		-2
1	-1	
		0
2	-1	
		2
3	1	
		4
4	5	
		6
5	11	

By using this property of polynomials, it follows that, for an nth-order polynomial, second differences can be represented by a polynomial of order $n - 2$, third differences by a polynomial of order $n - 3$, etc., until the nth differences are represented by a polynomial of order zero (a constant). Thus, the nth differences of an nth-order polynomial are constant. It follows directly from this that the $(n + 1)$st and all higher differences are zero. Table 8.8 represents a complete difference table for the previous example.

Table 8.8

x	$f(x) = x^2 - 3x + 1$	D^1	D^2	D^3	D^4	D^5
0	1					
		-2				
1	-1		2			
		0		0		
2	-1		2		0	
		2		0		0
3	1		2		0	
		4		0		
4	5		2			
		6				
5	11					

It is also possible to generate a polynomial by starting with a constant difference and forming a complete difference table. Consider the problem of forming a cubic polynomial with constant third differences equal to 4. A table is formed, starting with three entries in the D^3 column. In order to determine the second differences, one of them must be known a priori. In this example, it is assumed that, at the first point in the table x_1,

$$f(x_1) = \Delta f(x_1) = \Delta^2 f(x_1) = 0$$

The table is formed from right to left, as shown in Table 8.9. The given quantities used to start the table are underlined. If $x_1 = 1$ and $h = 1$,

Table 8.9

x	$f(x)$	D^1	D^2	D^3
x_1	$\underline{0}$			
		$\underline{0}$		
x_2	0		$\underline{0}$	
		0		$\underline{4}$
x_3	0		4	
		4		$\underline{4}$
x_4	4		8	
		12		$\underline{4}$
x_5	16		12	
		24		
x_6	40			

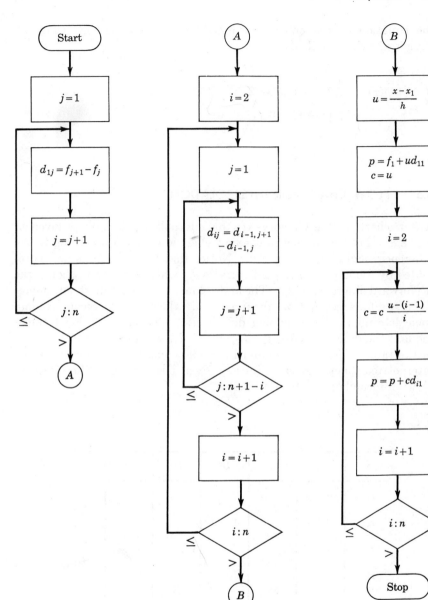

Fig. 8.8

the equation of the cubic polynomial can be obtained by using the Gregory-Newton method.

$$f(x) = p(x) = \frac{4}{3!}(x-1)(x-2)(x-3)$$
$$= \tfrac{2}{3}(x^3 - 6x^2 + 11x - 6)$$

8.5   FLOW CHART FOR INTERPOLATION

A flow chart for the Gregory-Newton interpolation formula is given in Fig. 8.8.   The terms used in the flow chart are defined according to the differences indicated in Fig. 8.9.   Notice that the differences are represented by a doubly subscripted variable d_{ij}, where the i and j signify the ith forward difference of x_j.   The first column in the flow chart determines the first differences d_{1j} from the functional values f_j.   The second column then determines the remaining differences.   The interpolation formula requires the differences $d_{11}, d_{21}, d_{31}, \ldots, d_{n1}$, as indicated by the arrow in the table.   The third column in the flow chart then determines the interpolated functional value for any given x according to the Gregory-Newton formula given by Eq. (8.8).

x	$f(x)$	D	D^2	D^3		D^n
x_1	f_1					
		d_{11}				
x_2	f_2		d_{21}			
		d_{12}		d_{31}		
x_3	f_3		d_{22}			
		d_{13}		d_{32}		d_{n1}
x_4	f_4		d_{23}		$\cdots$	
$\cdot$	$\cdot$	d_{14}	$\cdot$	d_{33}		
$\cdot$	$\cdot$	$\cdot$	$\cdot$	$\cdot$	$\cdots$	
$\cdot$	$\cdot$	$\cdot$	$\cdot$	$\cdot$		
x_n	f_n	$\cdot$	$d_{2(n-1)}$	$\cdot$		
		d_{1n}				
x_{n+1}	f_{n+1}					

Fig. 8.9

Table 8.10

Formula	Form	Result
Gregory-Newton forward	$p(x) = a_1 + a_2(x - x_1) + a_3(x - x_1)(x - x_2) + a_4(x - x_1)(x - x_2)(x - x_3) + \cdots$	$p(u) = f(x_1) + \dfrac{\Delta f(x_1)}{1!} u + \dfrac{\Delta^2 f(x_1)}{2!} u(u - 1) + \dfrac{\Delta^3 f(x_1)}{3!} u(u - 1)(u - 2) + \cdots$
Gregory-Newton backward	$p(x) = a_1 + a_2(x - x_{n+1}) + a_3(x - x_{n+1})(x - x_n) + a_4(x - x_{n+1})(x - x_n)(x - x_{n-1}) + \cdots$	$p(u) = f(x_{n+1}) + \dfrac{\nabla f(x_{n+1})}{1!} u + \dfrac{\nabla^2 f(x_{n+1})}{2!} u(u + 1) + \dfrac{\nabla^3 f(x_{n+1})}{3!} u(u + 1)(u + 2) + \cdots$
Gauss forward	$p(x) = a_1 + a_2(x - x_i) + a_3(x - x_i)(x - x_{i+1}) + a_4(x - x_i)(x - x_{i+1})(x - x_{i-1}) + a_5(x - x_i)(x - x_{i+1})(x - x_{i-1})(x - x_{i+2}) + \cdots$	$p(u) = f(x_i) + \dfrac{\delta f(x_i)}{1!} u + \dfrac{\delta^2 f(x_i)}{2!} u(u - 1) + \dfrac{\delta^3 f(x_i)}{3!} u(u - 1)(u + 1) + \dfrac{\delta^4 f(x_i)}{4!} u(u - 1)(u + 1)(u - 2) + \cdots$
Gauss backward	$p(x) = a_1 + a_2(x - x_i) + a_3(x - x_i)(x - x_{i-1}) + a_4(x - x_i)(x - x_{i-1})(x - x_{i+1}) + a_5(x - x_i)(x - x_{i-1})(x - x_{i+1})(x - x_{i-2}) + \cdots$	$p(u) = f(x_i) + \dfrac{\delta_b f(x_i)}{1!} u + \dfrac{\delta_b^2 f(x_i)}{2!} u(u + 1) + \dfrac{\delta_b^3 f(x_i)}{3!} u(u + 1)(u - 1) + \dfrac{\delta_b^4 f(x_i)}{4!} u(u + 1)(u - 1)(u + 2) + \cdots$
Stirling central	$\ldots\ldots\ldots\ldots\ldots\ldots\ldots\ldots\ldots\ldots$	$p(u) = f(x_i) + \dfrac{\delta f(x_i)}{1!} u + \dfrac{\delta^2 f(x_i)}{2!} u^2 + \dfrac{\delta^3 f(x_i)}{3!} u(u - 1)(u + 1) + \dfrac{\delta^4 f(x_i)}{4!} u^2(u - 1)(u + 1) + \dfrac{\delta^5 f(x_i)}{5!} u(u - 1)(u + 1)(u - 2)(u + 2) + \cdots$

8.6   OTHER INTERPOLATING FORMULAS

It has been demonstrated that the Gregory-Newton interpolation formula may be based upon forward finite differences. Technically, then, a more appropriate name would be the Gregory-Newton *forward*-interpolation formula. Formulas can also be developed from backward and central differences, and these are presented here. The method of deriving these formulas is exactly the same as that used in Sec. 8.3 for the Gregory-Newton formula. The variations in the several formulas occur in the form of the interpolation polynomial. Table 8.10 gives the assumed forms and resulting formulas for several interpolation polynomials. In the central-difference formulas, a central point is represented by x_i. The Stirling formula is obtained by averaging the Gauss forward and Gauss backward formulas.

As an illustration, all the interpolation polynomials are used for the data on the rocket problem given on page 285. The differences are formed in Table 8.11.

Table 8.11

Time	*Speed*	*D*	*D²*	*D³*	*D⁴*	*D⁵*
0	0.0000					
		0.0824				
60	0.0824		0.1099	Path 1		
		0.1923		0.0733		
120	0.2747		0.1832		0.1029	
		0.3755	Path 4	0.1762		0.4644
180	0.6502		0.3594		0.5673	
		0.7349	Path 3	0.7435		
240	1.3851		1.1029	Path 2		
		1.8378				
300	3.2229					

Gregory-Newton forward: The coefficients of this formula are obtained by following path 1 in the table.

$$p(u) = 0.0000 + \frac{0.0824}{1!} u + \frac{0.1099}{2!} u(u-1) + \frac{0.0733}{3!} u(u-1)(u-2)$$

$$+ \frac{0.1029}{4!} u(u-1)(u-2)(u-3)$$

$$+ \frac{0.4644}{5!} u(u-1)(u-2)(u-3)(u-4)$$

$$p(u) = 0.0824u + 0.05495u(u-1) + 0.01222u(u-1)(u-2)$$
$$+ 0.004288u(u-1)(u-2)(u-3)$$
$$+ 0.003870u(u-1)(u-2)(u-3)(u-4)$$

At $t = 150$ sec

$$u = \frac{t - t_1}{h} = \frac{150 - 0}{60} = 2.5$$

$p(2.5) = 0.4365$ mps

which agrees with the answer given in Sec. 8.3.1.

Gregory-Newton backward: The coefficients of this formula are obtained by following path 2 in the table.

$$p(u) = 3.2229 + \frac{1.8378}{1!} u + \frac{1.1029}{2!} u(u + 1) + \frac{0.7435}{3!} u(u + 1)(u + 2)$$

$$+ \frac{0.5673}{4!} u(u + 1)(u + 2)(u + 3)$$

$$+ \frac{0.4644}{5!} u(u + 1)(u + 2)(u + 3)(u + 4)$$

$$= 3.2229 + 1.8378u + 0.5514u(u + 1) + 0.1239u(u + 1)(u + 2)$$
$$+ 0.02364u(u + 1)(u + 2)(u + 3)$$
$$+ 0.003870u(u + 1)(u + 2)(u + 3)(u + 4)$$

At $t = 150$ sec

$$u = \frac{t - t_{n+1}}{h} = \frac{150 - 300}{60} = -2.5$$

$p(-2.5) = 0.4362$ mps

Gauss forward: The coefficients of this formula are obtained by following path 3 in the table. The central point is taken as $t = 180$.

$$p(u) = 0.6502 + \frac{0.7349}{1!} u + \frac{0.3594}{2!} u(u - 1) + \frac{0.7435}{3!} u(u - 1)(u + 1)$$

$$+ \frac{0.5673}{4!} u(u - 1)(u + 1)(u - 2)$$

$$= 0.6502 + 0.7349u + 0.1797u(u - 1) + 0.1239u(u - 1)(u + 1)$$
$$+ 0.02364u(u - 1)(u + 1)(u - 2)$$

At $t = 150$ sec

$$u = \frac{t - t_i}{h} = \frac{150 - 180}{60} = -0.5$$

$p(-0.5) = 0.4418$ mps

Note that only four differences could be used in this case.

Gauss backward: The coefficients of this formula are obtained by following path 4 in the table.

$$p(u) = 0.6502 + \frac{0.3755}{1!} u + \frac{0.3594}{2!} u(u+1) + \frac{0.1762}{3!} u(u+1)(u-1)$$
$$+ \frac{0.5673}{4!} u(u+1)(u-1)(u+2)$$
$$+ \frac{0.4644}{5!} u(u+1)(u-1)(u+2)(u-2)$$
$$= 0.6502 + 0.3755u + 0.1797u(u+1) + 0.02937u(u+1)(u-1)$$
$$+ 0.02364u(u+1)(u-1)(u+2)$$
$$+ 0.003870u(u+1)(u-1)(u+2)(u-2)$$

At $t = 150$ sec

$$u = \frac{t - t_i}{h} = \frac{150 - 180}{60} = -0.5$$

$$p(-0.5) = 0.4364 \text{ mps}$$

Stirling central: The coefficients of this formula are obtained by averaging the Gauss forward and Gauss backward formulas. The fifth differences are dropped, since the Gauss forward formula does not have a fifth difference.

$$p(u) = 0.6502 + \frac{0.7349 + 0.3755}{2} u + 0.1797u^2$$
$$+ \frac{0.1239 + 0.02937}{2} u(u-1)(u+1) + 0.2364u^2(u-1)(u+1)$$
$$= 0.6502 + 0.5552u + 0.1797u^2 + 0.07664u(u-1)(u+1)$$
$$+ 0.02364u^2(u-1)(u+1)$$

At $t = 150$ sec

$$u = \frac{t - t_i}{h} = \frac{150 - 180}{60} = -0.5$$

$$p(u) = 0.4416 \text{ mps}$$

8.6.1   Uniqueness

The laws of algebra state that there is only one unique nth-order polynomial which passes through $n + 1$ points. Therefore, the various interpolation formulas presented in this section are actually only different forms of the same polynomial. It follows, then, that any interpolated functional value should be the same regardless of which formula is used. This is illustrated in the examples just given. The three formulas which used six terms gave functional values at 150 sec of 0.4365, 0.4362, and 0.4364. The other two formulas, which used only five terms and thus represented slightly different interpolating polynomials, gave functional values of 0.4418 and 0.4416. The slight discrepancies in these results are due to round-off errors.

8.7 LAGRANGIAN INTERPOLATION

In the introduction of this chapter it was demonstrated that the determination of an interpolating polynomial involves solving a system of simultaneous equations for the coefficients of the polynomial. The methods of Gregory-Newton, Gauss, and Stirling assume the interpolating polynomial in such a way that the system of simultaneous equations is in triangular form. Lagrangian interpolation assumes the polynomial in such a way that the resulting system of simultaneous equations is in diagonal form. The solution can then be written down immediately. This is accomplished by assuming the interpolation polynomial in the form

$$
\begin{aligned}
p(x) = a_1 &\frac{(x - x_2)(x - x_3)(\cdots)(x - x_{n+1})}{(x_1 - x_2)(x_1 - x_3)(\cdots)(x_1 - x_{n+1})} \\
&+ a_2 \frac{(x - x_1)(x - x_3)(\cdots)(x - x_{n+1})}{(x_2 - x_1)(x_2 - x_3)(\cdots)(x_2 - x_{n+1})} \\
&\cdots\cdots\cdots\cdots\cdots\cdots\cdots \\
&+ a_{n+1} \frac{(x - x_1)(x - x_2)(\cdots)(x - x_n)}{(x_{n+1} - x_1)(x_{n+1} - x_2)(\cdots)(x_{n+1} - x_n)}
\end{aligned} \tag{8.13}
$$

The fact that this is an nth-order polynomial is easily ascertained. The numerator of each term is itself an nth-order polynomial, while the denominator is a constant. The coefficients a_1, a_2, . . . , a_{n+1} are obtained in the usual manner of requiring $p(x)$ to satisfy the functional values $f(x_1)$, $f(x_2)$, . . . , $f(x_{n+1})$ at the points x_1, x_2, . . . , x_{n+1}. This results in

$$f(x_1) = a_1$$
$$f(x_2) = a_2$$
$$\cdots\cdots$$
$$f(x_{n+1}) = a_{n+1}$$

Lagrange's interpolation formula can then be written as

$$
\begin{aligned}
p(x) = f(x_1) &\frac{(x - x_2)(x - x_3)(\cdots)(x - x_{n+1})}{(x_1 - x_2)(x_1 - x_3)(\cdots)(x_1 - x_{n+1})} \\
&+ f(x_2) \frac{(x - x_1)(x - x_3)(\cdots)(x - x_{n+1})}{(x_2 - x_1)(x_2 - x_3)(\cdots)(x_2 - x_{n+1})} \\
&\cdots\cdots\cdots\cdots\cdots\cdots\cdots \\
&+ f(x_{n+1}) \frac{(x - x_1)(x - x_2)(\cdots)(x - x_n)}{(x_{n+1} - x_1)(x_{n+1} - x_2)(\cdots)(x_{n+1} - x_n)}
\end{aligned} \tag{8.14}
$$

For the values given in Table 8.12, Lagrange's interpolation formula can be immediately written as

$$p(x) = 0.34202 \frac{(x - 25)(x - 30)}{(20 - 25)(20 - 30)} + 0.42262 \frac{(x - 20)(x - 30)}{(25 - 20)(25 - 30)}$$

$$+ 0.50000 \frac{(x - 20)(x - 25)}{(30 - 20)(30 - 25)}$$

$$= 0.0068404(x - 25)(x - 30) - 0.016905(x - 20)(x - 30)$$

$$+ 0.010000(x - 20)(x - 25)$$

For $x = 22$, $p(x) = 0.37465$, which agrees with the result given by the Gregory-Newton formula in Sec. 8.2.

Table 8.12

x	$f(x)$
20	0.34202
25	0.42262
30	0.50000

8.7.1   Equal Spacing

The Lagrangian formula is simplified somewhat if equal spacing between the points $x_1, x_2, \ldots, x_{n+1}$ is assumed. If we let h be the interval, the interpolation formula becomes

$$p(x) = f(x_1) \frac{(x - x_2)(x - x_3)(\cdots)(x - x_{n+1})}{(-h)(-2h)(\cdots)(-nh)}$$

$$+ f(x_2) \frac{(x - x_1)(x - x_3)(\cdots)(x - x_{n+1})}{(h)(-h)(\cdots)[-(n - 1)h]}$$

$$\cdot \cdot \cdot \cdot \cdot \cdot \cdot \cdot \cdot \cdot \cdot \cdot \cdot \cdot \cdot \cdot \cdot$$

$$+ f(x_{n+1}) \frac{(x - x_1)(x - x_2)(\cdots)(x - x_n)}{(nh)[(n - 1)h](\cdots)(h)}$$

$$= \frac{f(x_1)}{(-1)^n n! \, h^n} (x - x_2)(x - x_3)(\cdots)(x - x_{n+1})$$

$$+ \frac{f(x_2)}{(-1)^{n-1} 1! \, (n - 1)! \, h^n} (x - x_1)(x - x_3)(\cdots)(x - x_{n+1})$$

$$\cdot \cdot \cdot \cdot \cdot \cdot \cdot \cdot \cdot \cdot \cdot \cdot \cdot \cdot \cdot \cdot \cdot \cdot \cdot \cdot$$

$$+ \frac{f(x_{n+1})}{n! \, h^n} (x - x_1)(x - x_2)(\cdots)(x - x_n)$$

Making the substitution

$$x = x_1 + uh$$

produces

$$p(x) = \frac{f(x_1)}{(-1)^n n!} (u - 1)(u - 2)(\cdots)(u - n)$$

$$+ \frac{f(x_2)}{(-1)^{n-1} 1! \, (n-1)!} u(u - 2)(\cdots)(u - n)$$

$$\cdots\cdots\cdots\cdots\cdots\cdots\cdots\cdots\cdots$$

$$+ \frac{f(x_{n+1})}{n!} u(u - 1)(u - 2)(\cdots)[u - (n - 1)] \quad (8.15)$$

The data from the rocket problem given in Table 8.1 are repeated here.

Time, sec	Speed, mps
0	0.0000
60	0.0824
120	0.2747
180	0.6502
240	1.3851
300	3.2229

Passing a fifth-order polynomial through these six points gives the following Lagrangian interpolation formula:

$$p(u) = 0.0000 + \frac{0.0824}{1! \, 4!} u(u - 2)(u - 3)(u - 4)(u - 5)$$

$$+ \frac{0.2747}{-2! \, 3!} u(u - 1)(u - 3)(u - 4)(u - 5)$$

$$+ \frac{0.6502}{3! \, 2!} u(u - 1)(u - 2)(u - 4)(u - 5)$$

$$+ \frac{1.3851}{-4! \, 1!} u(u - 1)(u - 2)(u - 3)(u - 5)$$

$$+ \frac{3.2229}{5!} u(u - 1)(u - 2)(u - 3)(u - 4)$$

$$p(u) = 0.003433u(u - 2)(u - 3)(u - 4)(u - 5)$$
$$- 0.02289u(u - 1)(u - 3)(u - 4)(u - 5)$$
$$+ 0.05418u(u - 1)(u - 2)(u - 4)(u - 5)$$
$$- 0.05771u(u - 1)(u - 2)(u - 3)(u - 5)$$
$$+ 0.02686u(u - 1)(u - 2)(u - 3)(u - 4)$$

At the time $t = 150$ sec

$$u = \frac{t - t_1}{h} = \frac{150 - 0}{60} = 2.5$$

$$p(2.5) = 0.4364 \text{ mps}$$

which agrees with the results given in the example in Sec. 8.3.1.

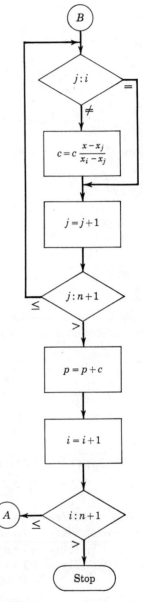
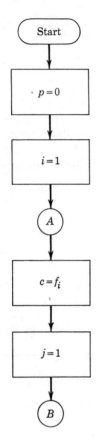

Fig. 8.10

Lagrangian interpolation certainly has the advantage that the coefficients of the interpolating polynomial are easily determined.   However, the resulting formula is somewhat complicated and tedious to apply. Further, the easily constructed difference table, which is very useful in the other interpolation methods, cannot be utilized.

The form of the Lagrangian formula given by Eq. (8.14) is somewhat easier to program than the latter form given by Eq. (8.15).   Further, it is more general in that it is applicable to either equal or unequal intervals and the points x_1, x_2, . . . , x_{n+1} need not be in order.   The flow chart is presented in Fig. 8.10.   The x_i's and the f_i's represent the given tabulated values of the independent and dependent variables, respectively.   The inner loop evaluates the ith term in the formula, while the outer loop sums these terms.   The interpolated functional value corresponding to any given x is given by p.

8.8   UNEQUAL INTERVALS

The approach used to pass an nth-order polynomial through $n + 1$ points, in which the spacing between x_1, x_2, . . . , x_{n+1} is not even, is generally

Table 8.13

x	$f(x)$
20	0.34202
30	0.50000
35	0.57358

the same as used previously.   To illustrate, consider the data in Table 8.13.   Now, assume the interpolating parabola to be in the form

$$p(x) = a_1 + a_2(x - 20) + a_3(x - 20)(x - 30)$$

Satisfying the parabola at the three tabular points gives

$$0.34202 = a_1$$
$$0.50000 = a_1 + a_2(30 - 20)$$
$$0.57358 = a_1 + a_2(35 - 20) + a_3(35 - 20)(35 - 30)$$

Solving the system of equations by successive substitution produces

$$a_1 = 0.34202 \qquad a_2 = 0.015798 \qquad a_3 = -0.0000722$$

and the interpolating parabola becomes

$$p(x) = 0.34202 + 0.015798(x - 20) - 0.0000722(x - 20)(x - 30)$$

At $x = 22$, $p(22) = 0.37477$. The given function is $\sin x$, and the true value of $\sin 22° = 0.37461$.

8.8.1   Divided Differences

As before, differences can be used advantageously in applying unequal-interval interpolation. Consider again a polynomial in the form of Eq. (8.6).

$$p(x) = a_1 + a_2(x - x_1) + a_3(x - x_1)(x - x_2)$$
$$+ a_4(x - x_1)(x - x_2)(x - x_3)$$
$$+ \cdots + a_{n+1}(x - x_1)(x - x_2)(\cdots)(x - x_n)$$

Following the usual procedure,

$$f(x_1) = a_1$$
$$f(x_2) = a_1 + a_2(x_2 - x_1)$$
$$f(x_3) = a_1 + a_2(x_3 - x_1) + a_3(x_3 - x_1)(x_3 - x_2)$$
$$f(x_4) = a_1 + a_2(x_4 - x_1) + a_3(x_4 - x_1)(x_4 - x_2)$$
$$+ a_4(x_4 - x_1)(x_4 - x_2)(x_4 - x_3)$$

The coefficients can now be determined by successive substitution and after a considerable amount of algebraic manipulation become

$$a_1 = f(x_1)$$

$$a_2 = \frac{f(x_2) - f(x_1)}{x_2 - x_1}$$

$$a_3 = \frac{[f(x_3) - f(x_2)]/(x_3 - x_2) - [f(x_2) - f(x_1)]/(x_2 - x_1)}{x_3 - x_1}$$

$$a_4 = \frac{\dfrac{\dfrac{f(x_4) - f(x_3)}{x_4 - x_3} - \dfrac{f(x_3) - f(x_2)}{x_3 - x_2}}{x_4 - x_2} - \dfrac{\dfrac{f(x_3) - f(x_2)}{x_3 - x_2} - \dfrac{f(x_2) - f(x_1)}{x_2 - x_1}}{x_3 - x_1}}{x_4 - x_1}$$

The pattern is apparent. Using the above equations, *forward divided differences* are defined as follows:

$$\Delta_d f(x_1) = a_2 = \frac{f(x_2) - f(x_1)}{x_2 - x_1}$$

$$\Delta_d^2 f(x_1) = a_3 = \frac{\Delta_d f(x_2) - \Delta_d f(x_1)}{x_3 - x_1} \tag{8.16}$$

$$\Delta_d^3 f(x_1) = a_4 = \frac{\Delta_d^2 f(x_2) - \Delta_d^2 f(x_1)}{x_4 - x_1}$$

The expression for the kth forward difference at the point x_1 is deduced as

$$\Delta_d^k f(x_1) = \frac{\Delta_d^{k-1} f(x_2) - \Delta_d^{k-1} f(x_1)}{x_{k+1} - x_1}$$

Table 8.14

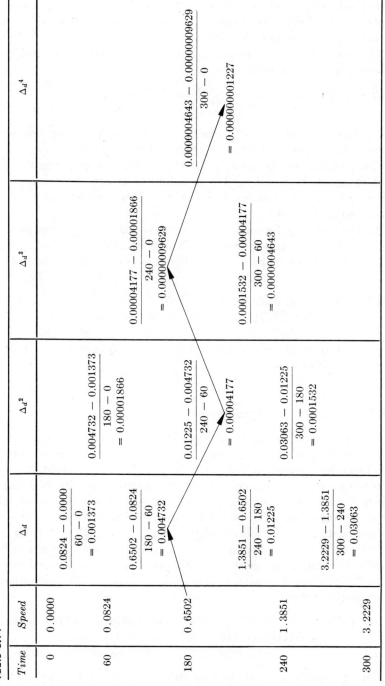

Time	Speed	Δ_d	$\Delta_d{}^2$	$\Delta_d{}^3$	$\Delta_d{}^4$
0	0.0000				
		$\dfrac{0.0824 - 0.0000}{60 - 0}$ $= 0.001373$			
60	0.0824		$\dfrac{0.004732 - 0.001373}{180 - 0}$ $= 0.00001866$		
		$\dfrac{0.6502 - 0.0824}{180 - 60}$ $= 0.004732$		$\dfrac{0.00004177 - 0.00001866}{240 - 0}$ $= 0.0000009629$	
180	0.6502		$\dfrac{0.01225 - 0.004732}{240 - 60}$ $= 0.00004177$		$\dfrac{0.0000004643 - 0.00000009629}{300 - 0}$ $= 0.0000000001227$
		$\dfrac{1.3851 - 0.6502}{240 - 180}$ $= 0.01225$		$\dfrac{0.0001532 - 0.00004177}{300 - 60}$ $= 0.0000004643$	
240	1.3851		$\dfrac{0.03063 - 0.01225}{300 - 180}$ $= 0.0001532$		
		$\dfrac{3.2229 - 1.3851}{300 - 240}$ $= 0.03063$			
300	3.2229				

Divided-difference tables can be easily constructed, and backward and central divided differences can also be determined, by use of the corresponding paths through the table.

The interpolating polynomial, known as *Newton's divided-difference formula*, can now be written as

$$p(x) = f(x_1) + [\Delta_d f(x_1)](x - x_1) + [\Delta_d^2 f(x_1)](x - x_1)(x - x_2)$$
$$+ \cdots + [\Delta_d^n f(x_1)](x - x_1)(x - x_2)(\cdots)(x - x_n) \quad (8.17)$$

Other interpolating formulas can be obtained by replacing the forward divided differences by backward or central divided differences.

Suppose that the speed at time $t = 120$ sec is not known in the rocket problem given in Sec. 8.1. Divided differences can then be utilized to determine an interpolating polynomial. The divided differences are presented in Table 8.14.

An interpolating formula, which corresponds to the Gauss backward formula given for equal intervals in Table 8.10, can be obtained by following the path shown in Table 8.14. The interpolated value at $t = 120$ sec is

$$p(120) = 0.6502 + 0.004732(120 - 180)$$
$$+ 0.00004177(120 - 180)(120 - 60)$$
$$+ 0.00000009629(120 - 180)(120 - 60)(120 - 240)$$
$$+ 0.000000001277(120 - 180)(120 - 60)(120 - 240)(120 - 0)$$
$$p(120) = 0.3211 \text{ mps}$$

8.8.2   Sheppard's Zigzag Rule

As mentioned previously, a divided-difference interpolation polynomial can be deduced which corresponds to each of the equal-interval interpolation formulas discussed in Sec. 8.6. Sheppard's rule states that the coefficients of an interpolation polynomial can be determined by following any zigzag path through the table. Each coefficient is obtained by taking either a forward or a backward difference. The kth term of the polynomial contains $k - 1$ factors of the form $x - x_i$. The $(k + 1)$st term includes all these factors plus one more, which is either forward or backward depending upon whether the difference $\Delta_d^k f(x_i)$ is forward or backward. To illustrate, consider the path in Table 8.15. The corresponding interpolation formula is

$$p(t) = 0.6502 + 0.004732(t - 180) + 0.00004177(t - 180)(t - 60)$$
$$+ 0.0000004643(t - 180)(t - 60)(t - 240)$$
$$+ 0.000000001227(t - 180)(t - 60)(t - 240)(t - 300)$$

One can readily see that the formulas of Newton and Gauss are special cases of Sheppard's rule.

Table 8.15

$Time$	$Speed$	Δ_d	$\Delta_d{}^2$	$\Delta_d{}^3$	$\Delta_d{}^4$
0	0.0000				
		0.001373			
60	0.0824		0.00001866		
		0.004732		0.00000009629	
180	0.6502		0.00004177		0.000000001227
		0.01225		0.0000004643	
240	1.3851		0.0001532		
		0.03063			
300	3.2229				

8.8.3 Graphical Representation

If we consider again the three points representing functional values and the interpolating polynomial passing through them, divided differences can be interpreted as follows: In Fig. 8.11 the slope of the line BC represents the first forward divided difference at point B. Similarly, the backward and central divided differences at point B are represented by the slopes of the lines AB and AC, respectively.

The flow chart given in Fig. 8.8 can be slightly modified to produce a flow chart for divided differences. In this figure only two changes are

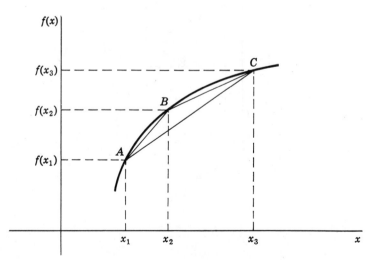

Fig. 8.11

needed in the first two columns.   They are

$$d_{1j} = \frac{f_{j+1} - f_j}{x_{j+1} - x_j}$$

$$d_{ij} = \frac{d_{i-1,j+1} - d_{i-1,j}}{x_{i+j} - x_j}$$

The third column in Fig. 8.8 applies only to equal intervals and must be replaced by the column shown in Fig. 8.12.

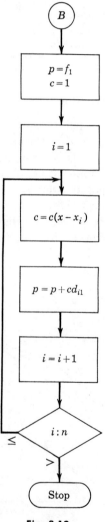

Fig. 8.12

8.8.4 Lagrangian Interpolation

The Lagrangian interpolation formula [Eq. (8.14)] can be applied even if the intervals are unequal. In addition, the flow chart given in Fig. 8.10 for the Lagrangian formula applies to any type of spacing. To illustrate, consider again the rocket problem as presented in the example in Sec. 8.8.1. Direct substitution into the Lagrangian formula gives an interpolated functional value. For instance, at $t = 120$ sec,

$$p(120) = 0.0000 + 0.0824 \frac{(120 - 0)(120 - 180)(120 - 240)(120 - 300)}{(60 - 0)(60 - 180)(60 - 240)(60 - 300)}$$
$$+ 0.6502 \frac{(120 - 0)(120 - 60)(120 - 240)(120 - 300)}{(180 - 0)(180 - 60)(180 - 240)(180 - 300)}$$
$$+ 1.3851 \frac{(120 - 0)(120 - 60)(120 - 180)(120 - 300)}{(240 - 0)(240 - 60)(240 - 180)(240 - 300)}$$
$$+ 3.2229 \frac{(120 - 0)(120 - 60)(120 - 180)(120 - 240)}{(300 - 0)(300 - 60)(300 - 180)(300 - 240)}$$
$$= 0.3211 \text{ mps}$$

8.9 INVERSE INTERPOLATION

As the name implies, inverse interpolation is the process of determining the value of the independent variable x corresponding to a given functional value $f(x)$. To illustrate the process, consider the data in Table 8.16. To obtain a value for the unknown x, linear interpolation may be used.

Table 8.16

x	$f(x)$
20	0.34202
?	0.39875
30	0.50000

$$\frac{x - 20}{30 - 20} = \frac{0.39875 - 0.34202}{0.50000 - 0.34202}$$
$$x = 23.59$$

If the above function represents $\sin x$, the actual value of x is 23.50.

In general, there are two methods of performing inverse interpolation. The first consists in constructing a standard interpolating polynomial, setting it equal to the given functional value $f(x)$, and solving the resulting polynomial equation for x. In the second method, the dependent and independent variables are simply interchanged. The two methods will

now be illustrated by examples.   In linear interpolation the two methods
are equivalent.

8.9.1   Polynomial Equation

In Table 8.17, assume that it is required to obtain a value for the missing

Table 8.17

x	$f(x)$
20	0.34202
?	0.39875
25	0.42262
30	0.50000

entry.   The equation for the parabola passing through the three known
points is given in Eq. (8.4) as

$$p(x) = -0.0000644x^2 + 0.01902x - 0.01258$$

Substitution of the given functional value produces

$$0.39875 = -0.0000644x^2 + 0.01902x - 0.01258$$

which when solved by the quadratic formula gives the two roots

$$x = 23.52, 272$$

The first root is the required answer and compares favorably with the
linear-interpolation result of 23.59 given earlier and the actual value of
23.50.   The second root lies outside the range of interpolation.
If higher-order interpolating polynomials are employed, the roots of
the equation may be found by the methods given in Chap. 6.

8.9.2   Interchanging Dependent and Independent Variables

This method has the advantage that it reduces the problem to ordinary
interpolation.   Usually the spacing of the dependent variable is uneven,
and use of divided differences or the Lagrangian method is necessary.
To illustrate, consider again the function given in Table 8.17.   Interchanging columns produces Table 8.18, and the Lagrangian polynomial

Table 8.18

x	$f(x)$
0.34202	20
0.39875	?
0.42262	25
0.50000	30

can be written down directly as

$$p(x) = 20 \frac{(x - 0.42262)(x - 0.50000)}{(0.34202 - 0.42262)(0.34202 - 0.50000)}$$
$$+ 25 \frac{(x - 0.34202)(x - 0.50000)}{(0.42262 - 0.34202)(0.42262 - 0.50000)}$$
$$+ 30 \frac{(x - 0.34202)(x - 0.42262)}{(0.50000 - 0.34202)(0.50000 - 0.42262)}$$

Substituting into this formula the value $x = 0.39875$ gives

$$f(0.39875) = 23.497$$

which is very close to arcsin $0.39875 = 23.500$.

8.9.3   Comparison of the Two Methods

Both methods presented here require two steps: first, the determination of an interpolating polynomial; second, the evaluation of the polynomial for a given value of the dependent variable. With the first method, evaluation of the polynomial requires finding its principal root, while, with the second method, only a direct substitution is needed. Thus, it appears that the method of interchanging dependent and independent variables is easier to apply. This is partially offset by the fact that it probably requires an interpolation with unequal spacing. However, the real issue is not which method is easier to apply, but which method gives a more accurate answer. In the preceding example interchanging variables produced a slightly more accurate result than did the root-determination method. However, this is not always the case, and this question is deferred to Sec. 8.11, where errors in interpolation are discussed.

8.10   EXTRAPOLATION

Given a set of points $x_1, x_2, \ldots, x_{n+1}$, which are ordered in such a way that $x_1 < x_2 < \cdots < x_{n+1}$, *interpolation* is concerned with determining functional values $f(x)$ for arguments x which lie within the interval $x_1 \leq x \leq x_{n+1}$. *Extrapolation*, on the other hand, utilizes the same interpolating polynomial to obtain functional values $f(x)$ for arguments x which lie outside the interval $x_1 \leq x \leq x_{n+1}$. Consider again the tabulated function given in Table 8.3, whose interpolating polynomial is

x	$f(x)$
20	0.34202
25	0.42262
30	0.50000

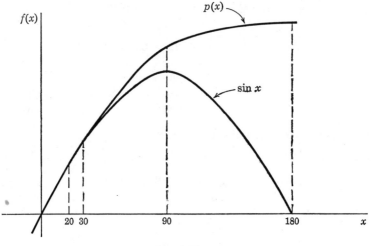

Fig. 8.13

determined by Eq. (8.5) as

$$p(x) = 0.34202 + 0.01612(x - 20) - 0.0000644(x - 20)(x - 25)$$

An extrapolated value for $f(32)$ is obtained by direct substitution.

$$p(32) = 0.34202 + 0.01612(32 - 20) - 0.0000644(32 - 20)(32 - 25)$$
$$= 0.53005$$

On the assumption that $f(x) = \sin x$, the actual value of $\sin 32°$ is 0.52992.

It is quite apparent that extrapolation will give very erroneous results if x is taken too far outside the interval $x_1 \leq x \leq x_{n+1}$. Table 8.19 compares actual values of $\sin x$ with extrapolated values obtained from $p(x)$ in the preceding example. Figure 8.13 shows how the interpolating polynomial diverges from the true function outside the interval $20 \leq x \leq 30$.

Table 8.19

x	$\sin x$	$p(x)$
0	0.00000	−0.01258
32	0.52992	0.53005
45	0.70711	0.71282
60	0.86603	0.89666
90	1.00000	1.17740
180	0.00000	1.32410

In predicting the future, interpolation, of course, is not possible, and extrapolation is necessary. The population of the United States at each of the last three censuses is given in Table 8.20. An interpolation parabola passing through the three given points estimates the population in 1970 as 215,661,406. A linear extrapolation from 1950 and 1960 produces a 1970 estimate of 207,320,552.

Table 8.20

Year	*Population*
1940	131,669,275
1950	151,325,798
1960	179,323,175

8.11   ERROR ANALYSIS

When an interpolating polynomial $p(x)$ is used to approximate a function $f(x)$, it is important to know the error involved in the approximation. *Error* is defined as follows:

$$\text{Error} = f(x) - p(x) \tag{8.18}$$

Quite obviously, the error at the points $x_1, x_2, \ldots, x_{n+1}$ is zero, since at these points $p(x) = f(x)$ by definition. It remains, then, to determine the error at intermediate points.

Consider, first, the linear case, where the interpolating polynomial has the form

$$p(x) = a_1 x + a_2$$

and satisfies the conditions

$$p(x_1) = f(x_1) \qquad p(x_2) = f(x_2)$$

In order to determine an expression for the error at any point $\bar{x}$ which is within the interval $x_1 \leq \bar{x} \leq x_2$, a function $F(x)$ is constructed in the form

$$F(x) = f(x) - p(x) - \frac{f(\bar{x}) - p(\bar{x})}{(\bar{x} - x_1)(\bar{x} - x_2)} (x - x_1)(x - x_2)$$

In this function it is important to note that x is the independent variable, while $\bar{x}$ is a constant representing the particular point at which the error is to be determined. The function $F(x)$ has the properties

$$F(x_1) = F(x_2) = F(\bar{x}) = 0$$

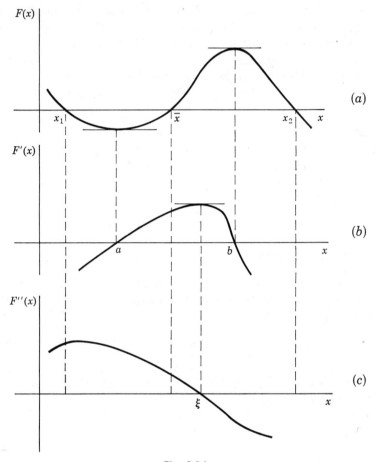

Fig. 8.14

and is illustrated graphically in Fig. 8.14a. Taking the derivative of $F(x)$ gives

$$F'(x) = f'(x) - p'(x) - \frac{f(\bar{x}) - p(\bar{x})}{(\bar{x} - x_1)(\bar{x} - x_2)} (2x - x_1 - x_2)$$

which is shown graphically in Fig. 8.14b. Notice that $F'(x)$ must have at least two zeros in the interval $x_1 \le x \le x_2$; these are designated by a and b. Taking the second derivative of $F(x)$ gives

$$F''(x) = f''(x) - p''(x) - 2 \frac{f(\bar{x}) - p(\bar{x})}{(\bar{x} - x_1)(\bar{x} - x_2)}$$

Since $p(x)$ is a linear function, $p''(x) = 0$, and the above expression becomes

$$F''(x) = f''(x) - 2\,\frac{f(\bar{x}) - p(\bar{x})}{(\bar{x} - x_1)(\bar{x} - x_2)}$$

$F''(x)$ is shown graphically in Fig. 8.14c and must have at least one zero in the interval (a,b), which is designated by ξ. Thus,

$$F''(\xi) = 0 = f''(\xi) - 2\,\frac{f(\bar{x}) - p(\bar{x})}{(\bar{x} - x_1)(\bar{x} - x_2)}$$

Upon solving for $f(\bar{x}) - p(\bar{x})$, the error at $\bar{x}$ is obtained.

$$\text{Error} = f(\bar{x}) - p(\bar{x}) = \frac{(\bar{x} - x_1)(\bar{x} - x_2)f''(\xi)}{2}$$

To illustrate the application of the formula for error, consider the linear interpolation of $\sin x$ in the example in Sec. 8.1.2. If x is given in degrees,

$$f(x) = \sin x$$
$$f'(x) = \frac{\pi}{180}\cos x$$
$$f''(x) = -\left(\frac{\pi}{180}\right)^2 \sin x$$

and the error at $x = 22$ is

$$\text{Error} = \frac{(22 - 20)(22 - 30)(\pi/180)^2(-\sin \xi)}{2}$$

A problem now arises in that ξ is not known. However, upper and lower bounds of the error can be obtained by using the maximum and minimum values of $\sin x$ in the interval $20 \leq x \leq 30$. The maximum and minimum values are obtained when $x = 30$ and 20, respectively, and are 0.50000 and 0.34202. Substitution of these values into the expression for the error gives the bounds

Maximum error $= 0.00122$
Minimum error $= 0.00083$

The actual error given in Sec. 8.1.2 is 0.00102, which lies within the above range.

An expression for the error in an nth-order interpolating polynomial can be obtained by generalizing the method used for the linear case.

The function $F(x)$ takes the form

$$F(x) = f(x) - p(x)$$
$$- \frac{f(\bar{x}) - p(\bar{x})}{(\bar{x} - x_1)(\bar{x} - x_2)(\cdot \cdot \cdot)(\bar{x} - x_{n+1})} (x - x_1)(x - x_2)(\cdot \cdot \cdot)(x - x_{n+1})$$

Taking the $(n + 1)$st derivative and remembering that the $(n + 1)$st derivative of $p(x)$ is zero, we get

$$F^{n+1}(x) = f^{n+1}(x) - \frac{f(\bar{x}) - p(\bar{x})}{(\bar{x} - x_1)(\bar{x} - x_2)(\cdot \cdot \cdot)(\bar{x} - x_{n+1})} (n + 1)!$$

As before, there must exist a point ξ such that $F^{n+1}(\xi) = 0$, and the expression for the error at $\bar{x}$ becomes

$$\text{Error} = f(\bar{x}) - p(\bar{x}) = \frac{(\bar{x} - x_1)(\bar{x} - x_2)(\cdot \cdot \cdot)(\bar{x} - x_{n+1})f^{n+1}(\xi)}{(n + 1)!} \quad (8.19)$$

Again, the value of ξ is generally not known, and the usual procedure is to determine the maximum value of $f^{n+1}(x)$ in the interpolation interval in order to obtain an upper bound on the error. In this discussion x has been assumed to be inside the interpolation interval. The results also apply to extrapolation, and in such a case ξ would lie within the extrapolation interval.

If equal-interval interpolation is being used, the error can be put into a different form by making the usual substitution

$$\bar{x} = x_1 + uh$$

The error then becomes

$$\text{Error} = \frac{u(u - 1)(u - 2)(\cdot \cdot \cdot)(u - n)h^{n+1}f^{n+1}(\xi)}{(n + 1)!} \quad (8.20)$$

For the parabolic interpolation given in the example in Sec. 8.2, $u = \frac{2}{5}$, $n = 2$, $h = 5$, $f^{n+1}(\xi) = -(\pi/180)^3 \cos \xi$. The maximum of $\cos x$ in the interval occurs at $x = 20$ and is 0.93969. Substitution into the expression for error gives the maximum error as -0.00004, which agrees with the results in Sec. 8.2.

8.11.1   Convergence

Suppose that it is desired to determine the maximum error of an equal-interval interpolation formula used in an interval $x = a$ to $x = b$. To obtain an upper bound of the error, the product of the factors u, $u - 1$, $u - 2$, . . . is replaced by n^{n+1}. Upon noting that $nh = b - a$, the upper bound of the error becomes

$$|\text{Error}| \leq \frac{(b - a)^{n+1}|f^{n+1}(x)|_{\max}}{(n + 1)!} \quad (8.21)$$

If $f^m(x)$ is bounded, $|f^m(x)| \leq M$, for all values of m. The error is then

$$|\text{Error}| \leq M \frac{(b - a)^{n+1}}{(n + 1)!}.$$

and since $(n + 1)!$ approaches infinity faster than $(b - a)^{n+1}$, the error approaches zero as n approaches infinity. Thus, the maximum error in using an interpolating polynomial within a specified interval can be reduced to any desired minimum value by taking a large enough n. This is, of course, assuming that the function $f(x)$ and all its derivatives are bounded in the interval. Under such conditions, the interpolating polynomial $p(x)$ is said to converge to the function $f(x)$ within the interval $a \leq x \leq b$.

As an illustration, suppose that it is desired to form an interpolating polynomial to represent $\sin x$ between $20°$ and $30°$ to an accuracy of five decimal places. Starting with $n = 1$, a maximum error can be obtained for each successive value of n until the error is sufficiently small.

$$n = 1: \quad |\text{Error}| \leq \frac{10^2 (\pi/180)^2 \sin 30°}{2} = 0.0076155$$

$$n = 2: \quad |\text{Error}| \leq \frac{10^3 (\pi/180)^3 \cos 20°}{6} = 0.0008327$$

$$n = 3: \quad |\text{Error}| \leq \frac{10^4 (\pi/180)^4 \sin 30°}{24} = 0.0000193$$

$$n = 4: \quad |\text{Error}| \leq \frac{10^5 (\pi/180)^5 \cos 20°}{120} = 0.0000013$$

Thus, a quartic polynomial guarantees five-place accuracy.

An alternative procedure in reducing the error in an interpolating polynomial is to hold n constant and reduce the interval (a,b) until the desired error bound is reached. It is quite obvious from Eq. (8.21) that any desired accuracy can be obtained by making $b - a$ small enough.

Intuitively, one would expect that, the higher the order of the interpolating polynomial, the less the error. While the error expression bears this out for large values of n, it may not be true in particular instances. For example, if $b - a = 10$,

$$\frac{10^3}{3!} = \frac{1000}{6} \qquad \frac{10^4}{4!} = \frac{2500}{6} \qquad \cdots$$

and the factor $10^{n+1}/n!$ does not start to decrease until $n = 11$. In many cases, however, the derivative $f^n(x)$ decreases as n increases. This aids convergence and is often the predominant factor in the error term. Such a case is illustrated in the example given above.

One may falsely conclude that reducing the scale of the independent variable, thereby reducing $b - a$, reduces the error. Reducing the scale, however, increases the derivative $f^{n+1}(x)$ by the same factor as $(b - a)^{n+1}$ is decreased, and thus the error remains the same. In the example given in Sec. 8.11, if x is measured in radians and $n = 2$,

$$|\text{Error}| \leq \frac{(10\pi/180)^3 |\cos x|_{\max}}{6}$$

which is the same as before.

8.11.2   Order of Error

The expression for the error in interpolation can be put into a slightly different form by replacing $b - a$ by nh.

$$|\text{Error}| \leq \frac{n^{n+1} |f^{n+1}(x)|_{\max}}{(n + 1)!} h^{n+1} \tag{8.22}$$

Treating h as the only variable, we customarily refer to this as an *error of order h^{n+1}*. Interpreting the order of error consists in examining how quickly the error decreases as h decreases. For example, if h is halved, the error is decreased by a factor of 4 if the order of error is h^2 and by 8 if the order of error is h^3. In general, an nth-order interpolating polynomial possesses an error of order h^{n+1}. Note that the order of error describes the *rate* at which error is reduced as h is reduced but gives no indication of the actual amount of error, which depends upon the actual values of n, h, and $|f^{n+1}(x)|_{\max}$.

8.11.3   Derivative Not Known

In some instances, it is not possible to determine the error in an interpolating polynomial because the $(n + 1)$st derivative is not known. The rocket problem, considered several times in this chapter, is an example of such a case. Thus, the error involved in the interpolating polynomials derived for the speed of the rocket is indeterminate. If the interpolating polynomial is passed through only some of the given points, an indication of the error can be obtained by checking the closeness of the interpolating polynomial to the remaining points. In the example in Sec. 8.8.1, a fourth-order interpolating polynomial passing through five of the given six points is presented. The interpolated speed at 120 sec is found to be 0.3211 mps, compared with the given value of 0.2747 mps.

In Sec. 8.4.6 it is shown that the nth differences of an nth-order polynomial are constant. Thus, it might be concluded that, if the nth differences of a tabulated function are constant, the function is an nth-order polynomial. This is not necessarily true. For example, the data in Table 8.21 indicate that

$$f(x) = 2x$$

However, an infinite number of polynomials of higher order, and other functions, can be found that will satisfy the given data. One such function is

$$f(x) = -\tfrac{1}{3}x^3 + 2x^2 - \tfrac{2}{3}x$$

Table 8.21

x	$f(x)$	D^1	D^2
0	0		
		4	
2	4		0
		4	
4	8		

To summarize, the following two statements can be made about an nth-order interpolating polynomial with regard to error:

1. If the $(n + 1)$st derivative of the function $f(x)$ is known, upper and lower bounds for the error can be determined.
2. If the $(n + 1)$st derivative of $f(x)$ is not known, the error cannot be determined. If $n + 1$ points of the function are known, and no other information is given, passing an nth-order interpolating polynomial through all the points is intuitively the most reasonable approach.

8.11.4   Inverse Interpolation

A comparison of the accuracy of the two methods of inverse interpolation presented in Sec. 8.9 will now be made. Let a general function be expressed as

$$y = f_1(x)$$

and its inverse as

$$x = f_2(y)$$

If $f_1(x)$ can be accurately represented by an nth-order interpolating polynomial, $f_2(y)$ may or may not also be represented accurately by an nth-order polynomial. Consider again the quadratic polynomial representing sin x between 20 and 30° given by Eq. (8.4). The upper bound of the error in this case is computed (Sec. 8.11) as -0.0004. The error in the quadratic interpolation of $x = \arcsin y$ (Sec. 8.9.2) will now be determined. The error is

$$\text{Error} = \frac{(y - y_1)(y - y_2)(y - y_3)f_2'''(\xi)}{6}$$

The third derivative of the inverse sine is

$$f_2'''(y) = \frac{180}{\pi} \frac{2y^2 + 1}{(1 - y^2)^{\frac{5}{2}}}$$

Using values of y to determine the upper bound on the error produces

$$|\text{Error}| \leq \left| \frac{(0.50000 - 0.34202)(0.34202 - 0.42262)(0.34202 - 0.50000)(180/\pi)[2(0.5)^2 + 1]}{6(1 - 0.25)^{\frac{5}{2}}} \right|$$
$$\leq |0.06|$$

These error analyses show that the interpolating parabolas for sin x and arcsin y give accuracy to within 4 and 6, respectively, in the fourth significant digit.

In some instances the inverse-interpolation method where the variables are interchanged gives very erroneous results. A check of the error formulas of Sec. 8.11 and above shows that the errors in parabolic interpolations of sin x and arcsin y near 90° approach 0 and ∞, respectively.

Generally speaking, the inverse-interpolation method in which the variables are interchanged is usually easier to apply, since it does not require the extraction of a root of a polynomial. Care must be taken, however, since the error may be larger than tolerable. If this is the case, either the other method must be employed or an inverse-interpolating polynomial of higher-order must be formed.

In cases where derivatives are not known and errors cannot be determined, no comparison of the two methods can be made.

8.12   APPLICATIONS

Engineers often use and study processes for which the physical laws either are not fully understood or cannot be expressed analytically. In such cases experimental data are often employed to determine relationships between the variables and parameters of the problem. Passing an interpolating polynomial through a set of experimentally determined points tacitly assumes that these points are exact. All experimental data contain measurement errors, and an interpolating polynomial would incorporate these errors. However, if the data represent averages taken from a large number of experiments, polynomial interpolation is a plausible method of obtaining relationships between the physical parameters of the problem. Methods of accounting for measurement errors are available and are discussed in Chap. 12.

Engineers constantly use mathematical tables for such functions as logarithms, sines, cosines, tangents, exponentials, etc. Usually these tables are given at such finely spaced intervals that linear interpolation is quite sufficient. However, if a computer subroutine is being designed to evaluate such a function, the storage of a complete table would take too much space to be practical. Thus, interpolating polynomials could be utilized in such a situation, but other methods are more commonly used. Sometimes the obtaining of a solution to a mathematical equation arising from a physical problem can be greatly simplified by employing an interpolating polynomial. An example of this is presented in the following applications.

8.12.1 Varistors

In Chap. 6, an application is presented which considers electrical resistors (varistors) that do not conform to Ohm's law. The characteristics of a varistor can be determined experimentally and represented by an interpolation formula of the form

$$I = a_2E + a_3E(E - E_2) + a_4E(E - E_2)(E - E_3) + \cdots$$

Consider the problem of a varistor and linear resistor in series as shown in Fig. 6.29, where R_2 represents the varistor and the expressions E_A and E_B replace E_1 and E_2 in the figure to avoid confusion with the experimentally determined points E_1 and E_2.

$$R_1[a_2E_B + a_2E_B(E_B - E_2) + a_4E_B(E_B - E_2)(E_B - E_3) + \cdots] + E_B = E \tag{8.23}$$

Once E_B is known, E_A and I can easily be determined (see Sec. 6.11.2).

8.12.2   Replacement of a Trigonometric Function
by an Interpolation Polynomial

In Chap. 6 an equation is developed for the angle of rotation of a circular bar to which a heavy lever is attached.   The form of this equation is

$$\cos\,\theta = k_1\theta$$

where k_1 is a constant.   The equation can be solved by the iterative techniques of Chap. 6, but if the approximate range in which θ lies is known, an interpolating parabola produces a ready solution.   Upon assuming the interpolating parabola in the form

$$\cos\,\theta \approx a_1\theta^2 + a_2\theta + a_3$$

the above equation becomes

$$a_1\theta^2 + a_2\theta + a_3 = k_1\theta$$

the solution of which is

$$\theta = \frac{-(a_2 - k_1) \pm \sqrt{(a_2 - k_1)^2 - 4a_1a_3}}{2a_1} \tag{8.24}$$

The sign in front of the radical should be such as to give $0 \leq \theta \leq 90°$. This solution should be substituted into the original equation as a check on the accuracy of the result.

8.12.3   Physical and Thermodynamic Properties of Materials

Engineering calculations in the areas of heat transfer and fluid dynamics require data on the physical and thermodynamic properties of the material being handled.   These data are usually determined experimentally and presented in tabular form.   For convenient use in a computer, either a polynomial may be fitted to the data and the coefficients stored, or else a part of the tabular values may be stored and then interpolated as needed.

As an example, suppose that a heat-balance calculation involves the latent heat of vaporization of water at a temperature in the range of 200 to 400°F.   It is not sufficiently accurate to use an average value over this range, since there is a variation of nearly 20 per cent.   By using a second-degree polynomial, the data can be fitted within less than 0.1 per cent.

PROBLEMS

8.1   Using the graph of Fig. 8.1, determine the speed of the rocket at the following times: 90, 250, and 270 sec.

◆**8.2** The torque T transmitted to the engine shaft by the piston, connecting rod, and crank mechanism shown was measured at various angles θ, and the cor-

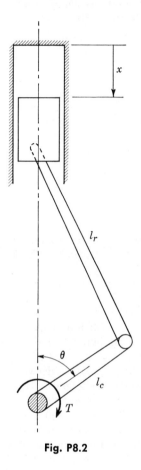

Fig. P8.2

responding values are given in the table below. By interpolation determine the torque for the following values of θ: 15, 45, 135, 225, and 345°.

θ, deg...	0	30	60	90	120	150	180	210	240	270	300	330	360
T, in.-lb.	0	23,100	10,500	10,000	7900	3700	0	−3000	−6000	−6000	−2100	−6800	0

8.3 The following data were taken from a standard tensile test of a steel specimen (see Prob. 3.55). Stress σ is defined as the tensile force P applied to the ends of the specimen, divided by the cross-sectional area of the central portion of the bar. Strain ϵ is defined as the elongation per unit length as measured by the attached gage. By interpolation determine the stress at the following strains: 400×10^{-6}, 900×10^{-6}, 1500×10^{-6}, and 2000×10^{-6} in./in.

σ, psi	ϵ, in./in.
0	0
5,000	167×10^{-6}
10,000	333×10^{-6}
15,000	500×10^{-6}
20,000	667×10^{-6}
25,000	833×10^{-6}
30,000	1000×10^{-6}
35,000	1310×10^{-6}
40,000	1730×10^{-6}
45,000	2580×10^{-6}

◆8.4   Using linear interpolation, determine the sine of 22.5° from the values $\sin 22° = 0.37461$ and $\sin 23° = 0.39073$. Check your answer in a trigonometric table.

8.5   Using linear interpolation, obtain a value of $\cos 22°$ from the values $\cos 20° = 0.93969$ and $\cos 30° = 0.86603$. What is the error?

◆8.6   With the data given in Table 8.1, determine the speed of the rocket at the following times by using linear interpolation: 90, 210, 250, and 270 sec.

8.7   Using the interpolating parabola derived in Sec. 8.2 for $\sin x$, determine the values of $\sin x$ for the following values of x: 21, 24, 27, and 28°.

8.8   Using the interpolating parabola derived in Sec. 8.2 for the speed of the rocket, determine the speed at the following times: 90, 100, and 160 sec.

◆8.9   Construct an interpolating parabola for the speed of a rocket, using the values given in Table 8.1 at times 120, 180, and 240 sec. Evaluate the speed at 150 sec, and compare with the result in the text.

8.10   Determine an interpolating parabola for $\cos x$, using the values $\cos 20° = 0.93969$, $\cos 25° = 0.90631$, and $\cos 30° = 0.86603$. Evaluate the parabola at $x = 22°$, and compare with the true value and with the linearly interpolated value obtained in Prob. 8.5.

◆8.11   A complete stress-strain diagram of a steel specimen (Prob. 8.3) to failure is shown. Construct an interpolating parabola passing through the three points A, B, and C. Evaluate the polynomial to obtain the stress at a strain of 0.3.

Point	Description	Strain, in./in.	Stress, psi
A	Yield point	0.02	40,000
B	Ultimate strength	0.20	62,000
C	Fracture point	0.38	44,000

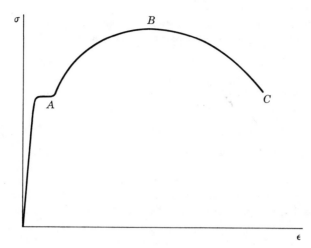

Fig. P8.11

8.12 The pressure p acting on the piston in Prob. 8.2 is a function of the volume behind the piston. This volume is proportional to x, which is in turn related to θ through geometry. Thus, the pressure is a function of θ. During the power stroke ($0° \leq \theta \leq 180°$) the pressure was determined at the points given in the table below. By interpolation determine the pressure when $\theta = 45$ and $135°$.

θ, deg.....	0	30	60	90	120	150	180
p, psi.....	842	310	100	43	24	17	15

8.13 During the compression stroke ($180° \leq \theta \leq 360°$) of the piston in Prob. 8.12, the pressure p was determined at the points given in the table below. By interpolation determine the pressure when $\theta = 225$ and $315°$.

θ, deg.....	180	210	240	270	300	330	360
p, psi.....	0	1	4	14	42	145	407

8.14 Construct an interpolating formula for $\sin x$, using functional values at $x = 20, 25$, and $30°$, by assuming a parabola in the form

$$p(x) = a_1 + a_2(x - 30) + a_3(x - 30)(x - 25)$$

Also, calculate $p(22°)$, and compare with the result given in Sec. 8.2.

◆**8.15** Construct an interpolating formula for sin x, using functional values at $x = 20$, 25, and 30°, by assuming a parabola in the form

$$p(x) = a_1 + a_2(x - 25) + a_3(x - 25)(x - 30)$$

Also, calculate $p(22°)$, and compare with the results given in Sec. 8.2 and Prob. 8.14.

8.16 Construct an interpolating formula for sin x, using functional values at $x = 20$, 25, and 30°, by assuming a parabola in the form

$$p(x) = a_1(x - 25)(x - 30) + a_2(x - 20)(x - 30) + a_3(x - 20)(x - 25)$$

Also, calculate $p(22°)$, and compare with the results given in Sec. 8.2 and Probs. 8.14 and 8.15.

8.17 Considering the parabolic interpolation of sin x, using functional values at $x = 20$, 25, and 30°, give some form of the parabola different from those used in Sec. 8.2 and in Probs. 8.14 to 8.16.   Do all these forms give the same value of $p(22°)$?

8.18 Construct a flow chart, and write a computer program which produces a linearly interpolated value for a general point x, given a set of tabulated points $[x_i, f(x_i)]$.

8.19 Construct a flow chart, and write a computer program which produces an interpolating parabola if given three points: $[x_1, f(x_1)]$, $[x_2, f(x_2)]$, $[x_3, f(x_3)]$.

8.20 Construct a flow chart, and write a computer program which produces a parabolically interpolated value for a general point x, given a set of tabular points $[x_i, f(x_i)]$.   Take two points before x and one after x, except when $x_1 \leq x \leq x_2$.

8.21 Using binomial coefficients, derive the expressions for $\Delta^5 f(x_1)$ and $\Delta^6 f(x_1)$.

8.22 Using the interpolating polynomial for the rocket problem given in Sec. 8.3.1, evaluate the speed of the rocket at the following times: 90, 250, and 270 sec.

◆**8.23** The powder pressure p in a gun barrel varies with the travel of the bullet, as shown.   The data from which the graph was constructed are given

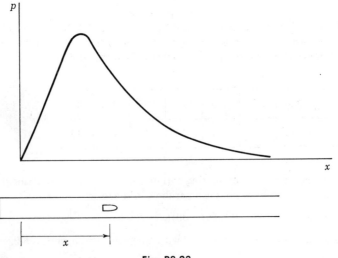

Fig. P8.23

below. By interpolation evaluate the pressure at the following values of x: 2, 7, and 12 in.

x, in......	0	5	10	15	20
p, psi.....	0	40,000	15,000	5000	1000

8.24 Construct a flow chart, and write a computer program which evaluates the binomial coefficient $\binom{i}{j}$.

8.25 Construct a flow chart, and write a computer program which produces a third-order interpolating polynomial if given four points: $[x_1,f(x_1)]$, $[x_2,f(x_2)]$, $[x_3,f(x_3)]$, $[x_4,f(x_4)]$.

8.26 Construct a flow chart, and write a computer program which produces an interpolated value from a third-order interpolating polynomial for a general point x, given a set of tabular points $[x_i,f(x_i)]$. Take two points on both sides of x, except when x is within the end intervals of the table.

8.27, ◆**8.28, 8.29** Construct complete difference tables with the given data. Then calculate $\Delta f(x_2)$, $\nabla^2 f(x_3)$, $\delta^2 f(x_3)$, $\delta^3 f(x_3)$.

Prob. 8.27		*Prob.* 8.28		*Prob.* 8.29	
x	$f(x)$	x	$f(x)$	x	$f(x)$
0	-10	2	100	-2	20
1	-5	4	50	-1	40
2	5	6	0	0	50
3	20	8	25	1	40
4	40	10	100	2	20

8.30 Construct a complete difference table for $\cos x$ with $0° \le x \le 60°$ at $10°$ increments. Then calculate $\Delta^3 f(0)$, $\nabla^2 f(40)$, $\delta^2 f(20)$, $\delta^3 f(30)$.

8.31 Determine algebraic expressions for $\Delta^5 f(x_i)$, $\nabla^5 f(x_i)$, $\delta^5 f(x_i)$, and $\delta^6 f(x_i)$.

8.32, ◆**8.33, 8.34** The entries in the table represent functional values of parabolas. Construct difference tables, and derive the equations of the parabolas and their differences.

Prob. 8.32		*Prob.* 8.33		*Prob.* 8.34	
x	$f(x)$	x	$f(x)$	x	$f(x)$
0	-4	-2	20	0	-3
1	-3	0	4	1	0
2	0	2	4	2	5
3	5	4	20	3	12
4	12	6	52	4	21

8.35 Generate a parabola by starting with three constant second differences equal to 2. Assume that $f(x_1) = \Delta f(x_1) = 0$, $x_1 = 0$, and $h = 1$. Determine the equation for $f(x)$.

◆**8.36** Generate a cubic polynomial by starting with four constant third differences equal to -3. Assume that $f(x_1) = -10$, $\Delta f(x_1) = 2$, $\Delta^2 f(x_1) = 0$, $x_1 = -1$, and $h = 2$. Determine the equation for $f(x)$.

8.37 Generate a fourth-order polynomial, starting with two constant fourth differences equal to 4. Assume that $f(x_1) = 1$, $\Delta f(x_1) = 2$, $\Delta^2 f(x_1) = 1$, $\Delta^3 f(x_1) = 0$, $x_1 = 0$, and $h = 1$. Determine the equation for $f(x)$.

8.38 A freely falling body has an acceleration of approximately 32 ft/sec². The distance s the body drops varies with the square of the time t. Construct a table which gives the distance s at increments of 1 sec from $t = 0$ to $t = 10$ sec by starting with constant second differences of 32. The initial conditions are $f(0) = 0$, $\Delta f(0) = 16$.

8.39 Construct a flow chart, and write a computer program which determines the first differences from a set of tabular data $[x_i, f(x_i)]$.

8.40 Construct a computer program which computes all the differences in a table consisting of four points: $[x_1, f(x_1)]$, $[x_2, f(x_2)]$, $[x_3, f(x_3)]$, $[x_4, f(x_4)]$.

The following data pertain to Probs. 8.41 to 8.44. A rocket is fired in the direction shown, and the x and y coordinates at various times t after blast-off are given in the accompanying table.

t, sec	x, ft	y, ft
0	0	0
100	80,000	300,000
200	200,000	700,000
300	380,000	1,200,000
400	500,000	1,000,000
500	550,000	600,000

◆**8.41** Construct an interpolation formula for $x(t)$.

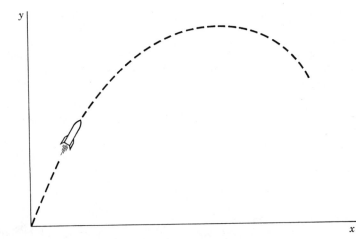

Fig. P8.41 to P8.44

8.42 Construct an interpolation formula for $y(t)$.

8.43 Construct an interpolation formula for $y(x)$.

◆**8.44** Determine $x(250)$ and $y(250)$ by using the interpolation formulas of Probs. 8.41 and 8.42. Then determine $y[x(250)]$ by using the interpolation formula of Prob. 8.43. Are the results the same? Should they be?

8.45 Write a computer program for the Gregory-Newton forward-interpolation formula, using the flow chart given in Fig. 8.8.

8.46 Construct a flow chart, and write a computer program for the Gregory-Newton backward-interpolation formula.

8.47 Construct a flow chart, and write a computer program for the Gauss forward-interpolation formula.

8.48 Construct a flow chart, and write a computer program for the Gauss backward-interpolation formula.

8.49 Construct a flow chart, and write a computer program for the Stirling central-interpolation formula.

8.50 Using the Lagrangian interpolation formula given in Sec. 8.7.1, calculate the speed of the rocket at the following times: 90, 210, 250, and 270 sec.

◆**8.51** Determine the Lagrangian interpolation formula for $\cos x$, $20° \leq x \leq 30°$, using functional values at $x = 20$, 25, and 30°. Evaluate the formula at $x = 22°$, and compare with the true value and values obtained in Probs. 8.5 and 8.10.

8.52 Write a computer program for Lagrangian interpolation, using the flow chart of Fig. 8.10.

8.53 Construct a flow chart, and write a computer program for equal-spacing Lagrangian interpolation.

8.54 Determine an interpolating parabola for $\cos x$, using the values $\cos 20° = 0.93969$, $\cos 30° = 0.86603$, and $\cos 35° = 0.81915$. Evaluate the parabola at $x = 22°$, and compare with the true value and values obtained in Probs. 8.5, 8.10, and 8.51.

8.55 Using the interpolation formula given in Sec. 8.8.1, determine the speed of the rocket at the following times: 90, 210, 250, and 270 sec.

8.56 Using the interpolation formula given in Sec. 8.8.2, determine the speed of the rocket at the following times: 90, 210, 250, and 270 sec.

◆**8.57** The velocity v of the water at various depths y in a channel was measured, and the results are presented in the table below. By interpolation determine the velocity of the water at depths of 4 ft and 7 ft.

y, ft......	0	1	3	6	8
v, ft/sec...	10	11	12	10	4

8.58 Derive the expressions for a_1, a_2, a_3, and a_4 given in Sec. 8.8.1.

8.59 Construct a computer program for divided-difference interpolation, using the flow charts given in Figs. 8.8 and 8.12.

In Probs. 8.60 to 8.65 determine the unknown argument for the given functional value, using linear interpolation.

8.60

x	sin x
20	0.34202
?	0.47500
30	0.50000

8.61

x	cos x
20	0.93969
?	0.90000
30	0.86603

8.62

Time, sec	Speed, mps
240	1.3851
?	2.0000
300	3.2229

♦8.63

θ, deg	Torque, in.-lb
120	7,900
?	5,000
150	3,700

8.64

Stress, psi	Strain, in./in.
20,000	667×10^{-6}
?	730×10^{-6}
25,000	833×10^{-6}

8.65

Depth, ft	Velocity, ft/sec
3	12
?	11.5
6	10

8.66 Using the interpolating parabola that is given in Sec. 8.9, determine arcsin 0.47500 by solving the quadratic equation.

8.67 Solve Prob. 8.66 by interchanging the dependent and independent variables.

◆**8.68** Using the interpolating parabola for cos x given in Prob. 8.10, determine arccos 0.90000 by solving the quadratic equation.

◆**8.69** Solve Prob. 8.68 by interchanging the dependent and independent variables.

8.70 Construct a flow chart, and write a computer program which performs inverse linear interpolation in a table of points $[x_i, f(x_i)]$.

8.71 Construct a flow chart, and write a computer program which performs inverse interpolation with a given interpolating parabola for a particular functional value $f(x)$.

8.72 Construct a flow chart, and write a computer program which performs inverse interpolation in a table of points $[x_i, f(x_i)]$ for a given functional value $f(x)$ by forming an interpolating parabola $p(x)$ with the three points nearest $f(x)$ and solving the resulting quadratic equation.

8.73 Plot a graph of cos x for $0° \le x \le 180°$. Then plot the interpolating parabola of Prob. 8.10, and compare the two curves.

8.74 Construct an interpolating parabola for the population data given in Sec. 8.10. Then estimate the population in 1965, 1980, and 2000.

◆**8.75** The gross sales of a small manufacturing company for three successive years were $1,000,000, $1,200,000, and $1,500,000. Assuming a similar rate of growth, estimate the gross sales in each of the next 3 years.

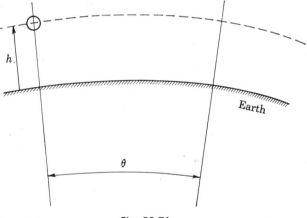

Fig. P8.76

8.76 Use the data given concerning the orbit of a satellite to predict the height h and angular orientation θ of the satellite at time $t = 1200$ sec. The time t is measured from the instant at which the satellite enters its orbit.

t, sec	h, miles	θ, deg
0	100	0
300	107	22
600	160	42
900	230	62

♦**8.77** Determine the upper bound on the error in Prob. 8.4.

8.78 Determine the upper bound on the error in Prob. 8.5.

8.79 Determine the upper bound on the error in Prob. 8.10.

8.80 Determine the order of a polynomial needed to represent cos x between 20° and 30° with an accuracy of five places.

8.81 Determine the equation of a third-order polynomial which will pass through the three points given below.

x	$f(x)$
1	-2
2	2
3	6

8.82 Determine the error bound of an interpolating parabola for $x = \arccos y$ where $20° \leq x \leq 30°$.

8.83 Determine an interval of convergence around $x = 25°$, so that an interpolating parabola will accurately represent sin x to four decimal places.

♦**8.84** Determine an interval of convergence around $x = 25°$, so that an interpolating parabola will accurately represent cos x to four decimal places.

8.85 Data on a varistor are given below. Construct a third-order interpolating polynomial passing through these points.

E, volts....	0	10	20	30
I, amp.....	0	0.35	0.48	0.60

♦**8.86** In Fig. 6.25, $W = 5000$ lb, and $r = 12$ in. A torque of 875 lb-in. is required to rotate the bar through 1°. Construct an interpolating parabola for $0° \leq \theta \leq 60°$, and solve the resulting quadratic equation for θ. Substitute this value back into the transcendental equation as a check of the result.

8.87 The output plate current from some types of electrical amplifiers comes in short pulses, as illustrated, where i is current in amperes and t is time. Taking $i_m = 0.5$ and $a = 0.004$ sec, construct an interpolating parabola which represents one current pulse.

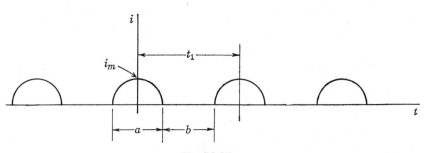

Fig. P8.87

8.88 The force F_R acting on a projectile, due to air resistance, varies with the velocity V. Results from wind-tunnel tests are given below. Construct an interpolating formula for F_R as a function of V.

V, ft/sec....	0	2000	4,000	6,000	8,000	10,000
F_R, lb......	0	3000	15,000	40,000	75,000	120,000

8.89 The data below present values of the force F_R due to water resistance as a function of the speed of a ship. Construct an interpolating polynomial to represent these data.

Speed, knots....	10	15	20	25	30
F_R, lb..........	17,000	30,000	56,000	100,000	165,000

8.90 The drag force D acting on a rocket is usually given in the form

$$D = C_D \rho A V^2$$

where ρ is the mass density of air, A the cross-sectional area of the rocket, V the speed of the rocket, and C_D an experimentally determined parameter which depends upon the speed V. Assuming that the functional dependence of C_D on V is given in tabular form, construct a flow chart, and write a computer program which determines the drag force D if ρ, A, and V are given. Use linear interpolation to determine C_D.

8.91 The latent heat of vaporization of water[1] at several temperatures is given in the following table:

°F	Btu/lb	°F	Btu/lb
200	977.9	310	902.6
210	971.6	320	894.9
220	965.2	330	887.0
230	958.8	340	879.0
240	952.2	350	870.7
250	945.5	360	862.2
260	938.7	370	853.5
270	931.8	380	844.6
280	924.7	390	835.4
290	917.5	400	826.0
300	910.1		

[1] From J. H. Keenan and F. G. Keyes, "Thermodynamic Properties of Steam," John Wiley & Sons, Inc., New York, 1937. Reprinted with permission of the copyright holders.

By interpolation, calculate the latent heat at the following temperatures: 225, 283, 357, and 410°F.

8.92   Aitken's iterated interpolation consists in two linear interpolations AB and AC and then the linear interpolation DE. Prove that this is equivalent to passing a parabola through the three points ABC. Note that the process can be applied to either equal or unequal spacing.

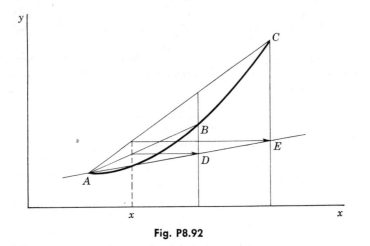

Fig. P8.92

8.93   The table below shows Aitken's method used to determine an interpolated value of 0.44 at $x = 0.2$. Apply Aitken's method to solve the two problems given in Sec. 8.2.

x	$f(x)$		
0	0		
1	3	0.6	
2	8	0.8	0.44

8.94   Apply Aitken's method to determine the speed of the rocket at $t = 150$ sec, using all the data given in Table 8.1.

8.95   Derive the Gregory-Newton backward formula.

8.96   Derive the Gauss forward formula.

8.97   Derive the Gauss backward formula.

8.98   Derive the Stirling formula.

9 NUMERICAL DIFFERENTIATION AND INTEGRATION

9.1 INTRODUCTION

The computing of derivatives and integrals of functions is often required in the solution of engineering problems. In mechanics the velocity of a body is defined as the time derivative of its displacement, and its acceleration is defined as the time derivative of its velocity. In a-c electrical circuits, the voltage across an inductance coil is proportional to the time derivative of the current flowing through the coil, and the voltage across a capacitor is proportional to the integral of the current flowing through the capacitor. The flow of heat along a rod is proportional to the derivative of the temperature with respect to the position along the rod. These represent only a few physical examples where differentiation and integration arise in connection with engineering problems. In a mechanics problem, for example, if the displacement of a body is given as a function of time, the velocity and acceleration of the body are obtained by direct differentiation. More often, however, the acceleration is given, and the velocity and displacement are determined by integration.

If a function is defined by an algebraic expression, its derivative can always be determined in terms of another algebraic expression. On the other hand, if a function is expressed only in tabular form, some approximate method must be employed. In fact, whether or not a derivative even exists for such a function should be considered. Generally, if a tabulated function arises from a physical situation, it seems reasonable to assume that its derivatives do exist. The general approach in determining the derivative of a tabulated function consists in replacing the function with an interpolating polynomial and then differentiating the polynomial. Successive higher-order derivatives are obtained by continued differentiation of the interpolating polynomial.

Basically, integration is concerned with determining the area under a given curve. If the curve can be expressed in terms of an algebraic expression, use of the fundamental theorem of integral calculus will often give an exact value for the area. Use of this theorem is not possible in the case of a function whose antiderivative is not known. Some algebraic expressions and functions which are defined only by a set of tabulated values fall into this category. The procedure in such a case is to replace the given function by an interpolating polynomial and then determine the area under the polynomial.

9.2   DIFFERENTIATION

The derivative $f'(x)$ of a function $f(x)$ at a specified point x_2 is defined as the limit as $h \to 0$ of

$$\frac{f(x_2 + h) - f(x_2)}{h} \tag{9.1}$$

The above expression represents the slope of the line AB given in Fig. 9.1. As $h \to 0$, the slope of line AB approaches the slope of the tangent line at point x_2. Thus, the slope of the tangent line at point x_2 is $f'(x_2)$.

Formulas for the derivatives of algebraic expressions are obtained by direct application of the above definition and can be found in any calculus text or book of mathematical tables.

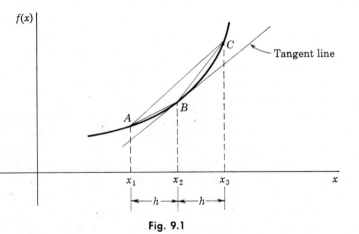

Fig. 9.1

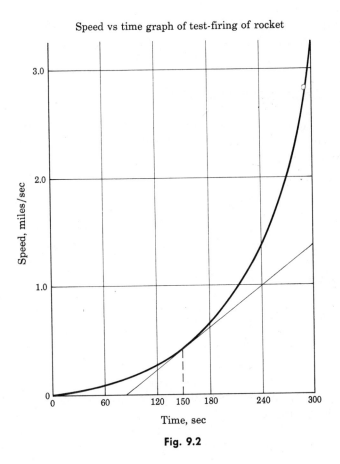

Fig. 9.2

9.2.1   Differentiation by Graphical Means

Figure 9.1 suggests a graphical method of determining the derivative of a function at a specified point.   This is accomplished by constructing a graph representing the function, drawing the tangent line at the given point, and measuring the slope of the line.   The procedure is illustrated in Fig. 9.2, where the velocity of a rocket is plotted against time from the data given in Sec. 8.1.1.   The value of the derivative at time $t = 150$ sec is taken from the graph as

$$\frac{1.38 - 0}{300 - 85} = 0.00642$$

Since the time derivative of velocity is acceleration, this result represents the acceleration of the rocket at time $t = 150$ sec, in units of miles per second per second.

9.3   NUMERICAL DIFFERENTIATION

An approximation for the derivative of a function at some point x_2 can be made by using a finite value of h in the definition of a derivative. This gives

$$f'(x_2) \approx \frac{f(x_2 + h) - f(x_2)}{h} \tag{9.2}$$

which is represented by the slope of the line BC in Fig. 9.1. This figure also illustrates other possibilities for approximating $f'(x_2)$. Using the slopes of lines AB and AC produces

$$f'(x_2) \approx \frac{f(x_2) - f(x_2 - h)}{h} \tag{9.3}$$

$$f'(x_2) \approx \frac{f(x_2 + h) - f(x_2 - h)}{2h} \tag{9.4}$$

These approximations correspond to replacing the function $f(x)$ by the three straight lines

$$BC = f(x_2) + \frac{f(x_2 + h) - f(x_2)}{h} (x - x_2)$$

$$AB = f(x_2) + \frac{f(x_2) - f(x_2 - h)}{h} (x - x_2)$$

$$AC = f(x_1) + \frac{f(x_2 + h) - f(x_2 - h)}{2h} (x - x_1)$$

As an example, consider the values of $\sin x$ shown in Table 9.1. The

Table 9.1

x	$f(x)$
20	0.34202
25	0.42262
30	0.50000

three approximations for $f'(x)$ at $x = 25$ are, respectively,

$$f'(25) \approx \frac{0.50000 - 0.42262}{5} = 0.015476$$

$$f'(25) \approx \frac{0.42262 - 0.34202}{5} = 0.016120$$

$$f'(25) \approx \frac{0.50000 - 0.34202}{10} = 0.015798$$

Compare these with the true value

$$f'(25) = \frac{\pi}{180} \cos 25° = 0.015818$$

Intuitively, from examining Fig. 9.1 and this example, it appears that the third approximation is more accurate than the other two.   This is true, since this approximation is equivalent to the result obtained by passing a parabola through the three points.   To prove this, assume that the parabola is in the form

$$p(x) = a_1 + a_2(x - x_1) + a_3(x - x_1)(x - x_2)$$

The coefficients a_1, a_2, and a_3, determined by the methods given in Chap. 8, are

$$a_1 = f(x_1)$$
$$a_2 = \frac{f(x_2) - f(x_1)}{h}$$
$$a_3 = \frac{f(x_3) - 2f(x_2) + f(x_1)}{2h^2}$$

Substituting these back into $p(x)$ and differentiating give

$$p'(x) = \frac{f(x_2) - f(x_1)}{h} + \frac{f(x_3) - 2f(x_2) + f(x_1)}{2h^2}(2x - x_1 - x_2)$$

At point x_2 this becomes

$$p'(x_2) = \frac{f(x_2) - f(x_1)}{h} + \frac{f(x_3) - 2f(x_2) + f(x_1)}{2h^2}h = \frac{f(x_3) - f(x_1)}{2h}$$

As another example, consider the data taken from the rocket problem given in Chap. 8 and presented aga'n in Table 9.2.   An approximation for the derivative at time $t = 150$ sec can be obtained by constructing a straight line passing through the two points at $t = 120$ and $t = 180$. The derivative is then represented by the slope of this line.

Table 9.2

Time, sec	Speed, mps
0	0.0000
60	0.0824
120	0.2747
180	0.6502
240	1.3851
300	3.2229

$$f'(150) \approx \frac{0.6502 - 0.2747}{60} = 0.006250$$

The graphical approximation of this derivative was found in Sec. 9.2.1 to be 0.00642.

9.3.1   Gregory-Newton Method

The methods presented thus far for obtaining the derivative of a tabulated function are equivalent to replacing the actual function by a straight line or parabola.   More accuracy can usually be obtained if the given function $f(x)$ is replaced by a polynomial $p(x)$ of higher order.   The derivative of this polynomial can then be used to represent $f'(x)$.   If $p(x)$ has the form

$$p(x) = a_1 x^n + a_2 x^{n-1} + \cdots + a_n x + a_{n+1}$$

the derivative is simply

$$p'(x) = na_1 x^{n-1} + (n-1)a_2 x^{n-2} + \cdots + a_n$$

Unfortunately, the standard interpolating polynomials are not in the above convenient form and thus cannot be differentiated so easily. Consider a polynomial in the Gregory-Newton form

$$p(x) = a_1 + a_2(x - x_1) + a_3(x - x_1)(x - x_2) \\ + \cdots + a_{n+1}(x - x_1)(\cdots)(x - x_n)$$

where $x_1, x_2, \ldots, x_{n+1}$ are the given points at which the function $f(x)$ is tabulated and $a_1, a_2, \ldots, a_{n+1}$ are defined in Sec. 8.3.   With a few more terms retained, the derivative of $p(x)$ is determined as follows:

$$\begin{aligned} p'(x) = \; & a_2 + a_3[(x - x_1) + (x - x_2)] \\ & + a_4[(x - x_1)(x - x_2) + (x - x_1)(x - x_3) + (x - x_2)(x - x_3)] \\ & + a_5[(x - x_1)(x - x_2)(x - x_3) + (x - x_1)(x - x_2)(x - x_4) \\ & \quad + (x - x_1)(x - x_3)(x - x_4) + (x - x_2)(x - x_3)(x - x_4)] \\ & + \cdots \end{aligned}$$

$$(9.5)$$

Equation (9.5) is quite general in that it is valid for any x and the subintervals (x_1,x_2), (x_2,x_3), $\ldots$, (x_n,x_{n+1}) are not necessarily equal.

To illustrate, consider the data in Table 9.3.   The interpolating parab-

Table 9.3

x	$f(x)$
20	0.34202
30	0.50000
35	0.57358

ola passing through these three points is given in Sec. 8.8 as

$$p(x) = 0.34202 + 0.015798(x - 20) - 0.0000722(x - 20)(x - 30)$$

Differentiating,

$$p'(x) = 0.015798 - 0.0000722[(x - 20) + (x - 30)]$$

At $x = 22$,

$$p'(22) = 0.015798 - 0.0000722(2 - 8) = 0.016231$$

Since the data in Table 9.3 correspond to sin x, the actual derivative is

$$f'(22) = \frac{\pi}{180} \cos 22° = 0.016183$$

9.3.2   Derivatives at Tabulated Points

Several terms in the general expression for the derivative drop out if the derivative is determined at one of the tabulated points x_i.   The most convenient point in the Gregory-Newton forward formula is the point x_1. The derivative at x_1 is obtained from the formula in Sec. 9.3.1.

$$\begin{aligned}
p'(x_1) = a_2 &+ a_3(x_1 - x_2) + a_4(x_1 - x_2)(x_1 - x_3) \\
&+ a_5(x_1 - x_2)(x_1 - x_3)(x_1 - x_4) \\
&+ \cdots + a_{n+1}(x_1 - x_2)(\cdot \cdot \cdot)(x_1 - x_{n+1}) \quad (9.6)
\end{aligned}$$

In the previous example, the derivative at $x = 20$ is approximated by

$$p'(20) = 0.015798 - 0.0000722(-10) = 0.16520$$

The true value of the derivative is

$$f'(20) = \frac{\pi}{180} \cos 20° = 0.016401$$

9.3.3   Equal Intervals

The Gregory-Newton interpolation formula is given in Sec. 8.3.1 in the form

$$p(u) = f(x_1) + \frac{\Delta f(x_1)}{1!} u + \frac{\Delta^2 f(x_1)}{2!} u(u-1)$$
$$+ \frac{\Delta^3 f(x_1)}{3!} u(u-1)(u-2) + \frac{\Delta^4 f(x_1)}{4!} u(u-1)(u-2)(u-3)$$
$$+ \cdots + \frac{\Delta^n f(x_1)}{n!} u(u-1)(\cdot\cdot\cdot)[u-(n-1)]$$

where

$$u = \frac{x - x_1}{h}$$

In order to differentiate $p(x)$, it is necessary to make use of the relation

$$p'(x) = p'(u) \frac{du}{dx} = \frac{1}{h} p'(u)$$

Thus,

$$p'(x) = \frac{1}{h} \left\{ \frac{\Delta f(x_1)}{1!} + \frac{\Delta^2 f(x_1)}{2!} [u + (u-1)] \right.$$
$$+ \frac{\Delta^3 f(x_1)}{3!} [u(u-1) + u(u-2) + (u-1)(u-2)]$$
$$+ \frac{\Delta^4 f(x_1)}{4!} [u(u-1)(u-2) + u(u-1)(u-3)$$
$$+ u(u-2)(u-3) + (u-1)(u-2)(u-3)]$$
$$\left. + \cdots \right\} \tag{9.7}$$

The formula becomes quite convenient for the derivative at point x_1, where $u = 0$.

$$p'(x_1) = \frac{1}{h} \left[\frac{\Delta f(x_1)}{1!} + \frac{\Delta^2 f(x_1)}{2!} (-1) + \frac{\Delta^3 f(x_1)}{3!} (-1)(-2) \right.$$
$$\left. + \frac{\Delta^4 f(x_1)}{4!} (-1)(-2)(-3) + \cdots \right]$$
$$= \frac{1}{h} \left[\Delta f(x_1) - \frac{\Delta^2 f(x_1)}{2} + \frac{\Delta^3 f(x_1)}{3} \right.$$
$$\left. - \frac{\Delta^4 f(x_1)}{4} + \cdots + \frac{(-1)^{n+1} \Delta^n f(x_1)}{n} \right] \tag{9.8}$$

As an example, consider the values of sin x and their differences given in Table 9.4. The derivative at $x = 20$ is approximated as

$$p'(20) = \frac{1}{5}\left(0.08060 + \frac{0.00322}{2}\right) = 0.016442$$

which compares favorably with the true value, 0.016401.

Table 9.4

x	$f(x)$	D	D^2
20	0.34202		
		0.08060	
25	0.42262		−0.00322
		0.07738	
30	0.50000		

9.3.4 Use of Other Interpolating Polynomials

As mentioned previously, the general form of the Gregory-Newton forward formula can be utilized to determine the derivative at any given point x, but it is particularly convenient at x_1. The derivative at the end point x_{n+1} of a set of equally spaced points $x_1, x_2, \ldots, x_{n+1}$ is easily obtained from the Gregory-Newton backward formula. The details are similar to those used in obtaining $p(x_1)$ previously, and the result is

$$p'(x_{n+1}) = \frac{1}{h}\left[\nabla f(x_{n+1}) + \frac{\nabla^2 f(x_{n+1})}{2} + \frac{\nabla^3 f(x_{n+1})}{3} \right.$$
$$\left. + \frac{\nabla^4 f(x_{n+1})}{4} + \cdots + \frac{\nabla^n f(x_{n+1})}{n} \right] \quad (9.9)$$

In Stirling's central-difference formula, all the even-order differences drop out of the resulting equation for the derivative $p'(x_i)$ at the central point x_i.

$$p'(x_i) = \frac{1}{h}\left[\delta f(x_i) - \frac{\delta^3 f(x_i)}{3!} 1^2 + \frac{\delta^5 f(x_i)}{5!}(1^2)(2^2) + \cdots \right] \quad (9.10)$$

The two formulas given here are applied in the following example.

Table 9.5 presents the differences for the data on the velocity of a rocket, which will be used to determine the derivatives at the times $t = 0$, 120, and 300 sec. The derivative at time $t = 0$ is best obtained by the Gregory-Newton forward formula.

Table 9.5

Time	Speed	D	D^2	D^3	D^4	D^5
0	0.0000					
		0.0824				
60	0.0824		0.1099			
		0.1923		0.0733		
120	0.2747		0.1832		0.1029	
		0.3755		0.1762		0.4644
180	0.6502		0.3594		0.5673	
		0.7349		0.7435		
240	1.3851		1.1029			
		1.8378				
300	3.2229					

$$p'(0) = \frac{1}{60}\left(0.0824 - \frac{0.1099}{2} + \frac{0.0733}{3} - \frac{0.1029}{4} + \frac{0.4644}{5}\right)$$
$$= 0.00198 \text{ mile/sec}^2$$

The derivative at time $t = 120$ sec can be obtained by using Stirling's central formula, the fifth differences being neglected.

$$p'(120) = \frac{1}{60}\left[\frac{0.1923 + 0.3755}{2} - \frac{(0.0733 + 0.1762)/2}{6}\right]$$
$$= 0.00438 \text{ mile/sec}^2$$

At time $t = 300$, the Gregory-Newton backward formula is most convenient.

$$p'(300) = \frac{1}{60}\left(1.8378 + \frac{1.1029}{2} + \frac{0.7435}{3} + \frac{0.5673}{4} + \frac{0.4644}{5}\right)$$
$$= 0.04786 \text{ mile/sec}^2$$

A flow chart for determining the derivative of $p'(x_1)$ can be easily constructed from the flow chart of Fig. 8.8 for the Gregory-Newton forward-interpolation formula. This flow chart is given in Fig. 9.3. As before, the first two columns evaluate the differences d_{ij}. The third column then calculates the derivative $p'(x_1)$.

9.4   HIGHER DERIVATIVES

Approximations for higher derivatives can be obtained by using finite values in the definition of these derivatives. Referring to Fig. 9.4, two

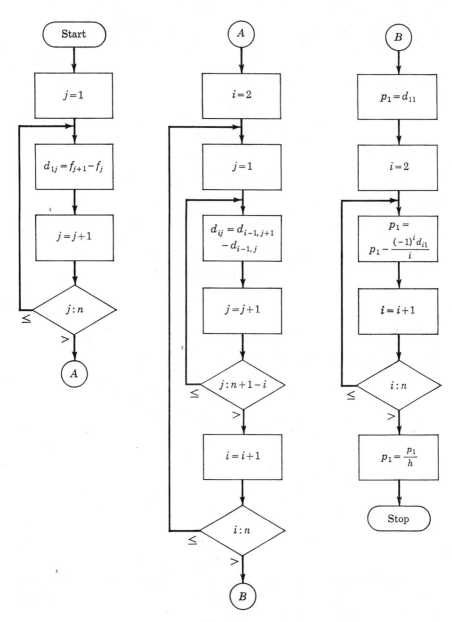

Fig. 9.3

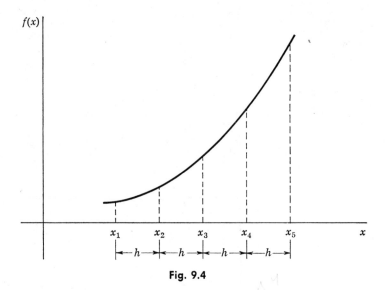

Fig. 9.4

approximations to $f''(x_3)$ are

$$f''(x_3) \approx \frac{f'(x_3 + h) - f'(x_3)}{h}$$
$$\approx \frac{[f(x_3 + 2h) - f(x_3 + h)]/h - [f(x_3 + h) - f(x_3)]/h}{h}$$
$$\approx \frac{f(x_3 + 2h) - 2f(x_3 + h) + f(x_3)}{h^2} \tag{9.11a}$$

$$f''(x_3) \approx \frac{f'(x_3) - f'(x_3 - h)}{h}$$
$$\approx \frac{[f(x_3) - f(x_3 - h)]/h - [f(x_3 - h) - f(x_3 - 2h)]/h}{h}$$
$$\approx \frac{f(x_3) - 2f(x_3 - h) + f(x_3 - 2h)}{h^2} \tag{9.11b}$$

Notice that the numerators of these expressions correspond to second forward and backward differences, respectively. An alternative expression using central differences is

$$f''(x_3) \approx \frac{f(x_3 + h) - 2f(x_3) + f(x_3 - h)}{h^2} \tag{9.11c}$$

which intuitively should be the most accurate. Applying these results to the data given in Table 9.2 produces the following three values for $f''(180)$:

$$f''(180) \approx \frac{3.2229 - 2(1.3851) + 0.6502}{60^2} = 0.0003064$$

$$f''(180) \approx \frac{0.6502 - 2(0.2747) + 0.0824}{60^2} = 0.00005089$$

$$f''(180) \approx \frac{1.3851 - 2(0.2747) + 0.0824}{60^2} = 0.00009983$$

Approximations for higher-order derivatives can be obtained in a similar fashion and the results expressed as

$$f^n(x) \approx \frac{D^n f(x)}{h^n} \tag{9.12}$$

where D represents forward, central, or backward differences. In the previous example

$$f^v(0) \approx \frac{0.4644}{60^5} = 0.5972 \times 10^{-9}$$

To obtain an approximation for the nth derivative of a function $f(x)$, at least an nth-order interpolating polynomial is necessary. The nth derivative of an nth-order interpolating polynomial is that found above. More accurate expressions for $f^n(x)$ are obtained if an interpolating polynomial of degree greater than n is used. The difficulty here, however, is that the forms of the interpolating polynomials are such that differentiation is rather awkward. Two methods of avoiding the straightforward (but tedious) differentiation are presented here.

Consider again the problem concerning the speed of a rocket given in Table 9.2. The Gregory-Newton forward-interpolation formula for this case is

$$p(u) = 0.0824u + 0.05495u(u - 1) + 0.01222u(u - 1)(u - 2)$$
$$+ 0.004288u(u - 1)(u - 2)(u - 3)$$
$$+ 0.003870u(u - 1)(u - 2)(u - 3)(u - 4)$$

Expanding and collecting terms give

$$p(u) = 0.003870u^5 - 0.03441u^4 + 0.9540u^3 - 0.1280u^2 + 0.1190u$$

which can now be easily differentiated any number of times. For example

$$p''(x) \Big]_{x=0} = \frac{p''(u)}{h^2} \Big]_{u=0} = \frac{-0.2560}{3600} = -0.0000711$$

A second method of obtaining the second derivative consists in constructing a table of values for $f'(x)$ and then numerically differentiating this function. In the previous section, derivatives at certain points in the data for the speed of a rocket were determined. By obtaining the derivatives at the remaining points, using the central-difference method, Table 9.6 may be constructed. Then,

$$p''(x)\Big]_{x=0} = \frac{1}{60}\left(0.00031 - \frac{0.00178}{2} - \frac{0.00028}{3} - \frac{0.00864}{4} - \frac{0.01395}{5}\right)$$
$$= 0.0000937$$

Table 9.6

Time	Speed	f'(x)	D	D²	D³	D⁴	D⁵
0	0.0000	0.00198					
			0.00031				
60	0.0824	0.00229		0.00178			
			0.00209		−0.00028		
120	0.2747	0.00438		0.00150		0.00864	
			0.00359		0.00838		−0.01395
180	0.6502	0.00797		0.00988		−0.00531	
			0.01347		0.00307		
240	1.3851	0.02144		0.01295			
			0.02642				
300	3.2229	0.04786					

9.5   GENERAL EXPRESSIONS FOR DERIVATIVES

Occasionally it is desired to express derivatives directly in terms of functional values. This is particularly true in the numerical solution of differential equations. Some expressions of this kind have already been developed in Secs. 9.3 and 9.4. The general expression depends upon the order of the derivative, the order of the interpolating polynomial, and the type of difference employed. Equal intervals occur in most applications. As an illustration, an approximation for $f'(x_1)$, obtained by retaining four terms in Eq. (9.7), is

$$f'(x_1) \approx \frac{1}{h}\left\{[f(x_2) - f(x_1)] - \frac{f(x_3) - 2f(x_2) + f(x_1)}{2}\right.$$
$$+ \frac{f(x_4) - 3f(x_3) + 3f(x_2) - f(x_1)}{3}$$
$$\left. - \frac{f(x_5) - 4f(x_4) + 6f(x_3) - 4f(x_2) + f(x_1)}{4}\right\}$$
$$= \frac{-3f(x_5) + 16f(x_4) - 36f(x_3) + 48f(x_2) - 25f(x_1)}{12h}$$

To obtain expressions for derivatives of higher order, the interpolating polynomials must be differentiated in general terms. For example, differentiating Eq. (9.8) produces

$$p''(x) = \frac{1}{h^2} \left\{ \frac{\Delta^2 f(x_1)}{2!} \right. \quad (2)$$

$$+ \frac{\Delta^3 f(x_1)}{3!} [u + (u-1) + u + (u-2) + (u-1) + (u-2)]$$

$$+ \frac{\Delta^4 f(x_1)}{4!} [u(u-1) + u(u-2) + (u-1)(u-2) + u(u-1)$$

$$+ u(u-3) + (u-1)(u-3) + u(u-2) + u(u-3)$$

$$+ (u-2)(u-3) + (u-1)(u-2) + (u-1)(u-3)$$

$$\left. + (u-2)(u-3)] + \cdots \right\}$$

At the point x_1, $u = 0$, and

$$p''(x_1) = \frac{1}{h^2} [\Delta^2 f(x_1) - \Delta^3 f(x_1) + \tfrac{11}{12} \Delta^4 f(x_1) + \cdots]$$

Substituting the expressions for the differences into this equation and retaining the terms shown produce an approximation for the second derivative from a fourth-order polynomial.

$$p''(x_1) = \frac{1}{12h^2} [11f(x_5) - 56f(x_4) + 114f(x_3) - 104f(x_2) + 35f(x_1)]$$

Other expressions for $p''(x_1)$ can be obtained by retaining more or fewer terms in the interpolating polynomial. Backward and central differences are conveniently used to obtain expressions for the second derivative at points other than x_1. Higher derivatives are determined in the same manner. Tables 9.7 to 9.9 present expressions for all derivatives up to the fourth, all orders up to the fourth of the Gregory-Newton and Stirling interpolating polynomials being used. For convenience, the notation y_i is employed in place of $f(x_i)$.

9.6   ERROR ANALYSIS IN NUMERICAL DIFFERENTIATION

An expression for the error involved in numerical differentiation can be obtained from the error in the interpolating polynomial which relates the true functional value $f(x)$ to the interpolated value $p(x)$ by

$$f(x) = p(x) + e(x)$$

Differentiating gives

$$f'(x) = p'(x) + e'(x)$$

Table 9.7 Forward differences about point x_1

Order of polynomial	p'	p''	p'''	p^{iv}
1	$\dfrac{y_2 - y_1}{h}$			
2	$\dfrac{-y_3 + 4y_2 - 3y_1}{2h}$	$\dfrac{y_3 - 2y_2 + y_1}{h^2}$		
3	$\dfrac{2y_4 - 9y_3 + 18y_2 - 11y_1}{6h}$	$\dfrac{-y_4 + 4y_3 - 5y_2 + 2y_1}{h^2}$	$\dfrac{y_4 - 3y_3 + 3y_2 - y_1}{h^3}$	
4	$\dfrac{-3y_5 + 16y_4 - 36y_3 + 48y_2 - 25y_1}{12h}$	$\dfrac{11y_5 - 56y_4 + 114y_3 - 104y_2 + 35y_1}{12h^2}$	$\dfrac{-3y_5 + 14y_4 - 24y_3 + 18y_2 - 5y_1}{2h^3}$	$\dfrac{y_5 - 4y_4 + 6y_3 - 4y_2 + y_1}{h^4}$

Table 9.8 Backward differences about point x_5

Order of polynomial	p'	p''	p'''	p^{iv}
1	$\dfrac{y_5 - y_4}{h}$			
2	$\dfrac{3y_5 - 4y_4 + y_3}{2h}$	$\dfrac{y_5 - 2y_4 + y_3}{h^2}$		
3	$\dfrac{11y_5 - 18y_4 + 9y_3 - 2y_2}{6h}$	$\dfrac{2y_5 - 5y_4 + 4y_3 - y_2}{h^2}$	$\dfrac{y_5 - 3y_4 + 3y_3 - y_2}{h^3}$	
4	$\dfrac{25y_5 - 48y_4 + 36y_3 - 16y_2 + 3y_1}{12h}$	$\dfrac{35y_5 - 104y_4 + 114y_3 - 56y_2 + 11y_1}{12h^2}$	$\dfrac{5y_5 - 18y_4 + 24y_3 - 14y_2 + 3y_1}{2h^3}$	$\dfrac{y_5 - 4y_4 + 6y_3 - 4y_2 + y_1}{h^4}$

Table 9.9 Central differences about point x_3

Order of polynomial	p'	p''	p'''	p^{iv}
1	$\dfrac{y_4 - y_2}{2h}$			
2	$\dfrac{y_4 - y_2}{2h}$	$\dfrac{y_4 - 2y_3 + y_2}{h^2}$		
3	$\dfrac{-y_5 + 8y_4 - 8y_2 + y_1}{12h}$	$\dfrac{y_4 - 2y_3 + y_2}{h^2}$	$\dfrac{y_5 - 2y_4 + 2y_2 - y_1}{2h^3}$	
4	$\dfrac{-y_5 + 8y_4 - 8y_2 + y_1}{12h}$	$\dfrac{-y_5 + 16y_4 - 30y_3 + 16y_2 - y_1}{12h^2}$	$\dfrac{y_5 - 2y_4 + 2y_2 - y_1}{2h^3}$	$\dfrac{y_5 - 4y_4 + 6y_3 - 4y_2 + y_1}{h^4}$

Thus, $e'(x)$ represents the error involved in using $p'(x)$ in place of $f'(x)$. It follows that the error in the nth derivative is $e^n(x)$. The expression for $e(x)$ is derived in Chap. 8 in the form

$$e(x) = \frac{(x - x_1)(x - x_2)(\cdot\,\cdot\,\cdot)(x - x_{n+1})f^{n+1}(\xi)}{(n + 1)!} \qquad (8.19)$$

where $x_1, x_2, \ldots, x_{n+1}$ represent the points through which the interpolating polynomial passes. Differentiating produces an expression for the error in the first derivative. In the derivation of Eq. (8.19) it may be seen that the value of ξ depends upon x. Thus,

$$\begin{aligned}
e'(x) = \frac{f^{n+1}(\xi)}{(n + 1)!} & [(x - x_1)(x - x_2)(\cdot\,\cdot\,\cdot)(x - x_n) \\
& + (x - x_1)(x - x_2)(\cdot\,\cdot\,\cdot)(x - x_{n-1})(x - x_{n+1}) \\
& + \cdot\,\cdot\,\cdot + (x - x_2)(x - x_3)(\cdot\,\cdot\,\cdot)(x - x_{n+1})] \\
& + \frac{(x - x_1)(x - x_2)(\cdot\,\cdot\,\cdot)(x - x_{n+1})}{(n + 1)!} \frac{d}{dx} [f^{n+1}(\xi)] \quad (9.13)
\end{aligned}$$

Because the derivative of $f^{n+1}(\xi)$ is difficult to determine, the above expression is not very useful. However, if x takes on one of the values x_1, $x_2, \ldots, x_{n+1}$, the second term vanishes and all the subterms except one vanish in the first term. For example, at x_1,

$$e'(x_1) = \frac{f^{n+1}(\xi)}{(n + 1)!} (x_1 - x_2)(x_1 - x_3)(\cdot\,\cdot\,\cdot)(x_1 - x_{n+1}) \qquad (9.14)$$

Notice that, unlike the interpolating polynomial, the approximation of the derivative contains an error at the points $x_1, x_2, \ldots, x_{n+1}$.

To illustrate, the error in the example in Sec. 9.3.3 is obtained as follows:

$$e'(20) = (-1)^2 \frac{5^2(\pi/180)^3(-\cos\xi)}{3}$$

$$|e'(20)| \leq 0.000042$$

The actual error is 0.000041.

If we let (a,b) be the interval of interpolation, the upper bound of error is

$$|e'(x)| \leq \frac{(b - a)^n |f^{n+1}(x)|_{\max}}{n!} + \frac{(b - a)^{n+1}}{(n + 1)!} \left| \frac{d}{dx} [f^{n+1}(\xi)] \right|_{\max}$$

Replacing $b - a$ by nh gives

$$|e'(x)| \leq \frac{n^n |f^{n+1}(x)|_{\max}}{n!} h^n + \frac{n^{n+1} |(d/dx)[f^{n+1}(\xi)]|_{\max}}{(n + 1)!} h^{n+1} \qquad (9.15)$$

Thus, the error is of an order h^n which is of an order one less than that of the interpolating polynomial. This implies that the error obtained in numerical differentiation is generally greater than that in interpolation. This is illustrated in the examples on the parabolic interpolation of sin x for $20° \leq x \leq 30°$.   In Sec. 8.11, the error in the interpolated value of sin $22°$ is about 4 in the fifth significant digit.   In Sec. 9.3.1 the error in the approximate derivative of sin $22°$ is about 5 in the fourth significant digit.

Each successively higher derivative of an interpolating polynomial contains an order of error one less than that of the preceding derivative. The nth derivative of an nth-order interpolating polynomial contains an error of order h.   This implies that, the higher the order of the derivative, the larger the error in numerical differentiation.   Derivatives of order greater than n obviously cannot be determined by use of an nth-order interpolating polynomial.

There exists a fallacy in all this discussion concerning error in numerical differentiation.   The purpose of numerical differentiation is to obtain derivatives of functions whose derivatives cannot be obtained analytically. However, to determine the error, the analytic expressions for the derivatives must be known.   Thus, in practical applications, the error in numerical differentiation can seldom be obtained.

9.7   THE FUNDAMENTAL THEOREM OF INTEGRAL CALCULUS

Consider an arbitrary function $f(x)$ shown graphically in Fig. 9.5 and suppose that it is desired to determine the area enclosed by the curves $y = 0, y = f(x), x = a,$ and $x = b$.   The fundamental theorem of integral calculus can be stated as follows:

$$\text{Area} = \int_a^b f(x)\ dx = F(b) - F(a) \tag{9.16}$$

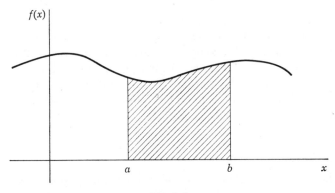

Fig. 9.5

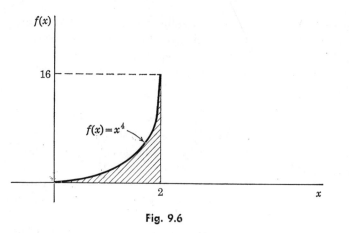

Fig. 9.6

where $F(b)$ and $F(a)$ represent the antiderivative of $f(x)$, indicated as $F(x)$, evaluated at $x = b$ and $x = a$, respectively, and the expression

$$\int_a^b f(x)\,dx$$

is the mathematical notation representing the area and is called a *definite integral*. The terms a and b are the limits of the integral; in particular, b is the upper limit, and a is the lower limit.

To illustrate the application of the theorem, the area under the curve $y = x^4$ between $x = 0$ and $x = 2$, shown in Fig. 9.6, is obtained as follows:

$$\text{Area} = \int_0^2 x^4\,dx = \left[\frac{x^5}{5}\right]_0^2 = \frac{2^5}{5} - \frac{0^5}{5}$$
$$= \tfrac{32}{5} = 6.4$$

9.8   NUMERICAL INTEGRATION

Often the fundamental theorem of integral calculus cannot be directly applied because either the antiderivative of $f(x)$ is not known or $f(x)$ is given in tabular form. The general procedure in such a case is to replace the given function $f(x)$ by an interpolating polynomial $p(x)$ and then apply the fundamental theorem to the integral containing $p(x)$. The accuracy of the result depends upon how closely $p(x)$ matches $f(x)$. This procedure is called *numerical integration*.

9.8.1   Trapezoidal Rule

One of the simplest methods of obtaining an approximate value of the area is to replace the actual area by a linear interpolating polynomial or

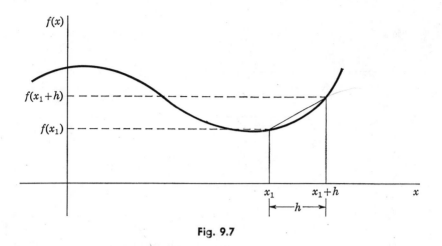

Fig. 9.7

trapezoid, as shown in Fig. 9.7.   Using the well-known formula for the area of a trapezoid gives

$$\text{Area} \approx \frac{f(x_1) + f(x_1 + h)}{2} h \tag{9.17}$$

The area under a curve between two limits $x = a$ and $x = b$ is approximately obtained by dividing the area into n trapezoids, as shown in Fig. 9.8, and applying the above relation to each trapezoid.   For convenience, the base of each trapezoid is made the same width and thus has a value

$$h = \frac{b - a}{n}$$

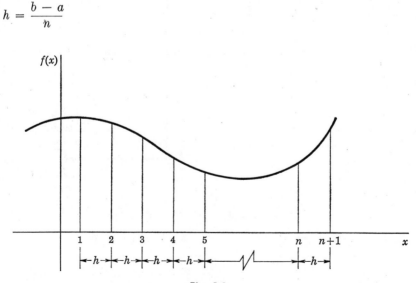

Fig. 9.8

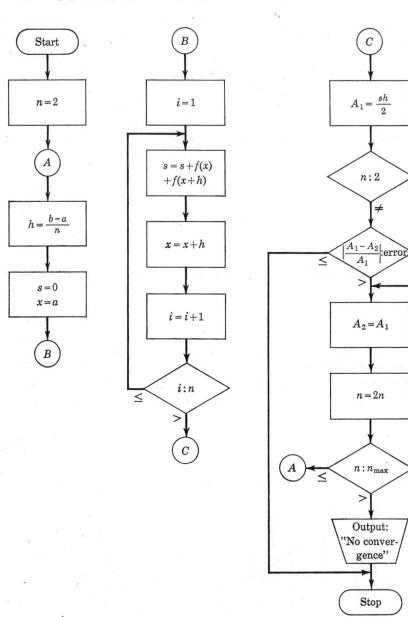

Fig. 9.9

To illustrate, the area shown in Fig. 9.6 is obtained approximately by being divided into four trapezoids. The values of the function $y = x^4$ at the necessary values of x are presented in Table 9.10. The value of h is $\frac{1}{2}$.

Table 9.10

x	y
0	0
$\frac{1}{2}$	$\frac{1}{16}$
1	1
$\frac{3}{2}$	$\frac{81}{16}$
2	16

$$\text{Area} \approx \frac{0 + \frac{1}{16}}{2}\frac{1}{2} + \frac{\frac{1}{16} + 1}{2}\frac{1}{2} + \frac{1 + \frac{81}{16}}{2}\frac{1}{2} + \frac{\frac{81}{16} + 16}{2}\frac{1}{2}$$

$$\approx 7\frac{1}{16} = 7.0625$$

In practice, it is convenient to factor out $h/2$, since then the calculations only involve taking a sum. A better result can be obtained by using a larger number of trapezoids.

Figure 9.9 presents a flow chart for evaluating an area under a curve between the limits a and b by using the trapezoidal rule. The flow chart determines successive approximations with decreasing values of h until a tolerable relative error is reached. The symbol n_{max} is a safeguard against infinite looping in case the procedure does not converge or there is a programming error.

9.8.2   Simpson's Rule

By using the trapezoidal rule, the area under a curve can usually be obtained to any degree of accuracy by taking a large enough number of divisions, n. Good accuracy can be obtained with a smaller value of n if the curve $f(x)$ is replaced by parabolic segments. The equation of a parabola is

$$y = ax^2 + bx + c$$

The parabola may be completely defined by being passed through three points, as shown in Fig. 9.10. The area under the parabola is approximately equal to the area under $f(x)$ and can be obtained by direct integra-

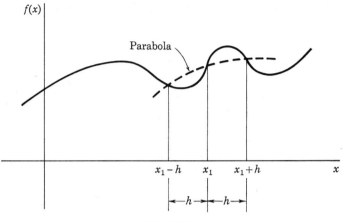

Fig. 9.10

tion as follows:

$$\text{Area} \approx \int_{x_1-h}^{x_1+h} (ax^2 + bx + c)\, dx = \left[\frac{ax^3}{3} + \frac{bx^2}{2} + cx\right]_{x_1-h}^{x_1+h}$$

$$= \frac{a}{3}[(x_1 + h)^3 - (x_1 - h)^3] + \frac{b}{2}[(x_1 + h)^2 - (x_1 - h)^2]$$
$$+ c[(x_1 + h) - (x_1 - h)]$$

$$= \frac{h}{3}[a(6x_1^2 + 2h^2) + 6bx_1 + 6c]$$

The result can be expressed in terms of functional values if the above expression is put into the form

$$\text{Area} = \frac{h}{3}[a(x_1 + h)^2 + b(x_1 + h) + c + 4(ax_1^2 + bx_1 + c)$$
$$+ a(x_1 - h)^2 + b(x_1 - h) + c]$$

$$= \frac{h}{3}[f(x_1 + h) + 4f(x_1) + f(x_1 - h)] \tag{9.18}$$

This relation is known as *Simpson's rule*. In order to obtain an approximate value of

$$\int_a^b f(x)\, dx$$

the interval $a \le x \le b$ is divided into an *even* number of increments and the above relation is applied to all successive pairs of increments.

As an illustration, again consider the area in Fig. 9.6. Upon taking four increments and using the values of $f(x)$ given in Table 9.10, Simpson's rule produces

$$\text{Area} \approx \frac{0.5}{3}\left[0 + 4\left(\frac{1}{16}\right) + 1\right] + \frac{0.5}{3}\left[1 + 4\left(\frac{81}{16}\right) + 16\right]$$

$$\approx \tfrac{77}{12} = 6.416$$

which is considerably more accurate than the result obtained by the trapezoidal rule.

In the examples presented so far, numerical integration was used to evaluate a known definite integral. The primary purpose of numerical integration, however, is to evaluate definite integrals of functions which do not possess antiderivatives in terms of known functions or of which the antiderivative is obtained only with great difficulty. The anti-derivatives of such functions as $\sqrt{1 + x^3}$, e^{-x^2}, e^x/x, $(\sin x)/x$, $1/(\ln x)$ cannot be expressed in terms of ordinary functions. An engineering problem is now presented in which numerical integration is applied.

Figure 9.11a represents an airplane wing which is acted upon by a lift force assumed to be uniformly distributed along the length of the wing at w kips (1 kip = 1 kilopound = 1000 lb) per linear foot. The deflection δ at the end of the wing is given by the integral

$$\int_0^L \frac{Mx}{EI}\,dx$$

M represents the internal moment in the wing, which is computed from Fig. 9.11b as

$$M = wx\,\frac{x}{2} = \frac{wx^2}{2}$$

E represents the modulus of elasticity of the material of which the wing

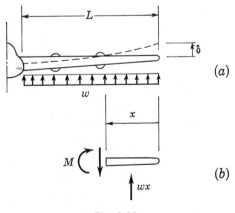

Fig. 9.11

is made. I (the moment of inertia) is a property of the shape of the wing cross section.   Since the shape of a wing generally varies with x, I is a function of x and is often complicated enough to make the determination of the antiderivative of Mx/EI quite difficult.

As a numerical example, let $w = 2$ kips/ft, $L = 40$ ft, and

$$E = 4.32 \times 10^6 \text{ kips/ft}^2 \text{ (steel)}$$

and let the moment of inertia I be given by the interpolating polynomial

$$I(x) = 30 \times 10^{-9}x^4 + 5.0 \times 10^{-6}x^3 + 0.30 \times 10^{-3}x^2$$
$$+ 8.0 \times 10^{-3}x + 0.1$$

The deflection is then given by

$$\delta = \frac{1}{4.32 \times 10^6}$$
$$\times \int_0^{40} \frac{x^3 \, dx}{30 \times 10^{-9}x^4 + 5.0 \times 10^{-6}x^3 + 0.3 \times 10^{-3}x^2 + 8.0 \times 10^{-3}x + 0.1}$$

The antiderivative of the function to be integrated can be found by using partial fractions but requires the determination of the factors of the denominator.   Thus, the application of numerical integration is worthwhile.   Values of the integrand are given in Table 9.11, and applying

Table 9.11

x	x^3/I
0	0
5	844
10	4,645
15	11,033
20	18,832
25	27,064
30	35,097
35	42,581
40	49,352

Simpson's rule for two interval spacings gives the following results:
    Four divisions:

$$\text{Area} \approx \frac{10}{3(4.32 \times 10^6)} \{[0 + 4(4645) + 18{,}832]$$
$$+ [18{,}832 + 4(35{,}097) + 49{,}352]\}$$
$$\approx 0.1900 \text{ ft}$$

Eight divisions:

$$\text{Area} \approx \frac{5}{3(4.32 \times 10^6)} \{[0 + 4(844) + 4645]$$
$$+ [4645 + 4(11,033) + 18,832]$$
$$+ [18,832 + 4(27,064) + 35,097]$$
$$+ [35,097 + 4(42,581) + 49,352]\}$$
$$\approx 0.1901 \text{ ft}$$

Since the two results are very close, the answer can be assumed to be quite accurate.

A flow chart for evaluating an area under a curve between the limits a and b by Simpson's rule is given in Fig. 9.12.

9.8.3   Use of Higher-order Polynomials

Simpson's rule is derived by replacing the actual curve by parabolic arcs. In a similar manner rules can be derived for higher-order polynomials. Table 9.12 gives results for all polynomials up to the fifth order.

Table 9.12

Order	Result
1	$(h/2)(y_1 + y_2)$ trapezoidal rule
2	$(h/3)(y_1 + 4y_2 + y_3)$ Simpson's rule
3	$(3h/8)(y_1 + 3y_2 + 3y_3 + y_4)$
4	$(2h/45)(7y_1 + 32y_2 + 12y_3 + 32y_4 + 7y_5)$
5	$(5h/288)(19y_1 + 75y_2 + 50y_3 + 50y_4 + 75y_5 + 19y_6)$

If a function is given in tabular form, it may be desirable to pass a high-order polynomial through all the given points and then determine the area under the polynomial. For example, the area under a fifth-order polynomial going through the six points for the rocket problem given in Table 9.2 can be obtained from the above table.

$$\text{Area} \approx \frac{5(60)}{288} [19(0) + 75(0.0824) + 50(0.2747) + 50(0.6502)$$
$$+ 75(1.3851) + 19(3.2229)]$$
$$\approx 227$$

The relationship between velocity and displacement is given by

$$v = \frac{ds}{dt}$$
$$s = \int_0^{300} v \, dt$$

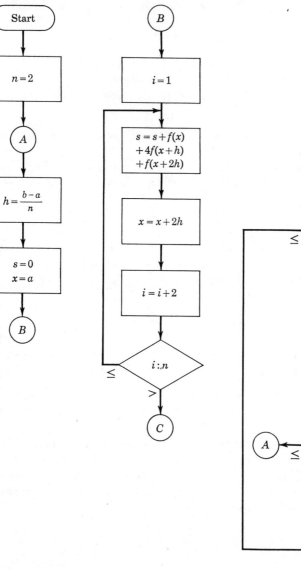
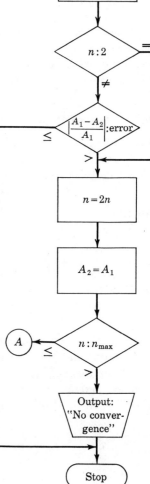

Fig. 9.12

and thus the above area represents the distance the rocket traveled during the 300 sec, i.e., 227 miles.

9.9   ERROR ANALYSIS IN NUMERICAL INTEGRATION

Using the trapezoidal rule to determine the area in Fig. 9.7 gives an error

$$E(h) = F(x_1 + h) - F(x_1) - \frac{h}{2}[f(x_1 + h) - f(x_1)]$$

where $F(x)$ represents the antiderivative of $f(x)$. If one thinks of x_1 as being fixed, the error is a function of h. It appears likely that, the smaller the value of h, the less the error. To determine the relationship between $E(h)$ and h, let h be a fixed quantity, and consider the function of t,

$$G(t) = E(t) - \frac{E(h)t^3}{h^3}.$$

This function has the property

$$G(0) = G(h) = 0$$

as illustrated in Fig. 9.13a. Substituting the value of $E(t)$ into the equation for $G(t)$ gives

$$G(t) = F(x_1 + t) - F(x_1) - \frac{t}{2}[f(x_1 + t) + f(x_1)] - \frac{E(h)t^3}{h^3}$$

Differentiation produces

$$G'(t) = f(x_1 + t) - \frac{t}{2}f'(x_1 + t) - \frac{1}{2}[f(x_1 + t) + f(x_1)] - \frac{3E(h)t^2}{h^3}$$

which is equal to 0 at $t = 0$ and $t = a$, as demonstrated in Fig. 9.13b. Differentiating again,

$$G''(t) = -\frac{t}{2}f''(x_1 + t) - \frac{6E(h)t}{h^3}$$

As demonstrated in Fig. 9.13c, $G''(t)$ equals 0 at some point ξ. Thus,

$$G'(\xi) = 0 = -\frac{\xi}{2}f''(x_1 + \xi) - \frac{6E(h)\xi}{h^3}$$

$$E(h) = -\frac{h^3 f''(x_1 + \xi)}{12}$$

Since the actual value of ξ is generally difficult to determine, an error bound is given by

$$|E(h)| \leq \frac{h^3}{12}|f''(x)|_{\max} \tag{9.19}$$

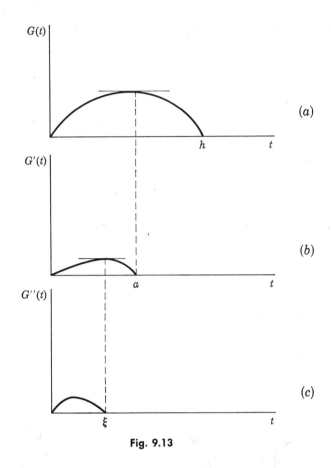

Fig. 9.13

It should be pointed out that Fig. 9.13a, b, and c implies that G, G', and G'' and therefore $f(x)$, $f'(x)$, and $f''(x)$ are all continuous functions.

Upon applying this error analysis to the example in Sec. 9.8.1, $h = \frac{1}{2}$, and

$$|f''(x)|_{\max} = |12x^2|_{\max} = 12(2^2) = 48$$

$$|\text{Error}| \leq \frac{\left(\frac{1}{2}\right)^3}{12} (48) = 0.5$$

This represents the error bound for one interval. Since four intervals are used in the example, the error bound is 2. The actual error is 0.6625.

The error in Simpson's rule for the area shown in Fig. 9.10 is

$$E(h) = F(x_1 + h) - F(x_1 - h) - \frac{h}{3} [f(x_1 - h) + 4f(x_1) + f(x_1 + h)]$$

Constructing the function

$$G(t) = E(t) - \frac{E(h)t^5}{h^5}$$

and differentiating three times,

$$G'''(t) = -\frac{t}{3}[f'''(x_1 + t) - f'''(x_1 - t)] - \frac{60E(h)t^2}{h^5}$$

With the same arguments used for the trapezoidal rule, $G'''(t) = 0$ at some point ξ, and it follows that

$$E(h) = -\frac{h^5[f'''(x_1 + \xi) - f'''(x_1 - \xi)]}{180\xi}$$

The bracketed term can be simplified as follows: In Fig. 9.14, $f'''(x)$ is graphically represented between $x_1 + \xi$ and $x_1 - \xi$. From the figure, one can see that there exists a point ξ_1 where the tangent to the curve is parallel to the line AB. Thus,

$$f^{iv}(\xi_1) = \frac{f'''(x_1 + \xi) - f'''(x_1 - \xi)}{2\xi}$$

$E(h)$ now becomes

$$E(h) = -\frac{h^5}{90}f^{iv}(\xi_1)$$

$$|E(h)| \leq \frac{h^5}{90}|f^{iv}(x)|_{\max} \tag{9.20}$$

Again, this holds only if $f(x)$, $f'(x)$, $f''(x)$, $f'''(x)$, and $f^{iv}(x)$ are all continuous.

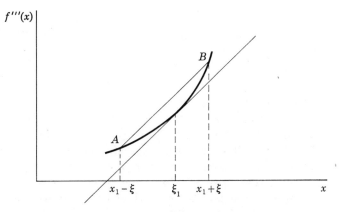

Fig. 9.14

The error in the example in Sec. 9.8.2 is bounded by

$$|Error| \leq 2\frac{(\frac{1}{2})^5}{90} \quad (24)$$

$$\leq 0.01667$$

The actual error is 0.016.

As in previous cases, the expression for the error contains a derivative of the function $f(x)$. If this derivative is not known, as is often the case with tabular data, an error bound cannot be determined.

It is informative to examine the order of error involved in interpolation, numerical differentiation, and numerical integration. In the case of an interpolating parabola, the order of error in interpolation is h^3, in numerical differentiation h^2, and in numerical integration h^5. Thus, the error involved in numerical integration tends to be less than that in interpolation, while numerical differentiation tends to magnify the error in interpolation.

9.10   APPLICATIONS

Many engineering problems require the use of numerical differentiation or integration. Three examples are presented here.

9.10.1   Buckling Load of a Column

Consider a bar subjected to a compressive force P, as shown in Fig. 9.15a. The stress σ in the bar is defined as

$$\sigma = \frac{P}{A}$$

where A is the cross-sectional area of the bar. Strain ϵ is defined as

$$\epsilon = \frac{\delta}{L}$$

where δ is the amount the bar shortens because of the load and L is the unstressed length of the bar. If the relation between σ and ϵ is linear, the material of which the bar is composed obeys Hooke's law, as illustrated in Fig. 9.15b, and is called an *elastic material*. Figure 9.16c represents a stress-strain law for a material which is not linear. A typical stress-strain law for structural steel is given in Fig. 9.15d. Notice that steel, as is common with many materials, obeys Hooke's law up to a certain point and then becomes nonlinear. The point σ_P at which the stress-strain

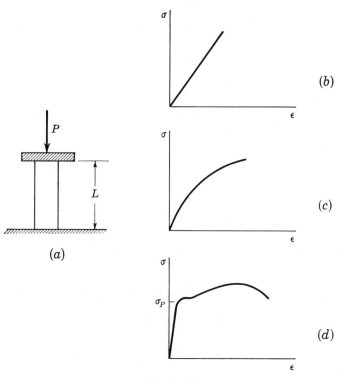

Fig. 9.15

relation is no longer linear is called the *proportional limit*, which has values between 35,000 and 120,000 for various steels.

A long, slender bar will fail by bending out laterally, as shown in Fig. 9.16a. This phenomenon, called *buckling*, can easily be observed by compressing a yardstick. Euler derived an equation for the load P which causes an elastic column to buckle.

$$P = \frac{\pi^2 EI}{L^2} \tag{9.21}$$

where I is the moment of inertia of the cross section. If a material is not elastic, the usual procedure is to replace the modulus of elasticity E by a tangent modulus, as shown in Fig. 9.16c. The buckling load is then

$$P = \frac{\pi^2 E_t I}{L^2} \tag{9.22}$$

Since the stress-strain diagram is obtained experimentally, the tangent modulus is obtained either graphically or numerically. Mathematically,

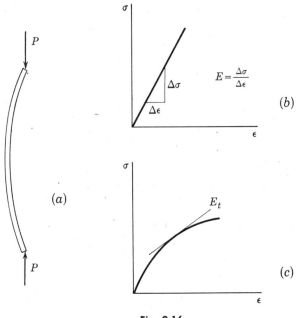

Fig. 9.16

of course,

$$E_t = \frac{d\sigma}{d\epsilon}$$

Another good computer problem also arises in this situation. Dividing the equation for the buckling load by the cross-sectional area A gives

$$\frac{P}{A} = \sigma = \frac{\pi^2 E_t(\sigma)I}{L^2 A}$$

Since E_t is a function of σ, some iterative method must be utilized to solve for σ. After σ is determined, the load P which causes buckling is obtained directly by $P = \sigma A$.

9.10.2   Amplifiers and Oscillators

Figure 9.17 shows a typical circuit containing an amplifier. Many types of amplifiers are used in such instruments as radio transmitters, television transmitters, measurement devices, and control devices. Some kinds of amplifiers (class B and class C) produce plate currents I_p in short bursts, as illustrated in Prob. 8.87. This current is periodic in time, with t_1 representing the period. In order to analyze circuits, it is usually necessary to express current in terms of an analytic function. In a problem such as this, it is convenient to assume this analytic expression in the form

$$I_p = I_0 + I_1 \cos \frac{2\pi t}{t_1} + I_2 \cos \frac{4\pi t}{t_1} + \cdots + I_n \cos \frac{2n\pi t}{t_1} \tag{9.23}$$

Notice that each term except the first in this series has a period which is a multiple of t_1. The accuracy of this representation generally depends upon the number of terms taken in the series. The standard method of obtaining a typical coefficient I_m is to multiply the equation by $\cos (2m\pi t/t_1) \, dt$ and integrate from $-t_1/2$ to $+t_1/2$.

$$\int_{-t_1/2}^{t_1/2} I_p \cos \frac{2m\pi t}{t_1} \, dt = I_0 \int_{-t_1/2}^{t_1/2} \cos \frac{2m\pi t}{t_1} \, dt$$

$$+ I_1 \int_{-t_1/2}^{t_1/2} \cos \frac{2\pi t}{t_1} \cos \frac{2m\pi t}{t_1} \, dt$$

$$+ I_2 \int_{-t_1/2}^{t_1/2} \cos \frac{4\pi t}{t_1} \cos \frac{2m\pi t}{t_1} \, dt$$

$$\cdots \cdots \cdots \cdots \cdots$$

$$+ I_n \int_{-t_1/2}^{t_1/2} \cos \frac{2n\pi t}{t_1} \cos \frac{2m\pi t}{t_1} \, dt$$

Every integral on the right side of this equation turns out to be zero except

$$\int_{-t_1/2}^{t_1/2} \cos \frac{2m\pi t}{t_1} \cos \frac{2m\pi t}{t_1} \, dt = \frac{t_1}{2}$$

Thus,

$$I_m = \frac{2}{t_1} \int_{-t_1/2}^{t_1/2} I_p \cos \frac{2m\pi t}{t_1} \, dt$$

Since I_p is zero over a portion of the cycle,

$$I_m = \frac{2}{t_1} \int_{-a/2}^{a/2} I_p \cos \frac{2m\pi t}{t_1} \, dt$$

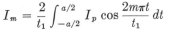

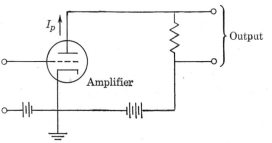

Fig. 9.17

The plate current I_p is generally obtained from experimental data, and thus the above integral must be evaluated numerically by some process such as Simpson's rule.

The constant term I_0 is called the *d-c plate current* and is given by

$$I_0 = \frac{2}{t_1} \int_{-a/2}^{a/2} I_p \, dt \tag{9.24}$$

The coefficient I_1 is called the *fundamental component of plate current* and is given by

$$I_1 = \frac{2}{t_1} \int_{-a/2}^{a/2} I_p \cos \frac{2\pi t}{t_1} \, dt \tag{9.25}$$

The voltage drop across an inductance coil is proportional to the time derivative of the current. If an inductance coil is in a line receiving the current I_p in pulses, the derivative dI_p/dt must be determined. One approach to this problem is numerically to differentiate the function shown in Prob. 8.87. The more common approach, however, is to differentiate the series representation of I_p.

$$\frac{dI_p}{dt} = -\frac{2\pi}{t_1} I_1 \sin \frac{2\pi t}{t_1} - \frac{4\pi}{t_1} I_2 \sin \frac{4\pi t}{t_1} - \cdots - \frac{2n\pi}{t_1} I_n \sin \frac{2n\pi t}{t_1}$$

The analysis of oscillators is similar to that of amplifiers. Oscillators are used in radios and television and for many industrial purposes.

The mathematical procedure used above in analyzing an amplifier is a special case of a general technique known as *Fourier analysis*. This technique also has applications in problems concerned with temperature distribution, vibrations, and deflections of elastic bodies.

9.10.3   Calculation of Fugacity

A thermodynamic quantity of interest in many chemical processes is the ratio of fugacity to the pressure, given by

$$\ln \frac{f}{p} = \frac{1}{RT} \int_{p^\circ}^{p} \left(v - \frac{RT}{pM} \right) dp$$

where

f = fugacity, atm
p = pressure, atm
p° = pressure at a standard state, often 1 atm
T = temperature, °R = °F + 460
v = specific volume, ft^3/lb
R = gas constant, 0.729 $(\text{ft}^3)(\text{atm})/(\text{lb mole})(°R)$
M = molecular weight

For an ideal gas $v = RT/pM$, and the value of the above integral is zero. For an actual gas the value of the integral at a pressure p is a measure of the deviation from ideality. The integral can be evaluated numerically from tabular values of v and p at a fixed temperature.

PROBLEMS

9.1 Construct the graph of Fig. 9.2, and determine the acceleration of the rocket at the following times: 0, 60, 120, 180, 240, and 300 sec.

9.2 Using the results of Prob. 9.1, construct an acceleration vs. time graph. Then, calculate the rate of change of acceleration at the following times: 60, 150, and 240 sec.

In Probs. 9.3 to 9.6, determine $f'(x)$ at the central point, using all three of the methods illustrated in Fig. 9.3. Compare with true values.

◆**9.3** $f(x) = \cos x: x = 20, 25,$ and $30°$

9.4 $f(x) = e^x: x = 1, 2,$ and 3

9.5 $f(x) = \ln x: x = 1, 2,$ and 3

9.6 $f(x) = x^2: x = 1, 2,$ and 3

9.7 Using the data given in Table 9.2, determine the acceleration of the rocket at the following times: 60, 120, 180, and 240 sec.

9.8 Using the data given in Table 9.2, determine the acceleration of the rocket at the following times: 30, 90, 210, and 270 sec.

9.9 Using the data given in Table 9.2, determine the acceleration of the rocket at the following times: 0 and 300 sec.

In Probs. 9.10 to 9.13, construct an interpolating parabola $p(x)$ passing through the given points, and then differentiate to obtain $p'(x)$. Compare results with the true values.

◆**9.10** $f(x) = \cos x: x = 20, 30,$ and $35°$. Determine $p'(x)$ for $x = 20, 22, 30,$ and $35°$.

9.11 $f(x) = e^x: x = 1, 2,$ and 4. Determine $p'(x)$ for $x = 1, 2, 3,$ and 4.

9.12 $f(x) = \ln x: x = 1, 2,$ and 4. Determine $p'(x)$ for $x = 1, 2, 3,$ and 4.

9.13 $f(x) = x^2: x = 1, 2,$ and 4. Determine $p'(x)$ for $x = 1, 2, 3,$ and 4.

9.14 In Eq. (9.5), determine the multiplier of the coefficient a_6.

9.15 In Eq. (9.7), determine the multiplier of the coefficient $\Delta^5 f(x_1)/5!$.

In Probs. 9.16 to 9.20, construct difference tables, and determine $p'(x_1)$. Compare with true values of the derivative.

◆**9.16** $f(x) = \cos x: x = 20, 25,$ and $30°$

9.17 $f(x) = \cos x: x = 20, 25, 30,$ and $35°$

9.18 $f(x) = e^x: x = 1, 2,$ and 3

9.19 $f(x) = \ln x: x = 1, 2,$ and 3

9.20 $f(x) = x^2: x = 1, 2,$ and 3

9.21 Derive the expression for $p'(x_{n+1})$ given in Sec. 9.3.4 from the Gregory-Newton backward-interpolation formula.

9.22 Derive the expression for $p'(x_i)$ given in Sec. 9.3.4 from the Stirling central-interpolation formula.

9.23 Referring to Prob. 8.41, determine the x component of the velocity of the rocket dx/dt at all the given points.

9.24 Referring to Prob. 8.42, determine the y component of the velocity of the rocket dy/dt at all the given points.

9.25 Combine the results of Probs. 9.23 and 9.24 to obtain the speed of the rocket at each point.

9.26 Referring to Prob. 8.43, determine dy/dx at all the given points.

9.27 Compare the results of Probs. 9.25 and 9.26. The slope of the velocity vector should be the same as dy/dx.

9.28 Referring to Prob. 8.57, determine dV/dy at all of the given points. The shear stress in the water is given by the equation

$$\tau = \mu \frac{dV}{dy}$$

where μ is called the *coefficient of viscosity*. Using $\mu = 0.235 \times 10^{-4}$, calculate the shear stress at all the given points.   The units of τ are pounds per square foot.

♦**9.29** Referring to Prob. 8.76, determine dh/dt at all the given points.

♦**9.30** Referring to Prob. 8.76, determine $d\theta/dt$ at all the given points.

♦**9.31** Use the results of Probs. 9.29 and 9.30 to determine the speed S of the satellite at all the given points.   The speed is given by the formula

$$S = \left\{ \left(\frac{dh}{dt}\right)^2 + \left[(R + h) \frac{d\theta}{dt} \right]^2 \right\}^{\frac{1}{2}}$$

where R is the radius of the earth, approximately 3960 miles.

9.32 Construct a flow chart, and write a computer program which produces a linearly interpolated value of the derivative for a general point x, given a set of tabulated points $[x_i, f(x_i)]$.

9.33 Construct a flow chart, and write a computer program which (1) produces a parabolically interpolated value for a general point x, given a set of tabulated points $[x_i, f(x_i)]$, and (2) produces a value for the derivative at x by differentiation of the parabola. For the interpolation take two points before x and one after x, except when $x_1 \le x \le x_2$.

9.34 Write a computer program for calculating the derivative $p'(x_1)$, using the flow chart of Fig. 9.3.

9.35 Construct a flow chart, and write a computer program for calculating the derivative $p'(x_{n+1})$, using backward differences.

9.36 Construct a flow chart, and write a computer program for calculating the derivative $p'(x_i)$, using central differences.

In Probs. 9.37 to 9.40, calculate three approximations of $f''(x)$ at the central point, using the three equations

$$f''(x) \approx \frac{\Delta^2 f(x)}{h^2} \qquad f''(x) \approx \frac{\nabla^2 f(x)}{h^2} \qquad f''(x) \approx \frac{\delta^2 f(x)}{h^2}$$

Compare with true values.

◆**9.37** $f(x) = \cos x \colon x = 15, 20, 25, 30,$ and $35°$

9.38 $f(x) = e^x \colon x = 1, 2, 3, 4,$ and 5

9.39 $f(x) = \ln x \colon x = 1, 2, 3, 4,$ and 5

9.40 $f(x) = x^2 \colon x = 1, 2, 3, 4,$ and 5

◆**9.41, 9.42–9.44** In Probs. 9.37 to 9.40, determine $f'''(x)$ and $f^{iv}(x)$ at the central point, using the equation

$$^n(x) = \frac{\delta^n f(x)}{h^2}$$

9.45 Using the interpolating polynomial for the speed of the rocket given in Sec. 9.4, determine the following quantities: $p'(180)$, $p''(180)$, $p'''(300)$, and $p^{iv}(0)$.

9.46 Referring to Prob. 8.41, determine the x component of the acceleration of the rocket d^2x/dt^2 at all the given points.

9.47 Referring to Prob. 8.42, determine the y component of the acceleration of the rocket d^2y/dt^2 at all the given points.

9.48 Combine the results of Probs. 9.46 and 9.47 to obtain the magnitude of the acceleration of the rocket at all the given points.

◆**9.49** Referring to Prob. 8.76, determine d^2h/dt^2 at all the given points.

◆**9.50** Referring to Prob. 8.76, determine $d^2\theta/dt^2$ at all the given points.

◆**9.51** Combine the results of Probs. 9.49 and 9.50 with those of Probs. 9.29 and 9.30 to determine the magnitude of the acceleration of the satellite at all the given points. The magnitude a is given by

$$a = \left\{ \left[\frac{d^2h}{dt^2} - (R + h) \left(\frac{d\theta}{dt} \right)^2 \right]^2 + \left[(R + h) \frac{d^2\theta}{dt^2} + 2 \left(\frac{dh}{dt} \right) \left(\frac{d\theta}{dt} \right) \right]^2 \right\}^{\frac{1}{2}}$$

where R is the radius of the earth, approximately 3960 miles.

9.52 Construct a flow chart, and write a computer program which calculates

$$f^n(x) \approx \frac{\delta^n f(x)}{h^n}$$

for any n from a set of tabulated points $[x_i, f(x_i)]$.

9.53 Construct a flow chart, and write a computer program which calculates $p''(x_1)$ from a set of tabulated points $[x_i, f(x_i)]$, using forward differences.

◆**9.54** Determine $p'(x_1)$ in terms of functional values, retaining five terms in the forward-difference formula for $p'(x_1)$.

9.55 Determine $p'(x_{n+1})$ in terms of functional values, retaining five terms in the backward-difference formula for $p'(x_{n+1})$.

9.56 Determine $p'(x_i)$ in terms of functional values, retaining five terms in the central-difference formula for $p'(x_i)$.

9.57 Derive an expression for $p''(x_{n+1})$ in terms of backward differences from a fourth-order interpolating polynomial. Then substitute functional values for the differences, and check the answers in Table 9.8.

9.58 Derive an expression for $p''(x_i)$ in terms of central differences from a fourth-order interpolating polynomial. Then substitute functional values for the differences, and check the answers in Table 9.9.

Assuming that the derivative is obtained from parabolic interpolation in Probs. 9.59 to 9.63, determine upper bounds of the error in $p'(x)$ for the intervals given. Assume that $d[f'''(\xi)]/dx$ is negligible.

♦9.59 $\cos x: 20° \leq x \leq 30°$
9.60 $e^x: 1 \leq x \leq 3$
9.61 $\ln x: 1 \leq x \leq 3$
9.62 $x^2: 1 \leq x \leq 3$
9.63 $x^3: 1 \leq x \leq 3$

In Probs. 9.64 to 9.68, determine the area under the given curves by numerical integration, and compare with the true area.

9.64 $f(x) = x^2: 0 \leq x \leq 2$
9.65 $f(x) = \sin x: 0° \leq x \leq 90°$
♦9.66 $f(x) = \cos x: 0° \leq x \leq 90°$
9.67 $f(x) = e^x: 0 \leq x \leq 2$
9.68 $f(x) = \ln x: 1 \leq x \leq 3$

In Probs. 9.69 to 9.73, determine the area under the given curve $f(x)$ between $x = 1$ and $x = 5$.

♦9.69 $f(x) = \sqrt{1 + x^3}$
9.70 $f(x) = e^{-x^2}$
9.71 $f(x) = e^x/x$
9.72 $f(x) = (\sin x)/x:$ in this case $\pi/2 \leq x \leq \pi$ rad
9.73 $f(x) = 1/(\ln x):$ in this case $2 \leq x \leq 5$

9.74   Referring to Prob. 8.2, determine the area under the T-θ curve during the power stroke ($0° \leq \theta \leq 180°$). This area represents the work done on the drive shaft.

9.75   Referring to Prob. 8.2, determine the area under the T-θ curve during the compression stroke ($180° \leq \theta \leq 360°$). This area is negative and represents the work done by the drive shaft to compress the gas behind the piston.

9.76   Combine the results of Probs. 9.74 and 9.75 to calculate the average horsepower transmitted to the drive shaft. The horsepower is calculated from the formula

$$\text{Horsepower} = \frac{\text{net work}}{6600(\text{time for one cycle})}$$

Assume an angular speed of 1600 rpm.

9.77   Referring to Prob. 8.11, determine the area under the stress-strain curve between points A and C. Assuming a straight line, determine the area under the curve between the origin and point A. The total area represents the energy per unit volume absorbed by the bar during loading.

9.78   Referring to Prob. 8.12, the work done by the gas on the piston during the power stroke is given by

$$\text{Work} = A \int p \, dx = A \int_0^\pi p \, \frac{dx}{d\theta} \, d\theta$$

where A is the cross-sectional area of the piston.   The factor $dx/d\theta$ is obtained from geometry,

$$x = l_r + l_c - \sqrt{l_r{}^2 - l_c{}^2 \sin^2 \theta} - l_c \cos \theta$$
$$\frac{dx}{d\theta} = \frac{l_c{}^2 \sin \theta \cos \theta}{\sqrt{l_r{}^2 - l_c{}^2 \sin^2 \theta}} + l_c \sin \theta$$

where l_c is the length of the crank and l_r the length of the connecting rod.   Taking a piston diameter of 6 in., $l_c = 5$ in., and $l_r = 14$ in., determine the work done by the gas during the power stroke.

9.79   Using the data of Prob. 9.78, determine the work done on the gas during the compression stroke.

◆**9.80**   Referring to Prob. 8.23, determine the area under the p-x curve.   Multiplying this area by the bore area gives the work done by the powder on the bullet.   With friction neglected, this work equals the kinetic energy of the bullet as it leaves the barrel.

$$\text{Work} = \tfrac{1}{2}mv^2$$

Using the following data, determine the muzzle velocity v of the bullet: diameter of bore $= 0.3$ in., mass of bullet $= 0.00007$.   The units of v are inches per second.

9.81   The quantity of water flowing in a channel per unit time is given by

$$Q = \int V \, dA$$

where V is the velocity of the water and dA is an element of area in the channel cross section.   Using the data of Prob. 8.57, determine the flow rate Q in the given channel.   The velocity is zero at $y = 9$ ft and can be assumed to vary linearly between $y = 8$ and 9 ft.

9.82   Derive the equation for the area under a third-order polynomial, and check your result in Table 9.12.

9.83   Derive the equation for the area under a fourth-order polynomial, and check your result in Table 9.12.

9.84   Derive the equation for the area under a fifth-order polynomial, and check your result in Table 9.12.

9.85   Derive the equation for the area under a sixth-order polynomial.

In Probs. 9.86 to 9.95, calculate upper bounds on the error in integration by both the trapezoidal rule and Simpson's rule, when using four intervals.

9.86 $f(x) = x^5: 0 \le x \le 2$
9.87 $f(x) = \sin x: 0 \le x \le \pi/2$
◆**9.88** $f(x) = \cos x: 0 \le x \le \pi/2$
9.89 $f(x) = e^x: 0 \le x \le 2$
9.90 $f(x) = \ln x: 1 \le x \le 3$
9.91 $f(x) = \sqrt{1 + x^3}: 1 \le x \le 5$
9.92 $f(x) = e^{-x^2}: 1 \le x \le 5$
9.93 $f(x) = e^x/x: 1 \le x \le 5$
9.94 $f(x) = (\sin x)/x: \pi/2 \le x \le \pi$
9.95 $f(x) = 1/(\ln x): 2 \le x \le 5$

9.96   Derive an expression for the error when using a cubic polynomial for numerical integration.

9.97   Derive an expression for the error when using a fourth-order polynomial for numerical integration.

9.98   Write a computer program for integration by the trapezoidal rule, using the flow chart in Fig. 9.9.

9.99   Write a computer program for integration by Simpson's rule, using the flow chart in Fig. 9.12.

9.100   Construct a flow chart, and write a computer program to evaluate

$$\int_c^d \int_a^b f(x)g(y) \, dx \, dy$$

9.101   Revise the flow chart in Fig. 9.12 so that numerical integration is accomplished even if there is an odd number of intervals.   Calculate the area under the last interval by the trapezoidal rule.   Write the corresponding computer program.

♦**9.102**   An American Standard I beam has a cross-sectional area of 5.0 in.² and a moment of inertia of 2.3 in.⁴   Determine the maximum length for which the column will not buckle, if a load of 200,000 lb is applied.   Use the stress-strain data given in Prob. 8.3.

9.103   An American Standard channel section has a cross-sectional area of 4.0 in.² and a moment of inertia of 1.5 in.⁴   Determine the maximum length for which the column will not buckle, if a load of 140,000 lb is applied.   Use the stress-strain data given in Prob. 8.3.

9.104   A Standard angle section has a cross-sectional area of 3.0 in.² and a moment of inertia of 1.0 in.⁴   Determine the maximum length for which the column will not buckle if a load of 135,000 lb is applied.   Use the stress-strain data given in Prob. 8.3.

9.105   If the column of Prob. 9.102 is 48 in. long, calculate the load P which causes buckling.

9.106   If the column of Prob. 9.103 is 23 in. long, calculate the load P which causes buckling.

9.107   If the column of Prob. 9.104 is 33 in. long, calculate the load P which causes buckling.

♦**9.108**   Taking $a = 0.004$ sec, $t_1 = 0.008$ sec, and $i_m = 0.5$ amp in the figure of Prob. 8.87, determine the coefficients I_0, I_1, and I_2.   Use these three terms to evaluate the plate current I_p at $t = 0$, 0.001, 0.002, 0.003, and 0.004 sec.

t, sec.......	-0.002	-0.001	0.000	0.001	0.002
I_p, amp....	0.0	0.4	0.5	0.4	0.0

9.109   Do Prob. 9.108 by retaining 17 terms in the trigonometric series.

9.110   Using the results of Prob. 9.109, calculate dI_p/dt at $t = 0$, 0.001, 0.002, 0.003, and 0.004 sec.

◆**9.111**   Pressure-volume data for ethylene at 100°F are given below.[1]   Determine the fugacity at a pressure of 10 atm.   The molecular weight of ethylene is 30.

p, atm	v, ft^3/lb	p, atm	v, ft^3/lb
1	14.51	50	0.201
2	7.217	60	0.144
4	3.571	80	0.082
6	2.356	100	0.057
10	1.382	150	0.046
15	0.897	200	0.042
20	0.652	250	0.039
30	0.405	300	0.038
40	0.279		

9.112   Construct a flow chart, and write a computer program for determining the fugacity of ethylene at 100°F at any pressure p ($1 \le p \le 300$ atm), using the data given in Prob. 9.111.

9.113   Hamming's method of finding formulas consists in assuming a linear combination of the known data and making the approximate equation exact for certain functions.[2]   To derive the trapezoidal rule, for example, if two functional values $f(x)$ and $f(x + h)$ are known, the formula is assumed in the form

$$a_1 f(x) + a_2 f(x + h) \approx \int_x^{x+h} f(x)\, dx$$

In order to solve for a_1 and a_2, the equation is made exact for the functions $f(x) = 1$ and $f(x) = x$.

$$f(x) = 1: \quad a_1 + a_2 = \int_x^{x+h} (1)\, dx = [x]_x^{x+h} = h$$

$$f(x) = x: \quad a_1 x + a_2(x + h) = \int_x^{x+h} x\, dx = \left[\frac{x^2}{2}\right]_x^{x+h} = \frac{h}{2}(h + 2x)$$

Solving for a_1 and a_2 gives

$$a_1 = \frac{h}{2} \qquad a_2 = \frac{h}{2}$$

and the trapezoidal rule is

$$\int_x^{x+h} f(x)\, dx \approx \frac{h}{2}[f(x) + f(x + h)]$$

Derive an expression for dy/dx from two given values $f(x)$ and $f(x + h)$.

---

[1] From R. H. Perry, C. H. Chilton, and S. D. Kirkpatrick (eds.), "Chemical Engineers' Handbook," McGraw-Hill Book Company, New York, 1963.   Reprinted with permission of the copyright holders.
[2] R. W. Hamming, "Numerical Methods for Scientists and Engineers," chap. 10, McGraw-Hill Book Company, New York, 1962.

9.114 Using Hamming's method, find an expression for dy/dx from three given values, $f(x)$, $f(x + h)$, and $f(x + 2h)$.

9.115 Derive Simpson's rule by using Hamming's method.

9.116 Using Hamming's method, find a linearly interpolated value of $f(x + \alpha h)$ from two known values $f(x)$ and $f(x + h)$.

9.117 Using Hamming's method, find an interpolated value of $f(x + \alpha h)$ from three known values $f(x)$, $f(x + h)$, and $f(x + 2h)$.

10 TAYLOR'S SERIES

10.1   INTRODUCTION

In Chap. 8 problems were considered in which data were given pertaining
to functional values prescribed at specified points x_1, x_2, . . . , x_{n+1}.
There is another commonly occurring problem in which data are specified
in terms of the derivatives of the function at a given point.   To illustrate,
consider the motion of a shell leaving a ship's gun, as shown in Fig. 10.1,
with a muzzle velocity of 2000 ft/sec.   Suppose that time is measured
from when the gun is fired and that it takes 1 sec for the shell to travel the
length of the gun barrel.   It is desired to calculate the vertical position
of the shell $y(t)$ as a function of time.   From the above description the
following data are known about the shell at the time $t = 1$ sec,

$$y(1) = 40 \text{ ft}$$

$$\frac{dy}{dt}(1) = \text{vertical component of shell velocity}$$
$$= 2000 \sin 30° = 1000 \text{ ft/sec}$$
$$\frac{d^2y}{dt^2}(1) = \text{acceleration of a freely falling body}$$
$$= -32.2 \text{ ft/sec}^2$$

Since three pieces of data are given, it is reasonable to assume the solution
in the form of a parabola.

$$y(t) = a_1 t^2 + a_2 t + a_3$$

The derivatives of $y(t)$ are, respectively,

$$\frac{dy}{dt} = 2a_1 t + a_2$$
$$\frac{d^2y}{dt^2} = 2a_1$$

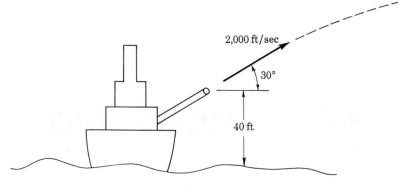

Fig. 10.1

and using the given data produces

$$40 = a_1 + a_2 + a_3$$
$$1000 = 2a_1 + a_2$$
$$-32.2 = 2a_1$$

Solving for a_1, a_2, and a_3 gives the solution as

$$y(t) = -16.1t^2 + 1032.2t - 976.1$$

The resulting simultaneous equations for a_1, a_2, and a_3 are solved in a straightforward manner by backward successive substitution. An even more convenient method of obtaining the coefficients is given by assuming the solution in the form

$$y(t) = a_1 + a_2(t - 1) + a_3(t - 1)^2$$

The derivatives are

$$\frac{dy}{dt}(t) = a_2 + 2a_3(t - 1)$$
$$\frac{d^2y}{dt^2}(t) = 2a_3$$

At time $t = 1$, the coefficients are determined by

$$40 = a_1$$
$$1000 = a_2$$
$$-32.2 = 2a_3$$

Thus, the solution is given by

$$y(t) = 40 + 1000(t - 1) - 16.1(t - 1)^2$$

It is seen that assuming $y(t)$ in this form removes the simultaneity of the equations for the coefficients.

The motion of the shell at other times can now be determined from the above relations. For example, at time $t = 4$ sec

Height of shell

$$y(4) = 40 + 1000(4 - 1) - 16.1(4 - 1)^2$$
$$= 2895.1 \text{ ft}$$

Vertical component of velocity of shell

$$\frac{dy}{dt}(4) = 1000 + 2(-16.1)(4 - 1)$$
$$= 903.4 \text{ ft/sec}$$

Vertical component of acceleration of shell

$$\frac{d^2y}{dt^2}(4) = -32.2 \text{ ft/sec}^2$$

10.2   TAYLOR'S SERIES

Consider now the general case where the values of a function and its first n derivatives are given at some point x_0 and it is desired to construct an nth-order polynomial with this information. Assume that the polynomial $T(x)$ has the form

$$T(x) = a_1 + a_2(x - x_0) + a_3(x - x_0)^2 + \cdots + a_{n+1}(x - x_0)^n \quad (10.1)$$

Differentiating n times gives

$$T'(x) = a_2 + 2a_3(x - x_0) + 3a_4(x - x_0)^2 + \cdots + na_{n+1}(x - x_0)^{n-1}$$
$$T''(x) = 2a_3 + 6a_4(x - x_0) + 12a_5(x - x_0)^2$$
$$+ \cdots + n(n - 1)a_{n+1}(x - x_0)^{n-2}$$
$$\cdots \cdots \cdots \cdots \cdots \cdots \cdots \cdots \cdots \cdots \cdots \cdots \cdots \cdots$$
$$T^n(x) = n(n - 1)(n - 2)(\cdots)(2)(1)a_{n+1}$$

Substituting the known values $f(x_0)$, $f'(x_0)$, $f''(x_0)$, $\ldots$, $f^n(x_0)$ into the above relations produces

$$f(x_0) = a_1$$
$$f'(x_0) = a_2$$
$$f''(x_0) = 2a_3$$
$$\cdots \cdots \cdots$$
$$f^n(x_0) = n!\, a_{n+1}$$

It is easily deduced that any coefficient a_i is simply $f^{i-1}(x_0)/(i - 1)!$.

Thus, the polynomial is expressed as

$$T(x) = f(x_0) + \frac{f'(x_0)}{1!}(x - x_0) + \frac{f''(x_0)}{2!}(x - x_0)^2$$

$$+ \cdots + \frac{f^n(x_0)}{n!}(x - x_0)^n \quad (10.2)$$

The above polynomial is commonly referred to as *Taylor's series*. For the special case where $x_0 = 0$,

$$T(x) = f(0) + \frac{f'(0)}{1!}x + \frac{f''(0)}{2!}x^2 + \cdots + \frac{f^n(0)}{n!}x^n \quad (10.3)$$

This polynomial is known as *Maclaurin's series*. Taylor's series can be put in a similar form by the substitution

$$x = x_0 + h$$

This produces

$$T(h) = f(x_0) + \frac{f'(x_0)}{1!}h + \frac{f''(x_0)}{2!}h^2 + \cdots + \frac{f^n(x_0)}{n!}h^n \quad (10.4)$$

As an example, the five-term Maclaurin's series and the five-term Taylor's series about the point $x = 1$ for the function $f(x) = \sqrt{1 + x^3}$ are now constructed. Table 10.1 gives the required coefficients. The

Table 10.1

i	$f^i(x)$	$f^i(0)$	$f^i(1)$
0	$\sqrt{1 + x^3}$	1	1.414
1	$\dfrac{3x^2}{2(1 + x^3)^{\frac{1}{2}}}$	0	1.061
2	$\dfrac{3x(x^3 + 4)}{4(1 + x^3)^{\frac{3}{2}}}$	0	1.327
3	$\dfrac{3(-x^6 - 20x^3 + 8)}{8(1 + x^3)^{\frac{5}{2}}}$	3	-0.8619
4	$\dfrac{9x^2(x^6 + 56x^3 - 80)}{16(1 + x^3)^{\frac{7}{2}}}$	0	-1.144

Maclaurin's series is then

$$T(x) = 1 + \frac{3}{3!}x^3 = 1 + \tfrac{1}{2}x^3$$

and the Taylor's series about $x = 1$ is

$$T(x) = 1.414 + \frac{1.061}{1!}(x-1) + \frac{1.327}{2!}(x-1)^2$$
$$- \frac{0.8619}{3!}(x-1)^3 - \frac{1.144}{4!}(x-1)^4$$
$$= 1.414 + 1.061(x-1) + 0.6635(x-1)^2$$
$$- 0.1436(x-1)^3 - 0.04767(x-1)^4$$

Table 10.2 compares actual values with values obtained from the above series.

Table 10.2

x	$\sqrt{1+x^3}$	$Maclaurin's$	$Taylor's$
0	1	1	1.112
$\frac{1}{2}$	1.061	1.062	1.064
1	1.414	1.5	1.414
2	3	5	2.947

Figure 10.2 represents a flow chart for evaluating, at the point x, an nth-order Taylor's series expanded about the point x_0. The functional and derivative values $f(x_0)$, $f'(x_0)$, $f''(x_0)$, . . . , $f^n(x_0)$ are represented by the variables f_0, f_1, f_2, . . . , f_n, respectively. The variable T represents the Taylor's-series evaluation.

10.3 REPRESENTATION OF FUNCTIONS, DIFFERENTIATION, AND INTEGRATION

In the previous section, a method of constructing a polynomial from information on the values of a function $f(x)$ and its derivatives at a specified point x_0 is given. This polynomial could then be utilized as an approximating polynomial. It could also be differentiated to represent the derivatives of $f(x)$ or integrated to obtain definite integrals of $f(x)$. A discussion of the errors involved in performing these operations is deferred to Sec. 10.4. It may be noted that the uses of Taylor's series are much the same as those of interpolating polynomials. Whether to use an

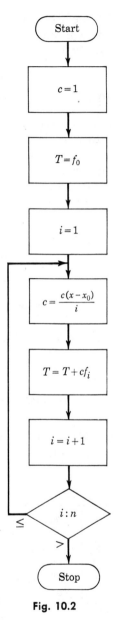

Fig. 10.2

interpolating polynomial or Taylor's series depends upon the particular problem under consideration. This point is taken up in more detail in Sec. 10.5.

10.3.1   Function Representation

In the previous section, two series were formed from the function $\sqrt{1 + x^3}$. Either of these could be used as an approximating polynomial for the given function. Table 10.2 illustrates the expected result, that the further x is from x_0, the larger the error.

As another example, a parabola for sin x in the neighborhood of $x = 25°$ is constructed. The necessary data are

$$f(x_0) = \sin 25° = 0.42262$$

$$f'(x_0) = \frac{\pi}{180} \cos 25° = 0.015818$$

$$f''(x_0) = -\left(\frac{\pi}{180}\right)^2 \sin 25° = -0.00012864$$

The Taylor's series about $x = 25$ is then

$$T(x) = 0.42262 + \frac{0.015818}{1!} (x - 25) - \frac{0.00012864}{2!} (x - 25)^2$$
$$= 0.42262 + 0.015818(x - 25) - 0.00006432(x - 25)^2$$

and Table 10.3 shows evaluations of this series and compares them with actual values of sin x.

Table 10.3

x	$T(x)$	sin x
20	0.34192	0.34202
22	0.37459	0.37461
25	0.42262	0.42262
30	0.50010	0.50000

It is interesting to compare the error at $x = 22°$ (0.00002) with the error (0.00004) obtained by an interpolating parabola going through the three points $x = 20, 25,$ and 30 given in Chap. 8.

10.3.2 Differentiation

Differentiation of Taylor's series is quite straightforward. With the series in the general form

$$T(x) = f(x_0) + \frac{f'(x_0)}{1!}(x - x_0) + \frac{f''(x_0)}{2!}(x - x_0)^2$$
$$+ \frac{f'''(x_0)}{3!}(x - x_0)^3 + \cdots + \frac{f^n(x_0)}{n!}(x - x_0)^n$$

the derivative is simply

$$T'(x) = f'(x_0) + \frac{f''(x_0)}{1!}(x - x_0) + \frac{f'''(x_0)}{2!}(x - x_0)^2$$
$$+ \cdots + \frac{f^n(x_0)}{(n-1)!}(x - x_0)^{n-1} \quad (10.5)$$

It is important to note that the polynomial for $T'(x)$ is itself Taylor's series. If we let $f'(x) = g(x)$ and $m = n - 1$, the derivative becomes

$$T'(x) = g(x_0) + \frac{g'(x_0)}{1!}(x - x_0) + \frac{g''(x_0)}{2!}(x - x_0)^2$$
$$+ \cdots + \frac{g^m(x_0)}{m!}(x - x_0)^m$$

As an example, consider the Taylor's series determined in Sec. 10.3.1 for $\sin x$ about the point $x = 25°$.

$$T(x) = 0.42262 + 0.015818(x - 25) - 0.00006432(x - 25)^2$$

Differentiating gives

$$T'(x) = 0.015818 - 0.00012864(x - 25)$$

Table 10.4 gives some results of this equation and compares them with

Table 10.4

x	$\dfrac{d}{dx}(\sin x)$	$T'(x)$	$p'(x)$
20	0.016401	0.016461	0.016442
22	0.016183	0.016204	0.016231
25	0.015818	0.015818	0.015798
30	0.015115	0.015175	0.015076

true values of the derivative of $\sin x$ and results obtained from the use of an interpolating parabola given in Chap. 9. The above expression for $T'(x)$ should be the Taylor's series for the function $(\pi/180) \cos x$. To demonstrate this, the two-term Taylor's series for this function is obtained as follows:

$$T(x) = \frac{\pi}{180}\left[\cos 25° + \frac{(\pi/180)(-\sin 25°)}{1!}(x - 25)\right]$$
$$= 0.015818 + 0.000012864(x - 25)$$

A similar result does not occur for an interpolating polynomial. This can be demonstrated with this example, by noting that the derivative of the interpolating polynomial for sin x is not an interpolating polynomial for $(\pi/180)$ cos x.

Higher derivatives are obtained by continued differentiation. The second derivative of the Taylor's series for sin x presented earlier is

$$T''(x) = 0.000012864$$

10.3.3   Integration

Taylor's series can be integrated term by term to give

$$\int T(x)\,dx = C + \frac{f(x_0)}{1!}(x - x_0) + \frac{f'(x_0)}{2!}(x - x_0)^2$$
$$+ \cdots + \frac{f^n(x_0)}{(n + 1)!}(x - x_0)^{n+1}$$

where C is an integration constant. Upon letting $F(x)$ be the anti-derivative of $f(x)$, $m = n + 1$, and noting that $C = F(x_0)$, the above equation becomes

$$\int T(x) = F(x_0) + \frac{F'(x_0)}{1!}(x - x_0) + \frac{F''(x_0)}{2!}(x - x_0)^2$$
$$+ \cdots + \frac{F^m(x_0)}{m!}(x - x_0)^m \quad (10.6)$$

Thus, the integration of the Taylor's series for a function $f(x)$ produces the Taylor's series for the antiderivative of $f(x)$.

In Chap. 9, numerical integration was utilized to determine the area shown in Fig. 9.6. The three-term Taylor's series corresponding to the function $f(x) = x^4$ about the point $x = 1$ is determined as follows:

$$f(1) = x^4\Big]_{x=1} = 1$$
$$f'(1) = 4x^3\Big]_{x=1} = 4$$
$$f''(1) = 12x^2\Big]_{x=1} = 12$$
$$T(x) = 1 + \frac{4}{1!}(x - 1) + \frac{12}{2!}(x - 1)^2$$
$$= 1 + 4(x - 1) + 6(x - 1)^2$$

If $f(x) = x^4$ is replaced by $T(x)$, an approximation to the area is obtained as follows:

$$\text{Area} \approx \int_0^2 [1 + 4(x - 1) + 6(x - 1)^2]\, dx$$
$$\approx [x + 2(x - 1)^2 + 2(x - 1)^3]_0^2$$
$$\approx 6$$

This result compares reasonably well with the actual area of 6.4 square units. Better accuracy is obtained by taking more terms in Taylor's series. Of course, if the fourth-order Taylor's series is constructed, it should reduce to the actual function $f(x) = x^4$. To verify this, consider the following:

$$f'''(1) = 24x \bigg]_{x=1} = 24$$
$$f^{iv}(1) = 24$$

$$T(x) = 1 + 4(x - 1) + 6(x - 1)^2 + \frac{24}{3!}(x - 1)^3 + \frac{24}{4!}(x - 1)^4$$
$$= 1 + 4(x - 1) + 6(x - 1)^2 + 4(x - 1)^3 + (x - 1)^4$$
$$= x^4 + (4 - 4)x^3 + (6 - 12 + 6)x^2 + (4 - 12 + 12 - 4)x$$
$$\qquad\qquad + (1 - 4 + 6 - 4 + 1)$$
$$= x^4$$

A more important utilization of Taylor's series is in the integration of a function for which the antiderivative is not known. To illustrate, suppose that it is desired to determine the area under the curve defined by $f(x) = \sqrt{1 + x^3}$ between $x = 0$ and $x = 1$. By using the Maclaurin's series corresponding to this function given in Sec. 10.2, the integration is

$$\int_0^1 \sqrt{1 + x^3}\, dx \approx \int_0^1 (1 + \tfrac{1}{2}x^3)\, dx$$
$$\approx \left[x + \frac{x^4}{8} \right]_0^1 = 1.125$$

If we use the Taylor's series about the point $x = 1$, given in Sec. 10.2, we have

$$\int_0^1 \sqrt{1 + x^3}\, dx \approx \int_0^1 [1.414 + 1.061(x - 1) + 0.6635(x - 1)^2$$
$$\qquad\qquad - 0.1436(x - 1)^3 - 0.04767(x - 1)^4]\, dx$$
$$\approx [1.414(x - 1) + 0.5305(x - 1)^2 + 0.2212(x - 1)^3$$
$$\qquad\qquad - 0.03590(x - 1)^4 - 0.009534(x - 1)^5]_0^1$$
$$\approx 1.131$$

Since Taylor's series and Maclaurin's series both give values that are higher than the true functional values, as shown in Table 10.2, the above results are probably a little high.

The trapezoidal and Simpson's rules can both be derived from the Taylor's series. To derive Simpson's rule, let $f(x)$ be represented by the Taylor's series

$$T(x) = a_1 + a_2(x - x_0) + a_3(x - x_0)^2$$

between the points $x = x_0 - h$ and $x = x_0 + h$. Integrating $T(x)$ produces

$$\int_{x_0-h}^{x_0+h} T(x)\, dx = \left[a_1(x - x_0) + \frac{a_2}{2}(x - x_0)^2 + \frac{a_3}{3}(x - x_0)^3 \right]_{x_0-h}^{x_0+h}$$

$$= a_1(h + h) + \frac{a_2}{2}(h^2 - h^2) + \frac{a_3}{3}(h^3 + h^3)$$

$$= \frac{h}{3}(6a_1 + 2a_3h^2)$$

Now, Simpson's rule is stated as

$$\text{Area} \approx \frac{h}{3}[f(x_0 - h) + 4f(x_0) + f(x_0 + h)]$$

$$\approx \frac{h}{3}[(a_1 - a_2h + a_3h^2) + 4a_1 + (a_1 + a_2h + a_3h^2)]$$

$$\approx \frac{h}{3}(6a_1 + 2a_3h^2)$$

It must be pointed out that the above analysis assumes that values of $f(x_0 - h)$ and $f(x_0 + h)$ are equal to the values obtained from the three-term Taylor's series at the points $x_0 - h$ and $x_0 + h$, respectively. At best this is only an approximation. In general, the area determined by a direct integration of the three-term Taylor's series will not exactly equal the area obtained by Simpson's rule.

10.4   ERROR ANALYSIS

The error $E(x)$ in Taylor's series can be defined in the usual way,

$$f(x) = T(x) + E(x) \tag{10.7}$$

and can be determined by considering the function

$$F(x) = f(x) - T(x) - \frac{f(\bar{x}) - T(\bar{x})}{(\bar{x} - x_0)^{n+1}}(x - x_0)^{n+1}$$

This function is illustrated in Fig. 10.3a. Notice that $F(x_0) = F(\bar{x}) = 0$. The error at the point $\bar{x}$ is determined as follows: Differentiating $F(x)$ gives

$$F'(x) = f'(x) - T'(x) - \frac{f(\bar{x}) - T(\bar{x})}{(\bar{x} - x_0)^{n+1}}(n + 1)(x - x_0)^n$$

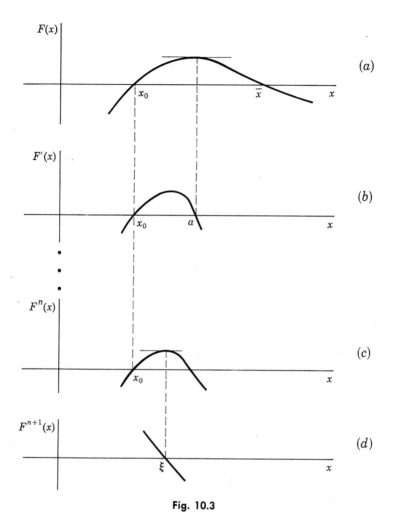

Fig. 10.3

This function possesses zeros at x_0 and some point a, as shown in Fig. 10.3b. If $F(x)$ is differentiated n times, each derivative possesses this property of having two zeros, one at x_0 and the other at some point between x_0 and $\bar{x}$, as shown in Fig. 10.3c.   Then, taking the $(n + 1)$st derivative gives

$$F^{n+1}(x) = f^{n+1}(x) - \frac{f(\bar{x}) - T(\bar{x})}{(\bar{x} - x_0)^{n+1}} (n + 1)!$$

which possesses a zero at some point ξ, as shown in Fig. 10.3d.   Thus,

$$F^{n+1}(\xi) = 0 = f^{n+1}(\xi) - \frac{f(\bar{x}) - T(\bar{x})}{(\bar{x} - x_0)^{n+1}} (n + 1)!$$

which can be solved for the error at $\bar{x}$,

$$E(\bar{x}) = f(\bar{x}) - T(\bar{x}) = \frac{f^{n+1}(\xi)}{(n+1)!} (\bar{x} - x_0)^{n+1} \qquad (10.8)$$

Letting $\bar{x} - x_0 = h$ gives

$$E(h) = \frac{f^{n+1}(\xi)}{(n+1)!} h^{n+1}$$

which is an error of order h^{n+1}.

Again, upon considering the equation

$$f(x) = T(x) + E(x)$$

substitution of the expressions for $T(x)$ and $E(x)$, Eqs. (10.2) and (10.8), gives

$$f(x) = f(x_0) + \frac{f'(x_0)}{1!} (x - x_0) + \frac{f''(x_0)}{2!} (x - x_0)^2$$
$$+ \cdots + \frac{f^n(x_0)}{n!} (x - x_0)^n + \frac{f^{n+1}(\xi)}{(n+1)!} (x - x_0)^{n+1} \quad (10.9)$$

The result is generally known as *Taylor's theorem*.

As an illustration, consider the three-term Taylor's series representing $\sin x$ about the point $x = 25°$ given in Sec. 10.3.1. The error is

$$\text{Error} = \frac{f'''(\xi)}{(3)!} h^3$$

If x is taken between 20 and 30°, the maximum h is 5 and the maximum $f'''(\xi)$ is $(\pi/180)^3 \cos 20°$. With these values, the maximum error is 0.000104, which compares with the actual maximum error of 0.00010, shown in Table 10.3.

It should be pointed out that it is tacitly assumed in this discussion of error that $f(x)$ and its first $n + 1$ derivatives do exist.

10.4.1   Differentiation and Integration

An expression for the error involved in differentiating or integrating Taylor's series is easily obtained, since the resulting series are also Taylor's series. Thus, the error in differentiating the nth-order Taylor's series is

$$\frac{f^n(\xi)}{n!} h^n$$

while in integrating it is

$$\frac{f^{n+2}(\xi)}{(n+2)!} h^{n+2}$$

10.4.2   Infinite Series

One can think of many functions which can be differentiated an infinite number of times. The functions used as examples, $\sin x$ and $\sqrt{1 + x^3}$, both possess this property. With this in mind, it is then possible to conceive of a Taylor's series as consisting of an infinite number of terms.

$$T(x) = f(x_0) + \frac{f'(x_0)}{1!} (x - x_0) + \frac{f''(x_0)}{2!} (x - x_0)^2 + \cdots$$

Such a series is said to *converge* if the sum of the infinite number of terms is a finite number. This sum may or may not equal $f(x)$. If it does, the series is said to *converge to* $f(x)$. In this case the equation can be written as

$$f(x) = f(x_0) + \frac{f'(x_0)}{1!} (x - x_0) + \frac{f''(x_0)}{2!} (x - x_0)^2 + \cdots$$

If the error term $E(x)$ approaches zero as the order n of the Taylor's series approaches infinity, the infinite Taylor's series will converge to $f(x)$. If $f^m(x)$ is bounded, $|f^m(x)| \leq M$, for all values of m, the error has the form

$$|\text{Error}| = \left| \frac{f^{n+1}(\xi)}{(n + 1)!} h^{n+1} \right| \leq M \frac{h^{n+1}}{(n + 1)!} \tag{10.10}$$

and since $(n + 1)!$ approaches infinity faster than h^{n+1}, as n approaches infinity, the error approaches zero. Thus, if a function $f(x)$ and all its derivatives are bounded in a given interval along the x axis, the infinite Taylor's series about a point x_0 in the interval will converge at any other point x in the interval. In the example $f(x) = \sin x$, all the derivatives of $\sin x$ are bounded for all values of x, and so any Taylor's series for $\sin x$ will converge. On the other hand, the Taylor's series for $\sqrt{1 + x^3}$ will not converge in any interval containing the point $x = -1$, since at this point all the derivatives of $\sqrt{1 + x^3}$ equal infinity. Notice that, for $x < -1$, the function becomes imaginary.

In practice, it is generally quite difficult to obtain the actual sum of an infinite series. Instead, a finite number of terms are summed, and the error involved is computed or estimated. As an illustration, suppose that it is desired to form the Taylor's series for $\sin x$ with x between 20 and 30°, which is accurate to five decimal places. In Sec. 10.4 it is established that the maximum error in a three-term series is no greater than 0.000104. The error in a four-term series is bounded by

$$|\text{Error}| \leq \left| \frac{[f^{iv}(\xi)]_{max} h^4}{4!} \right|$$

$$\leq \frac{(\pi/180)^4 \sin 30° \, 5^4}{24}$$

$$\leq 0.0000012$$

A four-term series is then sufficiently accurate.   The Taylor's series is

$$\sin x \approx 0.42262 + 0.015818(x - 25) - 0.00006432(x - 25)^2$$
$$- 0.00000083268(x - 25)^3$$

10.5   COMPARISON WITH INTERPOLATING POLYNOMIALS

Taylor's series and polynomial interpolation represent two methods of approximating a function.   In general, if a function $f(x)$ is represented by an nth-order polynomial using both these methods, the two polynomials are not the same.   This fact is easily ascertained by examining the different ways in which the two methods are formulated and is demonstrated in Table 10.5, which compares the interpolating parabola for $\sin x$ given in Chap. 8 with the three-term Taylor's series of $\sin x$ given in Sec. 10.3.1. A graph exaggerating the errors in this problem is shown in Fig. 10.4. As is expected, the Taylor's series has more accuracy in the neighborhood of $x = 25$, while the interpolating parabola has more accuracy near the end points $x = 20$ and $x = 30$.

10.5.1   Representation of Polynomials

The nth-order interpolating polynomial and the n-term Taylor's-series representation of an nth-order polynomial are both exact.   To illustrate

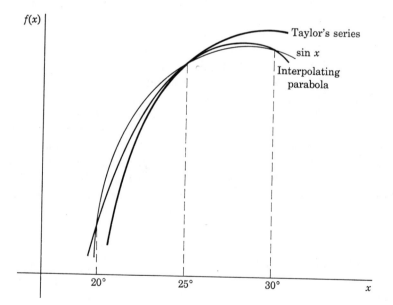

Fig. 10.4

Table 10.5

x	$\sin x$	*Interpolating parabola*	*Taylor's series, 3-term*
20	0.34202	0.34202	0.34192
21	0.35837	0.35840	0.35832
22	0.37461	0.37465	0.37459
23	0.39073	0.39077	0.39073
24	0.40674	0.40676	0.40674
25	0.42262	0.42262	0.42262
26	0.43837	0.43835	0.43838
27	0.45399	0.45396	0.45400
28	0.46947	0.46943	0.46949
29	0.48481	0.49478	0.48486
30	0.50000	0.50000	0.50010

this, consider the function $f(x) = x^2$. A short difference table for this function is given in Table 10.6.

Table 10.6

x	$f(x)$	D	D^2
0	0		
		1	
1	1		2
		3	
2	4		

Expanding in central differences about $x = 1$ gives

$$p(u) = 1 + 2u + u^2$$

where

$$u = \frac{x-1}{1} = x - 1$$

The interpolating polynomial in terms of x is thus

$$p(x) = 1 + 2(x - 1) + (x - 1)^2$$

The Taylor's series expanded about $x = 1$ is given by

$$T(x) = 1 + 2(x - 1) + (x - 1)^2$$

which agrees exactly with the interpolation parabola. Both of these are easily simplified to x^2.

10.5.2   Taylor's Series vs. Interpolation

The question naturally arises as to which method of approximating a function $f(x)$ is better, Taylor's series or polynomial interpolation. Quite obviously, if $f(x)$ is given only in tabular form, an interpolating polynomial should be used. On the other hand, if values of the function and its derivatives are prescribed at a certain point, it is better to apply Taylor's series. Often, however, enough information exists about a function so that either Taylor's series or an interpolating polynomial can be readily formed. In such instances, it cannot be said that one method is always better than the other. Taylor's series has the advantage of converging for most functions encountered in engineering practice. An interpolating polynomial can also be made to converge by utilizing more functional values for interpolation and decreasing the spacing between the points. Suppose that it is desired to construct a polynomial to approximate a function $f(x)$ in some interval $a \leq x \leq b$. The Taylor's series

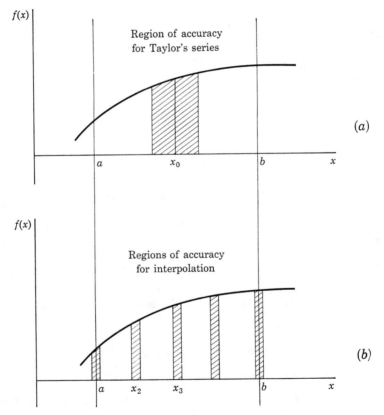

Fig. 10.5

about a point x_0 near the center of the interval produces a polynomial which is very accurate in the center of the interval and less accurate near the ends, as illustrated in Fig. 10.5a. The region of accuracy of an interpolating polynomial is illustrated in Fig. 10.5b. Any desired accuracy throughout the entire interval can be obtained with either type of series by taking enough terms, on the assumption that the successive derivatives of $f(x)$ do not approach infinity within the interval, causing the error term to become large.

It is possible to form an approximate Taylor's series from a tabulated function by replacing the derivatives appearing in the Taylor's series by their finite-difference approximations. For example, using forward differences with equal-interval spacing produces

$$T(x) \approx f(x_1) + \frac{\Delta f(x_1)}{1! \, h} (x - x_1) + \frac{\Delta^2 f(x_1)}{2! \, h^2} (x - x_1)^2 + \cdots$$

To illustrate, consider the function represented in Table 10.7. An approximate Taylor's series about $x = 25$, by use of a Gauss forward path, is

$$T(x) \approx 0.42262 + \frac{0.07738}{5} (x - 25) - \frac{0.00322}{(2)(5^2)} (x - 25)^2$$
$$\approx 0.42262 + 0.015476(x - 25) - 0.0000644(x - 25)^2$$

which compares approximately with the series given in Sec. 10.3.1.

Table 10.7

x	$f(x)$	D	D^2
20	0.34202		
		0.08060	
25	0.42262		-0.00322
		0.07738	
30	0.50000		

Similarly, if only a function and its derivatives are known at a given point, a difference table can be formed, from which an interpolating polynomial can be determined. Using the data given in Sec. 10.3.1, and again using a Gauss forward path, produces

x	$f(x)$	D	D^2
25	0.42262		$-0.00012864(25)$
		0.015818(5)	

Completing this table gives

20	0.34041		
		0.08221	
25	0.42262		−0.003216
		0.07909	
30	0.50171		

An approximation to the Newton-Gregory formula is then determined as follows:

$$p(x) \approx 0.34041 + \frac{0.08221}{5}(x - 20) - \frac{0.003216}{(2)(5^2)}(x - 20)(x - 25)$$
$$\approx 0.34041 + 0.016442(x - 20) - 0.00006432(x - 20)(x - 25)$$

This result agrees favorably with the interpolating polynomial given in Chap. 8.

10.6   APPLICATIONS

As with interpolating polynomials (Chap. 8), Taylor's series are used to represent functions, their derivatives, and their integrals. In the first application to follow, a simple approximate solution of a transcendental equation is obtained by replacing a trigonometric function by its Taylor's-series representation. The equation arises from the physical situation of a heavy lever twisting a circular bar. In the second application, a complicated integrand is replaced by Taylor's series in order to obtain the length of a suspension-bridge cable.

Another important use of both interpolating polynomials and Taylor's series occurs in the numerical solution of differential equations. As a third application, a numerical solution of the differential equation arising from the physical problem of an oscillating pendulum is obtained by using Taylor's series. This serves as an introduction to Chap. 11, which treats differential equations in greater detail.

10.6.1   Replacement of Trigonometric Functions by Taylor's Series

In Chap. 6 the equation

$$\cos \theta = K_1\theta$$

is developed for a heavy lever attached to a circular bar. In Chap. 8 an approximate solution to the above equation is obtained by replacing $\cos \theta$

by an interpolating parabola. An alternative procedure consists in replacing $\cos \theta$ by Taylor's series. For example, the first two terms of the Maclaurin's series for $\cos \theta$ are

$$\cos \theta \approx 1 - \frac{\theta^2}{2}$$

where θ is expressed in radians. Substituting this series into the above equation gives

$$1 - \frac{\theta^2}{2} = K_1 \theta$$

$$\theta^2 + 2K_1\theta - 2 = 0$$

$$\theta = \frac{-2K_1 + \sqrt{4K_1^2 + 8}}{2}$$

$$= -K_1 + \sqrt{K_1^2 + 2} \tag{10.11}$$

A more accurate solution can be obtained if the angle θ is approximately known beforehand. Then, Taylor's series expanded about this approximate value θ_0 can be utilized.

$$\cos \theta \approx a_0 + a_1(\theta - \theta_0) + a_2(\theta - \theta_0)^2$$

This produces

$$a_0 + a_1(\theta - \theta_0) + a_2(\theta - \theta_0)^2 = K_1 \theta$$

$$a_2\theta^2 + (a_1 - 2a_2\theta_0 - K_1)\theta + (a_0 - a_1\theta_0 + a_2\theta_0^2) = 0$$

The solution is then

$$\theta = \frac{-(a_1 - 2a_2\theta_0 - K_1) \pm \sqrt{(a_1 - 2a_2\theta_0 - K_1)^2 - 4a_2(a_0 - a_1\theta_0 + a_2\theta_0^2)}}{2a_2} \tag{10.12}$$

10.6.2   Analysis of a Suspension-bridge Cable

The cable of a suspension bridge as illustrated in Fig. 10.6a supports the weight of the roadbed by means of vertical wires called *suspenders*. The distance H is called the *sag*. In Fig. 10.6b, let T be the tension in the cable at any point x, T_c the tension at the center of the cable, and w the weight of the roadbed per unit length. Equilibrium of the portion of the cable shown gives

$$T \sin \theta = wx$$
$$T \cos \theta = T_c \tag{10.13}$$

Solving simultaneously and integrating, we get

$$\tan \theta = \frac{dy}{dx} = \frac{wx}{T_c}$$

$$y = \frac{wx^2}{2T_c} + C$$

The constant of integration C is zero, since, at $x = 0$, $y = 0$. Further, using the boundary condition that, at $x = L/2$, $y = H$ gives

$$H = \frac{wL^2}{8T_c}$$

$$T_c = \frac{wL^2}{8H}$$

Substituting these values back into the equation for y produces

$$y = \frac{4H}{L^2} x^2 \tag{10.14}$$

Thus, a suspension-bridge cable takes the form of a parabola. An important consideration is the length S of the cable, which is obtained by

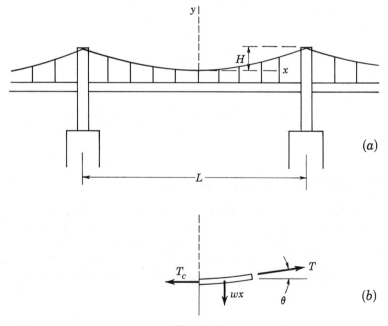

Fig. 10.6

the integration

$$S = \int_0^S ds$$

Figure 10.7 shows that the expression for ds is given by

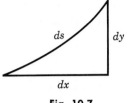

Fig. 10.7

$$ds = \sqrt{dx^2 + dy^2}$$

$$= dx \sqrt{1 + \left(\frac{dy}{dx}\right)^2}$$

Thus,

$$S = 2 \int_0^{L/2} \sqrt{1 + \left(\frac{dy}{dx}\right)^2}\, dx$$

$$= 2 \int_0^{L/2} \sqrt{1 + \left(\frac{8Hx}{L^2}\right)^2}\, dx$$

Although this function can be integrated directly, the result is fairly complicated and engineers have found it convenient to use the Taylor's-series representation of the integrand. Using three terms in the Maclaurin's series gives the following results:

$$\sqrt{1 + \left(\frac{8Hx}{L^2}\right)^2} \approx 1 + \frac{32H^2}{L^4} x^2 - \frac{512H^4}{L^8} x^4$$

$$S = 2 \int_0^{L/2} \left(1 + \frac{32H^2}{L^4} x^2 - \frac{512H^4}{L^8} x^4\right) dx$$

$$= L \left[1 + \frac{8}{3}\left(\frac{H}{L}\right)^2 - \frac{32}{5}\left(\frac{H}{L}\right)^4\right] \tag{10.15}$$

Experience with suspension-bridge cables has shown this to be a very accurate expression for the length.

10.6.3   Oscillations of a Pendulum

The forces acting on a simple pendulum are shown in Fig. 1.4. Newton's law of motion in the x direction gives

$$-W \sin \theta = ma_x$$

Since the mass moves along a circular path, the acceleration a_x is

$$a_x = L \frac{d^2\theta}{dt^2}$$

where t represents time. Mass m is related to weight W by

$$W = mg$$

where g is the acceleration of a freely falling body. Upon using these two relations, the equation of motion for the ball becomes

$$\frac{d^2\theta}{dt^2} + \frac{g}{L} \sin \theta = 0 \tag{10.16}$$

For small angles $\sin \theta \approx \theta$, and Eq. (10.16) reduces to the well-known equation for harmonic motion. If the pendulum is released from rest at an angle θ_0 at time $t = 0$, the solution of the equation is

$$\theta = \theta_0 \cos \sqrt{\frac{g}{L}} t \tag{10.17}$$

This is easily verified by substitution of Eq. (10.17) into Eq. (10.16), with $\sin \theta$ replaced by θ.

For large angles, the solution can be assumed in the form of Maclaurin's series

$$\theta(t) = \theta(0) + \frac{\theta'(0)}{1!} t + \frac{\theta''(0)}{2!} t^2 + \cdots$$

If the pendulum is started in the same manner as previously described,

$$\theta(0) = \theta_0$$
$$\theta'(0) = 0$$

Other coefficients are determined from Eq. (10.16) as follows:

$$\theta''(t) = -\frac{g}{L} \sin \theta$$

$$\theta''(0) = -\frac{g}{L} \sin \theta_0$$

$$\theta'''(t) = -\frac{g}{L} \cos \theta(\theta')$$

$$\theta'''(0) = 0$$

$$\theta^{iv}(t) = -\frac{g}{L} [- \sin \theta(\theta')^2 + \cos \theta(\theta'')]$$

$$\theta^{iv}(0) = -\frac{g}{L} \left[\cos \theta_0 \left(-\frac{g}{L} \sin \theta_0 \right) \right]$$

$$= \left(\frac{g}{L} \right)^2 \cos \theta_0 \sin \theta_0$$

$$\cdots \cdots \cdots \cdots \cdots \cdots \cdots$$

Thus, a three-term approximation to the solution of Eq. (10.16) is

$$\theta(t) = \theta_0 - \frac{g \sin \theta_0}{2L} t^2 + \frac{g^2 \sin \theta_0 \cos \theta_0}{24L^2} t^4 \tag{10.18}$$

As might be expected, this is accurate only for small values of t. Many more terms would be needed to represent θ for a long time, such as that for one complete swing of the pendulum.

The equation of motion for the pendulum is an example of a general type of mathematical equation known as a *differential equation*. While the Taylor's-series approach presented here is an example of a method of solving differential equations, better methods are available, some of which are presented in Chap. 11.

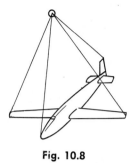

Fig. 10.8

The oscillation of a pendulum is one example of a problem taken from the general field of vibrations. The vibration analysis of buildings, bridges, ships, aircraft, rockets, and industrial machinery presents very important problems to the engineer. Figure 10.8 shows an airplane suspended as a pendulum. By setting the airplane into an oscillating motion, its moment of inertia can be determined by observing the period of its vibration. The moment of inertia of a body represents its resistance to rotational motion in much the same manner as its mass represents its resistance to linear motion.

PROBLEMS

10.1 The gun shown in Fig. 10.1 is replaced by a newer model in which the travel time of the shell in the barrel is 0.2 sec and the muzzle velocity is 5000 ft/sec. Determine the vertical position of the shell as a function of time, and calculate the height and vertical component of the velocity of the shell after 4 sec.

10.2 The gun shown in Fig. 10.1 is raised to an angle of 45° with the horizontal. In this position the end of the barrel is 45 ft above the surface of the water. Determine the vertical position of the shell as a function of time, and calculate the height and vertical component of the velocity of the shell after 4 sec.

♦10.3 If air resistance is neglected, the horizontal component of the acceleration of the shell shown in Fig. 10.1 is zero. Determine the horizontal position of the shell as a function of time, and calculate the horizontal position and the horizontal component of the velocity of the shell after 4 sec. What is the total speed of the shell after 4 sec?

10.4 A projectile is fired vertically upward from a point 10 ft above the ground and with a velocity of 100 ft/sec. Determine the maximum height the projectile will reach.

10.5 A second projectile is fired vertically upward from the same spot, but 2 sec after the projectile described in Prob. 10.4. At what time will the two projectiles collide?

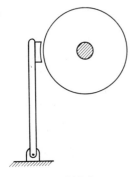

10.6   The flywheel shown rotates at a constant angular speed of 20 rad/sec for 5 sec.   The brake is then applied and causes a moment M of 80 lb-ft to act upon the flywheel.   Assuming that the relation between M and the angular deceleration α of the flywheel is

$$M = 40\alpha$$

determine the time at which the flywheel stops rotating.

Fig. P10.6

♦**10.7**   A pendulum is released from rest at $\theta = 0.8$ rad and at time $t = 0$ sec. The equation [Eq. (10.16)] governing the motion of the pendulum is

$$\frac{d^2\theta}{dt^2} + \frac{g}{L} \sin \theta = 0$$

Assuming a parabolic relationship, determine θ as a function of time if $L = 2$ ft. At what time will $\theta = 0$?   Why is this only an approximate solution of the problem?

  10.8   At the instant shown in Fig. 1.4, $t = 3$ sec, $\theta = 0.8$ rad, and $d\theta/dt = 2$ rad/sec.   Assuming a parabolic relationship, determine θ as a function of time if $L = 2$ ft.   At what time will $\theta = 0$?   Why is this only an approximate solution of the problem?

♦**10.9**   The mass m is released from rest at $x = 2$ ft and time $t = 0$ sec.   Assuming a parabolic relationship, determine x as a function of time if the spring con-

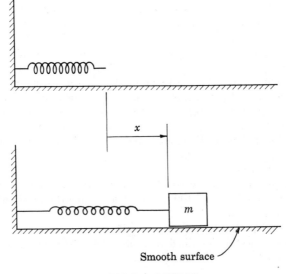

Smooth surface

Fig. P10.9 and P10.10

stant k is 5.0 lb/ft and $m = 0.25$ slug. At what time will $x = 0$? Why is this only an approximate solution of the problem?

10.10 At the instant shown, $t = 10$ sec, $x = 2$ ft, and the mass m (0.25 slug) is traveling to the right at the speed of 4 ft/sec. The spring constant k is 10 lb/ft. Assuming a parabolic relationship, determine x as a function of time. At what time will $x = 0$? Why is this only an approximate solution of the problem?

In Probs. 10.11 to 10.15, determine the Maclaurin's series for the given function.

◆**10.11** $f(x) = \cos x$
10.12 $f(x) = \tan x$
10.13 $f(x) = e^x$
10.14 $f(x) = e^{-x^2}$
10.15 $f(x) = (x - 1)/(x + 1)$

In Probs. 10.16 to 10.21, determine the three-term Taylor's series for $f(x)$ about the point x_0. Evaluate the series at x_1 and x_2, and compare with the true values.

◆**10.16** $f(x) = \cos x \colon x_0 = 25°$, $x_1 = 22°$, $x_2 = 45°$
10.17 $f(x) = e^x \colon x_0 = 2$, $x_1 = 1$, $x_2 = 3$
10.18 $f(x) = \ln x \colon x_0 = 2$, $x_1 = 1$, $x_2 = 3$
10.19 $f(x) = e^{-x^2} \colon x_0 = 2$, $x_1 = 1$, $x_2 = 3$
10.20 $f(x) = (\sin x)/x \colon x_0 = 3\pi/4$, $x_1 = \pi/2$, $x_2 = \pi$
10.21 $f(x) = 1/(\ln x) \colon x_0 = 3$, $x_1 = 1$, $x_2 = 5$

In Probs. 10.22 to 10.24, determine an approximation for the derivative $f'(x)$ of the given function by differentiating the corresponding Taylor's series found in Probs. 10.16 to 10.18. Evaluate $f'(x)$ at the point x_3. Compare with the true values and the results obtained from interpolating polynomials in Probs. 9.16, 9.18, and 9.19, respectively.

◆**10.22** $f(x) = \cos x \colon x_3 = 20°$
10.23 $f(x) = e^x \colon x_3 = 1$
10.24 $f(x) = \ln x \colon x_3 = 1$

◆**10.25** Differentiate the series for $f'(x)$ found in Prob. 10.22. How does the result compare with the second derivative of $\cos x$ for $20° \le x \le 30°$?

In Probs. 10.26 to 10.31, determine the area under the given curve $f(x)$ between x_1 and x_2 by integrating the corresponding series found in Probs. 10.16 to 10.21. Compare with true values (if possible) and with results obtained by numerical integration in Probs. 9.66, 9.67, and 9.70 to 9.73, respectively.

◆**10.26** $f(x) = \cos x \colon x_1 = 0$, $x_2 = 180°$
10.27 $f(x) = e^x \colon x_1 = 0$, $x_2 = 2$
10.28 $f(x) = e^{-x^2} \colon x_1 = 1$, $x_2 = 5$
10.29 $f(x) = e^x/x \colon x_1 = 1$, $x_2 = 5$
10.30 $f(x) = (\sin x)/x \colon x_1 = \pi/2$, $x_2 = \pi$
10.31 $f(x) = 1/(\ln x) \colon x_1 = 2$, $x_2 = 5$

10.32 Derive the third-order integration rule given in Table 9.12 by using Taylor's series.

10.33   Write a computer program for evaluating a Taylor's series, using the flow chart given in Fig. 10.2.

10.34   Construct a flow chart, and write a computer program which evaluates the derivative of a given Taylor's series at an arbitrary point x.

10.35   Construct a flow chart, and write a computer program which evaluates the integral of a given Taylor's series for the limits x_1 to x_2.

◆**10.36**   Determine the upper bound on the error in Prob. 10.16, assuming that x lies in the interval $20° \leq x \leq 30°$.   Compare with the result of Prob. 8.79.

10.37–10.51   Determine the upper bounds on the error in Probs. 10.17 to 10.31, respectively.

In Probs. 10.52 to 10.54, evaluate the Maclaurin's series of the given function $f(x)$ at the point x_0 found in Probs. 10.11 to 10.13, respectively, by summing the terms in the series until no significant change occurs in the third decimal place.

◆**10.52** $f(x) = \cos x : x_0 = 20°$

10.53 $f(x) = \tan x : x_0 = 20°$

10.54 $f(x) = e^x : x_0 = 2$

◆**10.55**   How many terms of a Taylor's-series expansion about $x = 25°$ are needed to represent $\cos x$ between 20 and 30° with an accuracy of five decimal places?   Compare with the result of Prob. 8.80.

10.56   Determine an interval of convergence around $x = 25°$, so that the three-term Taylor's series will accurately represent $\sin x$ to four decimal places. Compare with the result of Prob. 8.83.

◆**10.57**   Determine an interval of convergence around $x = 25°$, so that the three-term Taylor's series will accurately represent $\cos x$ to four decimal places.

10.58   Construct a flow chart, and write a computer program for evaluating Taylor's series.   The program should stop when the last term calculated is small compared with the sum of the previous terms.

10.59   Construct Taylor's series from the data given in Table 8.1.   Evaluate the series to determine the speed of the rocket when the time $t = 150$ sec, and compare with the results obtained by interpolation in Chap. 8.

◆**10.60**   Construct Taylor's series from the data given in Prob. 8.11.   Evaluate the series to obtain the stress at a strain of 0.3, and compare with the result obtained in Prob. 8.11.

10.61   Construct Taylor's series for the data given in Prob. 8.23.   Evaluate the series to obtain the pressures at $x = 2, 7$, and 12 in., and compare with the results of Prob. 8.23.

◆**10.62**   Referring to Fig. 6.25, $W = 5000$ lb, and $r = 12$ in.   A torque of 875 lb-in. is required to rotate the bar through 1°.   Determine an approximate solution by using Eq. (10.11), and compare with the actual solution found in Prob. 6.43.

10.63   Solve Prob. 10.62 by constructing the Taylor's series for $\cos \theta$ about $x = 30°$.

◆**10.64**   The cables of the Mackinac Straits bridge have a sag H of 350 ft.   The length L of the main span is 3800 ft.   Determine the length S of each cable.

10.65   Solve Prob. 10.64, using a four-term Taylor's series, and compare the results.

◆**10.66** Determine a four-term Maclaurin's-series solution of the pendulum problem, and calculate θ at $t = 0.25$ sec, using the following data: $\theta_0 = 0.8$ rad, $g = 32.2$ ft/sec^2, and $L = 2$ ft.

10.67 Determine a four-term Maclaurin's-series solution of the spring-mass system given in Prob. 10.9.   Calculate x at $t = 0.2$ sec, and compare with the true answer obtained from the relation

$$x = 2 \cos \sqrt{\frac{k}{m}}\, t$$

10.68 Derive Taylor's series, using Hamming's method.   (See Prob. 9.113.)

NUMERICAL SOLUTION OF ORDINARY DIFFERENTIAL EQUATIONS

11.1   INTRODUCTION

The mathematical description of many physical processes encountered in engineering problems is often conveniently made in terms of an equation containing differential quantities, i.e., a differential equation.   More often than not, however, the equation can be integrated analytically only if simplifying assumptions are made in its formulation from the physical process.   If that is done, the result is a mathematically exact answer to an approximation to the real problem.   An alternative is to perform a numerical integration of the more exact differential equation.   The result is now only an approximation to the solution of that equation, but the error in the integration can usually be kept within prescribed limits. Thus, numerical integration represents an important and practical technique in the solution of differential equations.

Only ordinary differential equations will be considered here, i.e., functions of only one independent variable.   The general form of such an equation of order n is

$$F(x,y,y', \ldots , y^{(n)}) = 0 \tag{11.1}$$

where x is the independent variable, y is an unknown function of x, $y(x)$, and the derivatives of y with respect to x are $y', y'', \ldots , y^{(n)}$.   It is, of course, the purpose of integration to find the function $y(x)$ that satisfies a given differential equation of the form of Eq. (11.1).

11.2   FORMULATION OF DIFFERENTIAL EQUATIONS

Facility in formulating or setting up a differential equation and recognizing its type requires a knowledge of the physical or chemical process being described, a certain amount of mathematical skill, and a considerable degree of experience.   A few examples will be given here to illustrate some of the basic principles that are involved.

Probably one of the problems most frequently giving rise to a differential equation is that in which a basic physical law is applied to a process occurring under variable conditions.   For example, with an incompressible fluid of density ρ, the increase in pressure Δp due to an increase in depth Δz is given by

$$\Delta p = \rho\,\Delta z \tag{11.2}$$

but if the density is not constant, the law holds true only for a differential depth.   As shown in Fig. 11.1, the increase in pressure due to an increase in depth dz is given by the differential equation

$$dp = \rho\,dz \tag{11.3}$$

Equation (11.3) is a first-order, ordinary differential equation, apparently very simple.   As written, however, it contains three variables, and one of them must be eliminated in terms of the other two before the equation can be integrated.   Since we are interested in finding p as a function of z, we must eliminate ρ.   This can be done if we have a relationship between the density and the pressure.   The type of the resulting differential equation and how it is integrated will depend to a large extent on the density function.   Several possibilities will be considered.

The problem can be made more specific if we apply the equation to the calculation of the atmospheric pressure at a height z above sea level.

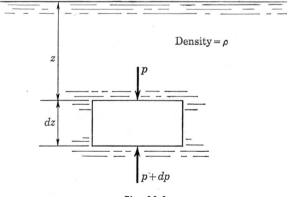

Fig. 11.1

Because of the change in direction, the equation becomes

$$dp = -\rho \, dz \qquad (11.4)$$

On the assumption that the atmosphere behaves like an ideal gas, the expression for density is given by the combined law of Charles and Boyle as

$$\rho = \frac{pM}{RT} \qquad (11.5)$$

where

p = pressure
M = molecular weight of air
R = a constant
T = absolute temperature

Substituting Eq. (11.5) into Eq. (11.4) gives

$$dp = -\frac{pM}{RT} \, dz \qquad (11.6)$$

Equation (11.6) still contains three variables and therefore cannot be integrated until a relationship between T and p or z is known. As the first possibility, let us assume that T is constant. The variables may now be separated to give

$$\frac{dp}{p} = -\frac{M}{RT} \, dz$$

and integration yields

$$\ln p = -\frac{M}{RT} z + C$$

where C is the constant of integration. To evaluate C, we may use the *initial condition* that p equals the sea-level pressure p_0 at $z = 0$ to get

$$\ln p_0 = C$$

Substituting this value for C, rearranging, and taking the antilogarithm of both sides, we obtain

$$p = p_0 e^{-(M/RT)z}$$

Instead of using indefinite integrals and evaluating a constant, we may, of course, use definite integrals. Thus,

$$\int_{p_0}^{p} \frac{dp}{p} = -\frac{M}{RT} \int_{0}^{z} dz$$

yields the same final result.

If T is not assumed to be constant, from the nature of the physical problem it must depend on z. Several possible functions of z could be hypothesized; one of the simplest would be the negative exponential form

$$T = T_0 e^{-az}$$

where T_0 is the temperature at sea level and a is a constant. Substituting this relation into Eq. (11.6) and integrating produce

$$\ln \frac{p}{p_0} = - \frac{M}{RT_0} \int_0^z e^{az}\, dz$$

which gives

$$p = p_0 e^{(M/RT_0 a)(e^{az}-1)}$$

In reality, the relationship between T and z is not simple. It must be determined experimentally, by measurements made from balloons and rockets, yielding data that can only be tabulated or else fitted by a complex equation. The variable T must be retained under the integral sign, the equation becoming

$$\ln \frac{p}{p_0} = - \frac{M}{R} \int_0^z \frac{dz}{T} \tag{11.7}$$

In this case, the problem is that of finding the area under the curve of the reciprocal of the temperature plotted vs. the height. This can be accomplished by one of the methods discussed in Chap. 9, such as Simpson's rule.

As a second example, consider the process shown in Fig. 11.2, in which a tank, initially containing liquid to a height h_0, is being filled at a rate F_1 and simultaneously emptied at a rate F_2. These rates may be (1) constant, (2) functions of time t, or (3) functions of the height of liquid in the

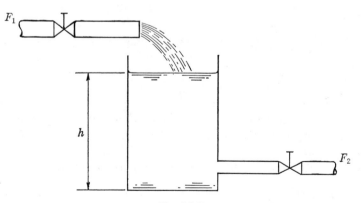

Fig. 11.2

tank, h.   All are physically attainable by means of control valves on the inlet and outlet lines, the valve settings depending on either the time or the liquid level.

The formulation of the differential equation to represent this process may be carried out by using the principle of a material balance.   Since the general method of making a balance is applicable to many types of problems involving material, energy, or momentum transfer, it merits explanation.   The basis for any of these processes is the equation

$$\text{Input} = \text{output} + \text{accumulation} \tag{11.8}$$

where input and output represent the *total amount* entering and leaving the system during some chosen time interval Δt, and accumulation represents the *net change* in the inventory of the system during that interval. If, instead of input and output *amounts*, we use *rates*, and the rates are constant, the equation becomes

$$\text{(Input rate)}(\Delta t) = \text{(output rate)}(\Delta t) + \text{accumulation}$$

If the rates are variable, it is necessary to take a differential time interval dt and to write the equation in differential form as

$$\text{(Input rate)}(dt) = \text{(output rate)}(dt) + d(\text{accumulation}) \tag{11.9}$$

Note that the accumulation is now also a differential quantity, since it is the change in inventory during a differential time interval.

Applying Eq. (11.9) to the tank system gives

$$F_1 \, dt = F_2 \, dt + dV \tag{11.10}$$

where the flow rates are assumed to be in volumetric units and V is the volume of liquid contained in the tank at time t.   For a tank of constant cross-sectional area A,

$$dV = A \, dh$$

Substituting into Eq. (11.10) and rearranging yield

$$\frac{dh}{dt} + \frac{F_2}{A} - \frac{F_1}{A} = 0 \tag{11.11}$$

Equation (11.11) is a first-order differential equation containing possibly as many as four variables.   On the assumption that the desired function is $h(t)$, a variety of equations may be written, depending on how F_1 and F_2 are eliminated.

If both rates are constant, the variables may be separated to produce

$$dh = \frac{1}{A} (F_2 - F_1) \, dt \tag{11.12}$$

Assuming that F_1 is constant and that F_2 is proportional to h in the form $F_2 = \alpha h$ gives

$$\frac{dh}{dt} + \frac{\alpha h}{A} - \frac{F_1}{A} = 0 \tag{11.13}$$

Equation (11.13) may be regarded as either linear or variables-separable.

Suppose, now, that F_1 is also variable, for example, as a cyclic function of time,

$$F_1 = \beta(1 + \cos \omega t)$$

where β and ω are constants. The differential equation becomes

$$\frac{dh}{dt} + \frac{\alpha h}{A} - \frac{\beta(1 + \cos \omega t)}{A} = 0 \tag{11.14}$$

Equation (11.14) is still linear and integrable by the standard methods for linear equations.

As a final set of conditions on the flow rates, we might assume that the outlet line behaves like an orifice, for which the flow rate is proportional to the square root of the liquid height. Letting the inlet rate be a cyclic function of time as above, we have

$$\frac{dh}{dt} + \frac{\alpha \sqrt{h}}{A} - \frac{\beta(1 + \cos \omega t)}{A} = 0 \tag{11.15}$$

where α is a new constant. The equation is now nonlinear. Since the variables cannot be separated, the methods of Chap. 9 are not applicable to its numerical integration. Methods to be developed in this chapter must be used.

As a third and final example, a process will be considered in which a *rate* function is involved. Many of the fundamental laws in engineering are concerned with rates, either physical, such as the rate of heat transfer through a piece of metal, or chemical, such as the rate at which a chemical reaction takes place. The example to be used here is that of radioactive decay. In words, the fundamental law states that the amount of a radioactive element A decaying per unit time is directly proportional to the total amount of A present. In equation form, it becomes

$$-\frac{dA}{dt} = kA \tag{11.16}$$

where

A = amount of element A present
k = first-order rate constant, sec^{-1}

An interesting point to be noted here is that a differential equation, which strictly applies to a *continuous* process, is being used to represent a process that occurs at discrete points in time, i.e., those points at which individual atoms decay. Actually, the rate is the *expected*, or *statistical average*, rate that would be found when a large number of atoms are observed for a long time interval. In practice, these conditions are usually fulfilled.

Equation (11.16) may readily be integrated to yield

$$A = A_0 e^{-kt} \tag{11.17}$$

where A_0 is the amount of A initially present. If we let A equal $A_0/2$, the *half-life* time may be calculated as

$$t_{\frac{1}{2}} = \frac{\ln 2}{k} \tag{11.18}$$

During this time interval, half the amount present at the beginning of the interval will decay. Equation (11.18) may be used to calculate k from an experimental value of $t_{\frac{1}{2}}$.

In a nuclear reactor, an isotope of uranium reacts with neutrons and undergoes fission into many elements, most of which are radioactive. As a greatly simplified example, it will be assumed that the four elements A, B, C, and D are formed by fission at constant rates R_A, R_B, R_C, and R_D. The radioactive decay scheme of these elements is shown below, where

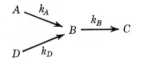

the respective rate constants for decay are indicated over the arrows. Element C is assumed to be stable.

Assuming that we wish to find the amount of each of the elements in the reactor as a function of time, we may combine a material balance (input = output + accumulation) with a rate equation. The resulting differential equations are:

$$\begin{aligned}
R_A\,dt &= k_A A\,dt + dA \\
R_B\,dt + k_A A\,dt + k_D D\,dt &= k_B B\,dt + dB \\
R_C\,dt + k_B B\,dt &= dC \\
R_D\,dt &= k_D D\,dt + dD
\end{aligned} \tag{11.19}$$

Rearranging to the standard form gives

$$\frac{dA}{dt} + k_A A - R_A = 0 \tag{11.20}$$

$$\frac{dB}{dt} + k_B B - (R_B + k_A A + k_D D) = 0 \tag{11.21}$$

$$\frac{dC}{dt} - (R_C + k_B B) = 0 \tag{11.22}$$

$$\frac{dD}{dt} + k_D D - R_D = 0 \tag{11.23}$$

As written, all the equations are linear and may be integrated analytically. The result of integrating Eqs. (11.20) and (11.23) may be substituted into Eq. (11.21) to eliminate variables A and D from that equation, and B may likewise be eliminated from Eq. (11.22). The work will not be carried out here. More complicated schemes of decay, or the assumption of a nonconstant rate of uranium fission, can easily lead to a system of equations that must be integrated numerically.

11.3   GENERAL COMMENTS ON APPROXIMATE INTEGRATION

A first-order ordinary differential equation may be written in the form

$$\frac{dy}{dx} = f(x,y) \tag{11.24}$$

On the assumption that the equation must be integrated approximately, the method to be used depends on the form of the function $f(x,y)$. If it is a function of x alone, or if the variables can be separated as

$$f_1(y) \, dy = f_2(x) \, dx$$

the methods of Chap. 9 can be used. Otherwise, a different approach is needed.

Since y is a function of x (granted that it is an unspecified function), $y = y(x)$, we may write

$$\frac{dy}{dx} = f(x,y(x)) = y'(x) \tag{11.25}$$

The right-hand side of Eq. (11.25) is, symbolically, now a function of only x. In order for the equation to be integrated, however, we must either insert an explicit expression for $y(x)$ in the function $f(x,y(x))$ or else replace $f(x,y(x))$ with a different, approximate function of x that is more easily handled. The former idea is the basis for Picard's method, the latter for the replacement of $y'(x)$ by a Taylor expansion or an interpolating polynomial. In the following sections, several of these methods will be discussed.

11.4   PICARD'S METHOD

In this method, an assumed function $y^{(0)}(x)$ is substituted into the right-hand side of Eq. (11.25). The equation may then be integrated between the limits of x_0 and x to give a new and better approximation $y^{(1)}(x)$. The process is repeated, according to the iteration

$$y^{(r+1)}(x) = y_0 + \int_{x_0}^{x} f(x, y^{(r)}(x)) \ dx \tag{11.26}$$

The sequence of functions $y^{(0)}$, $y^{(1)}$, . . . , $y^{(r+1)}$ will, under suitable restrictions, converge to $y(x)$. The conditions for convergence are presented in more advanced textbooks.[1] Essentially, $f(x,y)$ must be a single-valued continuous function in the region of the xy plane being considered and must also be bounded in that region.

As an example in the application of Picard's method, the equation

$$\frac{dy}{dx} = x^2 + y = f(x,y) \tag{11.27}$$

will be integrated, with an initial condition $x_0 = 1$, $y_0 = 1$. For the purpose of checking the accuracy of the approximate integration, the equation may be integrated analytically, yielding

$$y = 6e^{x-1} - x^2 - 2x - 2 \tag{11.28}$$

A convenient choice for $y^{(0)}(x)$ is the initial value y_0. Substituting Eq. (11.27) into Eq. (11.26) with $y^{(0)}(x)$ gives

$$y^{(1)}(x) = 1 + \int_{1}^{x} (x^2 + 1) \ dx$$

and

$$y^{(1)}(x) = -\frac{1}{3} + x + \frac{x^3}{3}$$

Continuing the iteration produces

$$y^{(2)}(x) = 1 + \int_{1}^{x} \left(x^2 - \frac{1}{3} + x + \frac{x^3}{3} \right) dx$$

$$y^{(2)}(x) = \frac{5}{12} - \frac{x}{3} + \frac{x^2}{2} + \frac{x^3}{3} + \frac{x^4}{12}$$

[1] See, for example, K. S. Kunz, "Numerical Analysis," pp. 171–175, McGraw-Hill Book Company, New York, 1957.

Similarly,

$$y^{(3)}(x) = \frac{3}{20} + \frac{5x}{12} + \frac{x^2}{6} + \frac{x^3}{6} + \frac{x^4}{12} + \frac{x^5}{60}$$

$$y^{(4)}(x) = \frac{23}{120} + \frac{3x}{20} + \frac{5x^2}{24} + \frac{7x^3}{18} + \frac{x^4}{24} + \frac{x^5}{60} + \frac{x^6}{360}$$

The result is a series which, in the limit, would be that obtained from Eq. (11.28) by expanding the exponential term in a Taylor's series. Although convergent, the series requires a large number of terms in order to obtain a reasonable number of significant figures. At $x = 1.10$, for example, $y^{(4)}(x) = 1.204$, compared with 1.221 for the rounded analytical value.

Picard's method is of great importance in the theoretical study of differential equations, but its application as a method of numerical integration is of limited utility. In addition to the low rate of convergence shown in the above example, the successive integrations are tedious and with more complex functions than that used here may become very difficult.

11.5 USE OF TAYLOR'S SERIES

The Taylor's-series expansion of a function $y(x)$ about a point x_0 is given by Eq. (10.9) as

$$y(x) = y(x_0) + (x - x_0)y'(x_0) + \frac{(x - x_0)^2}{2!} y''(x_0)$$

$$+ \cdots + \frac{(x - x_0)^N}{N!} y^{(N)}(\xi) \quad (11.29)$$

where $x_0 < \xi < x$. Given values of x_0 and $y(x_0)$ and an expression for $y'(x)$, it is possible, in theory, to differentiate $y'(x)$ to obtain the derivatives needed in Eq. (11.29). In general, $y'(x) = f(x,y)$, and the derivatives are given by partial differentiation and the use of the chain rule. Thus,

$$y''(x) = \frac{\partial f(x,y)}{\partial x} + \frac{dy}{dx} \frac{\partial f(x,y)}{\partial y}$$

$$= f_x + ff_y \quad (11.30)$$

where f_x indicates the partial derivative of $f(x,y)$ with respect to x and f_y the partial derivative with respect to y. Another differentiation yields

$$y'''(x) = f_{xx} + ff_{xy} + ff_{yx} + f^2 f_{yy} + f_y(f_x + ff_y)$$

$$= f_{xx} + 2ff_{xy} + f^2 f_{yy} + f_y(f_x + ff_y)$$

General expressions for the higher derivatives can be derived, but in using Taylor's series, it is simpler to apply Eq. (11.30) to the successive derivatives.

Upon referring to Eq. (11.27) again as an example, the derivatives are

$y'(x) = x^2 + y$ the given differential equation
$\qquad\qquad\qquad y'(x_0) = 1 + 1 = 2$
$y''(x) = 2x + y'$ $y''(x_0) = 2 + 2 = 4$
$y'''(x) = 2 + y''$ $y'''(x_0) = 2 + 4 = 6$
$y^{iv}(x) = y'''$ $y^{iv}(x_0) = 6$
$\cdots\cdots\cdots$

Then,

$$y(x) = 1 + (x - 1)(2) + \frac{(x - 1)^2}{2!}(4) + \frac{(x - 1)^3}{3!}(6)$$
$$+ \frac{(x - 1)^4}{4!}(6) + \cdots$$

or

$$y(x) = 1 + 2(x - 1) + 2(x - 1)^2 + (x - 1)^3 + \frac{(x - 1)^4}{4} + \cdots$$

At $x = 1.1$, the series converges very rapidly. The fifth term is 0.000025, and the sum of the first four terms is 1.221, agreeing exactly with the analytical value to four significant figures. At $x = 2.0$, the agreement is not so good. The first five terms of the series yield 6.250, whereas the analytical value is 6.310, rounded to four figures. More terms in the series are needed as x increases.

In this example, a Taylor expansion worked very well, since the differentiation was increasingly simple. With a different type of function, such as one involving negative fractional powers of x, the differentiation might not be trivial.

11.6 · EULER'S METHOD

In this method and those following, Eq. (11.25) is integrated in a stepwise fashion. If we start with the initial condition (x_0, y_0), a second point (x_1, y_1) may be calculated from

$$y_1 = y_0 + \int_{x_0}^{x_1} y'(x)\, dx$$

where a convenient approximation for $y'(x)$ is used over the interval (x_0, x_1). The calculation is continued by using (x_1, y_1) to get (x_2, y_2), a new approximation for $y'(x)$ being used. It may be continued in this way, yielding not an analytical expression for y in terms of x but a series of points defining a y-x curve. In general, then,

$$y_{n+1} = y_n + \int_{x_n}^{x_{n+1}} y'(x)\, dx \tag{11.31}$$

Two forms of approximation for $y'(x)$ have been widely used. In the first, $y'(x)$ is replaced by an interpolating polynomial that passes through the $N + 1$ values of y'_n, y'_{n-1}, . . . , y'_{n-N}. The value of y_{n+1} is thus determined by the already calculated values of y_n, y_{n-1}, . . . , y_{n-N}, since $y'_n = f(x_n,y_n)$, $y'_{n-1} = f(x_{n-1},y_{n-1})$, etc. Methods based on this approach are called *marching*, or *open*, methods.

In the second form of approximation, the expression for $y'(x)$ is given by an interpolating polynomial passing through the $N + 1$ values of y'_{n+1}, y'_n, . . . , y'_{n-N+1}. An *iteration* scheme is then necessary, since y_{n+1} appears on the left-hand side of Eq. (11.31) and is also needed in the calculation of y'_{n+1} on the right-hand side. Such methods are often referred to as *closed* methods.

The simplest function with which to replace $y'(x)$ in the calculation of y_{n+1} is the constant value $y'_n = f(x_n,y_n)$. The integration is then

$$y_{n+1} = y_n + \int_{x_n}^{x_{n+1}} f(x_n,y_n) \, dx = y_n + (x_{n+1} - x_n)f(x_n,y_n) \tag{11.32}$$

Letting h be the increment in x gives

$$y_{n+1} = y_n + hf(x_n,y_n) \tag{11.33}$$

which is known as *Euler's method*. The geometrical interpretation is very simple, as shown in Fig. 11.3. Starting at the given initial condition (x_0,y_0), the slope of the function $y(x)$ is calculated and then used to extrapolate the function as a straight line to (x_1,y_1). The process is repeated, the new slope at point 1 being used to extrapolate to point 2, and so on. The curve connecting these points is an approximation to $y(x)$. The difference between the true value of $y(x)$ and the curve will

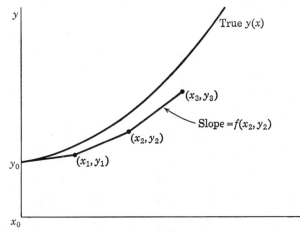

Fig. 11.3

depend on the function being integrated and the size of the increment h, as discussed below.

As an example of the application of Euler's method, Eq. (11.27) will be integrated from $x = 1$ to $x = 2$. An initial condition of $y = 1$ is specified. With the given function, Eq. (11.33) becomes

$$y_{n+1} = y_n + h(x_n^2 + y_n)$$

Thus, with $h = 0.1$,

$$y_1 = 1.0 + 0.1[(1.0)^2 + 1.0] = 1.2$$
$$y_2 = 1.2 + 0.1[(1.1)^2 + 1.2] = 1.441$$
$$y_3 = 1.441 + 0.1[(1.2)^2 + 1.441] = 1.7291$$

. . . .

The complete results are shown in Table 11.1, along with the values obtained from the analytical integration. It can be seen that the numerical values lie below the analytical values and that the error builds up as the integration progresses. The cause of this is the positive and increasing slope of the function, as discussed in Sec. 11.6.1. For a large interval of integration, the results would be very poor.

The process may be improved by taking a smaller value of h. The resulting values of y at the same x values, for $h = 0.01$, 0.001, 0.0001, 0.00001, and 0.000001, are also included in Table 11.1. There is a steady increase in accuracy as h is decreased, with about one more significant digit being obtained as h is cut by a factor of 10. In order to keep from building up excessive round-off error, it is necessary to do the calculations in double precision, particularly for the smallest h values. The difficulty is that the increment to be added to y is small, relative to the value of y, causing a loss in significant figures in the increment when it is added to the

Table 11.1†

x	Analytical	$h = 0.1$	$h = 0.01$	$h = 0.001$	$h = 0.0001$	$h = 0.00001$	$h = 0.000001$
1.1	1.22102551	1.20000000	1.21877897	1.22079930	1.22100287	1.22102324	1.22102528
1.2	1.48841655	1.44100000	1.48334214	1.48790551	1.48836541	1.48841143	1.48841604
1.3	1.80915284	1.72910000	1.80057198	1.80828853	1.80906635	1.80914420	1.80915198
1.4	2.19094818	2.07101000	2.17807104	2.18965089	2.19081836	2.19093520	2.19094689
1.5	2.64232762	2.47411100	2.62423725	2.64050481	2.64214520	2.64230938	2.64232580
1.6	3.17271279	2.94652210	3.14834716	3.17025724	3.17246705	3.17268823	3.17271034
1.7	3.79251624	3.49717431	3.76064784	3.78930397	3.79219476	3.79248409	3.79251303
1.8	4.51324556	4.13589174	4.47245846	4.50913355	4.51283403	4.51320441	4.51324145
1.9	5.34761865	4.87348092	5.29628237	5.34244215	5.34710058	5.34756685	5.34761349
2.0	6.30969096	5.72182901	6.24593111	6.30326052	6.30904738	6.30962661	6.30968453

† Digits printed in italics in the results of the numerical integrations lie outside the range of agreement with the analytical values.

larger value.   If, for example, eight significant figures are retained after each calculation and the relative value of the increment, $\Delta y / y$, is 10^{-5}, only three significant figures of the increment will be used in the addition.   In an extreme case, with h taken sufficiently small, the successive values of y would be constant.

The problem of round-off error can be somewhat alleviated by collecting a certain number of the calculated increments of y to form a larger, more significant increment that can then be added to y.   Thus, y_{n+1} may be represented as

$$y_{n+1} = y_0 + \sum_{i=0}^{n} \Delta y_i = y_0 + h \sum_{i=0}^{n} f(x_i, y_i)$$

Note that h has been taken outside the sum, which also helps to reduce the error.   In carrying out the integration, the sum is initially set to zero, corresponding to $y = y_0$.   Successive values of y are calculated by the above equation, with a fixed y_0, until the desired number of increments have been summed.   Then, y_0 is replaced by the current value of y, and the sum is reset to zero to begin accumulating another set of increments. A convenient point in the program at which to do this is at the time when an output line is printed.

11.6.1   Error Analysis

As was shown experimentally in the above example, the error that is accumulated during a numerical integration depends very strongly on the step size h.   To obtain a more quantitative idea of the dependence, consider the first step, starting with the initial values (x_0, y_0).   If we denote $Y(x_1)$ as the approximation obtained by Euler's method, we have

$$Y(x_1) = y_0 + hf(x_0, y_0)$$

The exact value, which would be obtained by integrating the differential equation, may be calculated by a Taylor expansion.   Letting it be $y(x_1)$ gives

$$y(x_1) = y_0 + hf(x_0, y_0) + \frac{h^2}{2!} f'(\xi)$$

where $f'(\xi)$ is the derivative of $f[x, y(x)]$ evaluated at $x = \xi$, $y = y(\xi)$, and $x_0 < \xi < x_1$.   Letting $e_1 = y(x_1) - Y(x_1)$ be the error due to the truncation of the Taylor's series, we obtain

$$e_1 = \frac{h^2}{2!} f'(\xi) \tag{11.34}$$

Equation (11.34) shows that the truncation error in Euler's method is proportional to h^2, or *of the order of h^2*, abbreviated as $O(h^2)$. An upper bound for the error can be calculated by substituting the maximum absolute value of $f'(x,y)$ in the interval (x_0,x_1) for $f'(\xi)$, provided that it can be determined. In the present example,

$$f(x,y) = x^2 + y$$
$$f'(x,y) = f_x + ff_y = 2x + x^2 + y$$

In the interval $(1.0, 1.1)$, the maximum absolute value of $f'(x,y)$ occurs at $x = 1.1$. By use of the data of Table 11.1, the value is 4.63102551, based on the analytical value for y at $x = 1.1$. Then,

$$e_1 = \frac{0.01}{2}(4.63102551) = 0.023155128$$

The actual error is 0.02102551. Ordinarily, of course, values of $y(x)$ are not known. In that case, the approximate value $Y(x)$ must be used in estimating the error in itself.

A change in h causes a change of $O(h^2)$ in the truncation error, as seen by Eq. (11.34). Hence, if h is decreased by a factor of 10, the error for a *single step* is correspondingly decreased by a factor of 100. Since ten times as many steps are needed to cover a given interval, however, the error at the end point is decreased by a factor of only 10. In general, then, the overall truncation error will vary as $1/h$.

A check of some of the results in Table 11.1 bears this out. For example, the errors at $x = 1.1$, as a function of h, are:

h	Error
0.1	0.02102551
0.01	0.00224654
0.001	0.00022621
0.0001	0.00002264

While the above analysis is of some importance, it does not indicate how the error will accumulate when more than one step is taken. A difficulty arises in the analysis of a multistep process in that the initial conditions for the second and subsequent steps are not the same for the numerical method as for the analytical method. Referring to Fig. 11.4, it can be seen that the second step in the numerical integration is being started from the point (x_1, Y_1), whereas the correct starting point would be (x_1, y_1). In general, the slope $f(x_1, y_1)$ will not be the same as $f(x_1, Y_1)$, causing the straight-line segment to be aimed somewhat closer toward the true curve (convergent) or somewhat farther away from it (divergent).

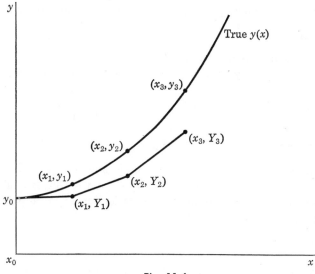

Fig. 11.4

Note that this effect is due to the error already accumulated and will cause it to decrease or to grow. It is in addition to the new truncation error in the second step. Stated more precisely, if the function $f(x,y)$ is such that curves of $y(x)$ for different initial conditions converge with increasing x, the error in any step will tend to be damped out in later steps. If, however, the curves diverge with increasing x, the error in any step will be propagated into later steps and will be amplified. For parallel curves, the error in the overall integration will just be the sum of the errors in the individual steps.

This can be demonstrated in the example, the integration of Eq. (11.27). The analytical integral may be written as

$$y = Ce^x - x^2 - 2x - 2$$

where C is the constant of integration. With initial conditions of (x_0, y_0),

$$C = (y_0 + x_0^2 + 2x_0 + 2)e^{-x_0}$$

and

$$y = (y_0 + x_0^2 + 2x_0 + 2)e^{x - x_0} - x^2 - 2x - 2$$

Differentiating with respect to y_0 at constant x_0 produces

$$\frac{dy}{dy_0} = e^{x - x_0}$$

For a small increment in y_0, Δy_0, the change in y is approximately

$$\Delta y = e^{x-x_0} \, \Delta y_0$$

and, for values of x greater than x_0, Δy increases with x. Thus, the overall error is greater than the sum of the individual errors indicated by Eq. (11.34).

11.7   MODIFIED EULER'S METHOD

Euler's method was an example of a simple *marching* type of numerical integration, because the value of y'_{n+1} was not used in calculating y_{n+1}. An *iterative* scheme may be developed by letting $y'(x)$ be calculated as the average of y'_n and y'_{n+1}. Thus,

$$y'(x) = \tfrac{1}{2}[f(x_n, y_n) + f(x_{n+1}, y_{n+1})]$$

Substituting into Eq. (11.31) and again letting h be the increment in x, we obtain

$$y_{n+1}^{(r+1)} \doteq y_n + \frac{h}{2}[f(x_n, y_n) + f(x_{n+1}, y_{n+1}^{(r)})] \tag{11.35}$$

where the superscript on y_{n+1} is an iteration index. In order to get started, $y_{n+1}^{(0)}$ may be calculated from the simple Euler equation, or it may be set equal to y_n. Successive values of $y_{n+1}^{(r)}$ are calculated from Eq. (11.35) and compared. When the relative change is as small as desired, the iteration is stopped.

Geometrically, the modified Euler's method consists again of a straight-line extrapolation from (x_n, y_n) to (x_{n+1}, y_{n+1}), but the slope that is used is the average of the slopes at the two points. Obviously, this should help to decrease the systematic error made by using only one of the slopes.

Substituting Eq. (11.27) into Eq. (11.35) gives

$$y_{n+1}^{(r+1)} = y_n + \frac{h}{2}(x_n{}^2 + y_n + x_{n+1}^2 + y_{n+1}^{(r)})$$

With $h = 0.1$, the iterations for y_2 progress as follows:

$$
\begin{aligned}
y_2{}^{(0)} &= 1.0 + 0.1(1.0^2 + 1.0) = 1.2 \qquad \text{starting value}\\
y_2{}^{(1)} &= 1.0 + 0.05(1.0^2 + 1.0 + 1.1^2 + 1.2) = 1.2205\\
y_2{}^{(2)} &= 1.0 + 0.05(1.0^2 + 1.0 + 1.1^2 + 1.2205) = 1.2215\\
y_2{}^{(3)} &= 1.0 + 0.05(1.0^2 + 1.0 + 1.1^2 + 1.2215) = 1.2216
\end{aligned}
$$

The results of the integration over the interval from $x = 1$ to $x = 2$, with $h = 0.1, 0.01, 0.001$, and 0.0001, are presented in Table 11.2.

Table 11.2

x	$Analytical$	$h = 0.1$	$h = 0.01$	$h = 0.001$	$h = 0.0001$
1.1	1.22102551	1.22157881	1.22103098	1.22102545	1.22102551
1.2	1.48841655	1.48963958	1.48842864	1.48841642	1.48841655
1.3	1.80915284	1.81118041	1.80917288	1.80915262	1.80915284
1.4	2.19094818	2.19393604	2.19097770	2.19094784	2.19094818
1.5	2.64232762	2.64645540	2.64236839	2.64232712	2.64232762
1.6	3.17271279	3.17818728	3.17276686	3.17271212	3.17271280
1.7	3.79251624	3.79957512	3.79258593	3.79251534	3.79251624
1.8	4.51324556	4.52216164	4.51333357	4.51324440	4.51324556
1.9	5.34761865	5.35870459	5.34772806	5.34761718	5.34761865
2.0	6.30969096	6.32330465	6.30982529	6.30968911	6.30969095

11.7.1   Error Analysis for Modified Euler's Method

An analysis of the single-step truncation error in this method is complicated by the fact that two equations are used for calculating y_{n+1}, the first for predicting a trial value and the second for improving that value by iteration. On the assumption, however, that the number of iterations is greater than two, it can be shown that the order of error reaches and remains at a fixed value. Considering the value of y_n to be exact and letting the sequence of approximations to y_{n+1} be $Y_{n+1}^{(0)}$, $Y_{n+1}^{(1)}$, $Y_{n+1}^{(2)}$, etc., and using the simple Euler's method to calculate $Y_{n+1}^{(0)}$, we have

$$Y_{n+1}^{(0)} = y_n + hy_n'$$

and for the subsequent iterations

$$Y_{n+1}^{(r+1)} = y_n + \frac{h}{2}(y_n' + Y_{n+1}'^{(r)})$$

or

$$Y_{n+1}^{(1)} = y_n + hy_n' + \frac{h^2}{2}y_n''$$

$$Y_{n+1}^{(2)} = y_n + hy_n' + \frac{h^2}{2}y_n'' + \frac{h^3}{4}y_n'''$$

$$Y_{n+1}^{(3)} = y_n + hy_n' + \frac{h^2}{2}y_n'' + \frac{h^3}{4}y_n''' + \frac{h^4}{8}y_n^{iv}$$

Further iterations result in changes in terms beyond that in h^3.

We may obtain the exact value of y_{n+1} by expansion in Taylor's series about (x_n, y_n). Thus,

$$y_{n+1} = y_n + hy_n' + \frac{h^2}{2}y_n'' + \frac{h^3}{6}y'''(\xi)$$

where $x_n < \xi < x_{n+1}$. Therefore, on the assumption that $y'''(x)$ does not change strongly in the interval (x_n, x_{n+1}), the principal part of the error is

$$e = y_{n+1} - Y_{n+1} = h^3 \left[\frac{y'''(\xi)}{6} - \frac{y'''_n}{4} \right] \approx -\frac{h^3}{12} y'''(\xi) \tag{11.36}$$

Taking the maximum absolute value of $y'''(x) = f''(x,y)$ in the interval $(1.0, 1.1)$ in the example being studied,

$$f''(x,y) = y + x^2 + 2x + 2$$

At $x = 1.1$, the value of $f''(x,y)$ is 6.63102551. Substitution into Eq. (11.36) gives an estimate of -0.00055258. The actual error, calculated from the data of Table 11.2, is -0.00055330.

11.7.2   Rate of Convergence of Modified Euler's Method

Another point to be considered in an iterative process is the *rate of convergence* to the solution. For the modified Euler's method, this can easily be found by comparing the equations for successive iterations. Thus, the differences are

$$Y^{(2)}_{n+1} - Y^{(1)}_{n+1} = \frac{h^3}{4} y'''_n$$

$$Y^{(3)}_{n+1} - Y^{(2)}_{n+1} = \frac{h^4}{8} y^{iv}_n$$

$$Y^{(4)}_{n+1} - Y^{(3)}_{n+1} = \frac{h^5}{16} y^v_n$$

$$\cdot \cdot \cdot \cdot \cdot \cdot \cdot \cdot \cdot \cdot \cdot \cdot$$

$$Y^{(r+1)}_{n+1} - Y^{(r)}_{n+1} = \frac{h^{r+2}}{2^{r+1}} y^{(r+2)}_n$$

On the assumption that the derivatives of y are bounded, then the method will converge, the rate depending on the actual values of the derivatives. The final value of Y is, however, still only an approximation to the exact solution of the differential equation. In the present example, the derivatives are constant above y'''_n, and the convergence is very rapid.

11.7.3   Choice of Step Size

Clearly, the step size h is of critical importance in a numerical integration. Too large a value causes the truncation error to be excessive; too small a value wastes computer time and increases round-off error. Furthermore, it is desirable to permit h to be variable from one step to the next, or at least every few steps, since the truncation error depends on derivatives that change as the integration progresses.

A valuable technique for aiding in the choice of an optimum step size has been presented by Richardson.[1] Two integrations are carried out over the same interval, which may be a subinterval of the total integration, the first with a step size of h and the second with a step size of $h/2$. If the single-step truncation error of the process is of order $r + 1$, the errors are given by

$$e_1 = y - Y^{(1)} = C_1 h^{r+1}$$
$$e_2 = y - Y^{(2)} = 2C_2 \left(\frac{h}{2}\right)^{r+1}$$

where $Y^{(1)}$ and $Y^{(2)}$ are the results of the numerical integrations. For small intervals, $C_1 \approx C_2$, and the equations may be solved to give

$$y - Y^{(2)} \approx \frac{Y^{(2)} - Y^{(1)}}{2^r - 1} \tag{11.37}$$

Equation (11.37) gives an estimate of the error without the evaluation of a derivative, as in Eq. (11.36), and is, accordingly, convenient and practical to use in a computer program.

With the example and data of Sec. 11.7, integration from $x = 1.0$ to 1.1, with step sizes of 0.1 and 0.05 produces, $Y^{(1)} = 1.22157881$ and $Y^{(2)} = 1.22116371$. For the modified Euler's method, the truncation error is $O(h^3)$, or $r = 2$. Hence, an estimate of the error in $Y^{(2)}$, found by substituting the above values into Eq. (11.37), is -0.00013837. The actual error is -0.00013820.

11.7.4   Flow Charts

A program for integration by the modified Euler's method is presented in the flow charts of Figs. 11.5 and 11.6.   The first flow chart constitutes a general program for reading the necessary parameters, controlling the step size, and printing the results.   It calls a subprogram for the actual integration; in this case the subprogram uses the modified Euler's method, but any other self-starting method could be used.

The input parameters include:

(x_0, y_0), initial conditions
x_f, upper limit of x
p, increment in x between output printings
h_0, initial choice of h; must equal p divided by an integral power of 2
$h_{\min}$, minimum allowable h
$h_{\max}$, maximum allowable h; not greater than p
e, permissible relative error in iteration
n, maximum number of iterations

[1] L. F. Richardson and J. A. Gaunt, The Deferred Approach to the Limit, *Trans. Roy. Soc. London*, **226A**: 300 (1927).

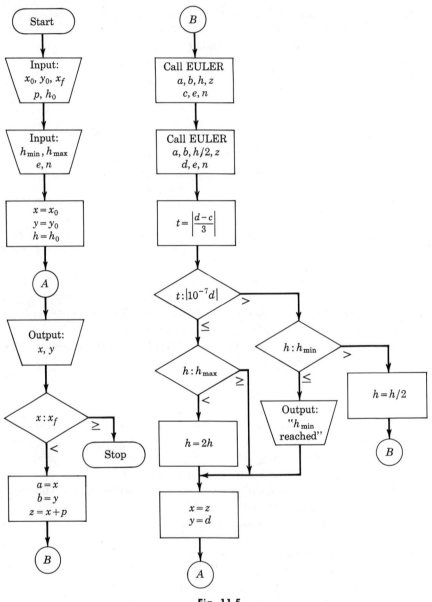

Fig. 11.5

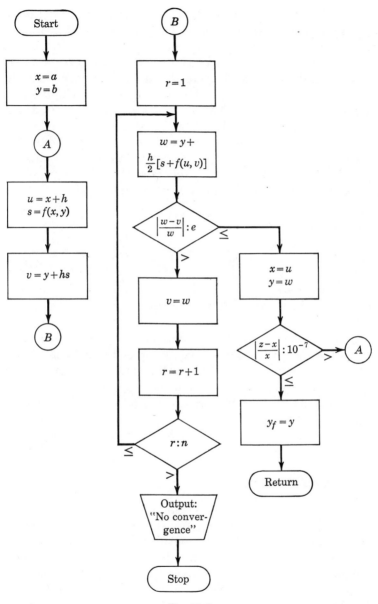

Fig. 11.6

In the main flow chart, the variables a and b are set equal to the current values of x and y, respectively, and the limit of the next integration before printing is stored as z. The subprogram is then called twice, first with a step size of h and then with a step size of $h/2$. It returns the two new values of y, called c and d. They are compared, according to the method of Eq. (11.37), and if they are sufficiently close, an attempt is made to double h for the next step. The values of x and y are updated, and the program loops back to point A. If the values of c and d are not so close as desired, an attempt is made to reduce h and to repeat the calculation. If h reaches h_{min} and the error is still too large, a message is printed to that effect, with the program continuing the integration.

The arguments of the subprogram include:

a, current value of x

b, current value of y

h, step size

z, current upper limit of x

e and n, defined above

y_f, value of y to be returned to main program

In the second flow chart the function $y'(x) = f(x,y)$ is defined by an arithmetic function statement. Then, in order to save computer time, the terms that do not depend on y_{n+1} in Eq. (11.35) are calculated before the iteration is begun and are saved as the single quantity s. The variables u, v, and w are used to designate x_{n+1}, $y_{n+1}^{(r)}$, and $y_{n+1}^{(r+1)}$, respectively. The iteration index is r.

11.8   ADAMS'S METHOD

In the methods described above, a constant value of the slope $f(x,y) = y'(x)$ is assumed over the interval (x_n, x_{n+1}). This facilitates integration of the right-hand side of Eq. (11.25) but is obviously not a good approximation for most functions unless h is chosen to be very small. A more sophisticated approach is to pass an interpolating polynomial through the N known points y_n', y_{n-1}', . . . , y_{n-N+1}' $(n > N)$ and then to *extrapolate* the resulting function over the interval of integration. (In order to get the integration started, the first N points must be calculated in some other way, such as by the modified Euler's method.)

A convenient form to use is the Gregory-Newton backward-interpolation formula, shown in Table 8.10. In the notation of Eq. (11.25), it becomes

$$y'(u) = y'_n + \frac{\nabla y'_n}{1!} u + \frac{\nabla^2 y'_n}{2!} u(u + 1) + \frac{\nabla^3 y'_n}{3!} u(u + 1)(u + 2)$$

$$+ \cdots + \frac{\nabla^N y'_n}{N!} u(u + 1)(u + 2)(\cdots)(u + N - 1) \quad (11.38)$$

where

$$y'_n = y'(x_n) = f(x_n, y_n)$$

$$u = \frac{x - x_n}{h}$$

Upon noting that $dx = h\, du$, the integration of Eq. (11.25) is now given by

$$y_{n+1} = y_n + \int_{x_n}^{x_{n+1}} y'(x)\, dx = y_n + h \int_0^1 y'(u)\, du$$

and

$$y_{n+1} = y_n + h \int_0^1 \left[y'_n + \frac{\nabla y'_n}{1!} u + \frac{\nabla^2 y'_n}{2!} u(u + 1) \right.$$

$$\left. + \frac{\nabla^3 y'_n}{3!} u(u + 1)(u + 2) + \cdots \right] du \quad (11.39)$$

The integration with respect to u in Eq. (11.39) is straightforward. The first seven terms yield, after simplification,

$$y_{n+1} = y_n + h \left(y'_n + \tfrac{1}{2} \nabla y'_n + \tfrac{5}{12} \nabla^2 y'_n + \tfrac{3}{8} \nabla^3 y'_n + \tfrac{251}{720} \nabla^4 y'_n \right.$$

$$\left. + \tfrac{95}{288} \nabla^5 y'_n + \frac{19{,}087}{60{,}840} \nabla^6 y'_n \right) \quad (11.40)$$

Equation (11.40) is often truncated so as to retain only the differences up through the third. Thus,

$$y_{n+1} = y_n + h(y'_n + \tfrac{1}{2} \nabla y'_n + \tfrac{5}{12} \nabla^2 y'_n + \tfrac{3}{8} \nabla^3 y'_n)$$

For programming purposes, a more useful form can be obtained by replacing the differences by their definitions in terms of y'_n, y'_{n-1}, y'_{n-2}, and y'_{n-3}, listed in Eq. (8.11). This gives the Adams's *four-point* formula,

$$y_{n+1} = y_n + \frac{h}{24} (55 y'_n - 59 y'_{n-1} + 37 y'_{n-2} - 9 y'_{n-3}) \quad (11.41)$$

Like the simple Euler's method, Eq. (11.41) has the disadvantage that it uses extrapolation, basing the calculation of y_{n+1} on values of $y'(x)$ that all lie to the left of the interval of integration. An iterative form, using y'_{n+1}, can be derived by passing an interpolating polynomial through the N points y'_{n+1}, y'_n, . . . , y'_{n-N+2}. Again, from the Gregory-Newton backward-interpolation formula from Table 8.10,

$$y'(u) = y'_{n+1} + \frac{\nabla y'_{n+1}}{1!} u + \frac{\nabla^2 y'_{n+1}}{2!} u(u+1) + \frac{\nabla^3 y'_{n+1}}{3!} u(u+1)(u+2)$$

$$+ \cdots + \frac{\nabla^N y'_{n+1}}{N!} u(u+1)(u+2)(\cdots)(u+N-1) \quad (11.42)$$

where

$$u = \frac{x - x_{n+1}}{h}$$

Integration of Eq. (11.25) with the substitution of $y'(u)$ for $y'(x)$ over the interval $(-1,0)$ in u, corresponding to (x_n, x_{n+1}), yields, after simplification,

$$y_{n+1} = y_n + h(y'_{n+1} - \tfrac{1}{2}\nabla y'_{n+1} - \tfrac{1}{12}\nabla^2 y'_{n+1} - \tfrac{1}{24}\nabla^3 y'_{n+1} - \tfrac{19}{720}\nabla^4 y'_{n+1} + \cdots)$$
$$(11.43)$$

Again, retaining only the terms through the third difference, and then replacing the differences by their definitions, we obtain

$$y_{n+1} = y_n + \frac{h}{24} (9y'_{n+1} + 19y'_n - 5y'_{n-1} + y'_{n-2}) \quad (11.44)$$

The complete algorithm includes the use of Eq. (11.41) as a *predictor* for $y^{(0)}_{n+1}$, following which Eq. (11.44) is used in iteration form. Thus,

$$y^{(0)}_{n+1} = y_n + \frac{h}{24} (55y'_n - 59y'_{n-1} + 37y'_{n-2} - 9y'_{n-3})$$

$$y^{(r+1)}_{n+1} = y_n + \frac{h}{24} (9y'^{(r)}_{n+1} + 19y'_n - 5y'_{n-1} + y'_{n-2})$$

where $y'_i = f(x_i, y_i)$.

Adams's method will now be used to calculate the fifth point in the integration of Eq. (11.27). First, the analytical values for y at $x = 1.0$, 1.1, 1.2, and 1.3 will be used to calculate $y'(x)$ at those points. Thus,

$$y'(x_n) = x_n{}^2 + y_n$$
$$y'_1 = (1.0)^2 + 1.0 = 2.0$$
$$y'_2 = (1.1)^2 + 1.22102551 = 2.43102551$$
$$y'_3 = (1.2)^2 + 1.48841655 = 2.92841655$$
$$y'_4 = (1.3)^2 + 1.80915284 = 3.49915284$$

With $h = 0.1$, the value for y_5 is

$$y_5^{(1)} = 1.80915284 + \frac{0.1}{24} [55(3.49915284) - 59(2.92841655)$$
$$+ 37(2.43102551) - 9(2.0)] = 2.19092273$$
$$y_5'^{(1)} = (1.4)^2 + 2.19092273 = 4.15092273$$
$$y_5^{(2)} = 1.80915284 + \frac{0.1}{24} [9(4.15092273) + 19(3.49915284)$$
$$- 5(2.92841655) + 2.43102551] = 2.19094928$$

Continuing the iteration,

$$y_5^{(3)} = 2.19095040$$
$$y_5^{(4)} = 2.19095034$$
$$y_5^{(5)} = 2.19095034$$

The accuracy is comparable with that obtained in the modified Euler's method with $h = 0.01$. The complete results of the integration, with $h = 0.1$, 0.01, 0.001, and 0.0001, are presented in Table 11.3. In all cases, the first four points were calculated from the analytical expression for $y(x)$. This was done so as to make it easier to compare the analytical results with the numerical results for the remainder of the integration. Ordinarily, of course, a numerical method would also be used to obtain the starting values.

Table 11.3

x	$Analytical$	$h = 0.1$	$h = 0.01$	$h = 0.001$	$h = 0.0001$
1.1	1.22102551		1.22102549	1.22102542	1.22102544
1.2	1.48841655		1.48841653	1.48841645	1.48841647
1.3	1.80915284		1.80915283	1.80915273	1.80915276
1.4	2.19094818	2.19095023	2.19094816	2.19094806	2.19094810
1.5	2.64232762	2.64233231	2.64232760	2.64232749	2.64232752
1.6	3.17271279	3.17272065	3.17271278	3.17271265	3.17271269
1.7	3.79251624	3.79252786	3.79251622	3.79251608	3.79251612
1.8	4.51324556	4.51326167	4.51324554	4.51324540	4.51324544
1.9	5.34761865	5.34764006	5.34761863	5.34761847	5.34761852
2.0	6.30969096	6.30971859	6.30969093	6.30969076	6.30969080

11.8.1   Error Analysis

The expression for the error term in the Gregory-Newton backward-difference interpolation formula is similar to that given in Eq. (8.20) for the forward differences. With the necessary sign changes to account for the change in direction, it is

$$\text{Error} = \frac{h^{N+1}}{(N+1)!} u(u+1)(u+2)(\cdots)(u+N)f^{N+1}(\xi) \tag{11.45}$$

where N is the order of the interpolating polynomial and ξ lies within the interval of interpolation (or extrapolation as used here).

Letting y_{n+1} be the exact value of $y(x)$ at x_{n+1}, and taking y'_n, y'_{n-1}, y'_{n-2}, and y'_{n-3} to be exact, we have from the Adams's four-point predictor formula

$$y_{n+1} = y_n + \frac{h}{24}(55y'_n - 59y'_{n-1} + 37y'_{n-2} - 9y'_{n-3}) + \tfrac{251}{720}h^5 y^{\text{v}}(\xi_1) \tag{11.46}$$

and from the corrector formula

$$y_{n+1} = y_n + \frac{h}{24}(9y'_{n+1} + 19y'_n - 5y'_{n-1} + y'_{n-2}) - \tfrac{19}{720}h^5 y^{\text{v}}(\xi_2) \tag{11.47}$$

where both ξ_1 and ξ_2 lie within the interval (x_{n-3}, x_{n+1}). The numerical coefficients in the error terms are just those derived in Eqs. (11.40) and (11.43).

Letting $Y_{n+1}^{(0)}$ be the approximation to y_{n+1} obtained by the Adams's predictor formula and Y_{n+1} that obtained by the corrector formula, the respective errors are

$$e_p = y_{n+1} - Y_{n+1}^{(0)} = \tfrac{251}{720}h^5 y^{\text{v}}(\xi_1) = O(h^5) \tag{11.48}$$

$$e_c = y_{n+1} - Y_{n+1} = -\tfrac{19}{720}h^5 y^{\text{v}}(\xi_2) = O(h^5) \tag{11.49}$$

If $y^{\text{v}}(x)$ does not change rapidly in the interval (x_{n-3}, x_{n+1}), $y^{\text{v}}(\xi_1) \approx y^{\text{v}}(\xi_2)$ and Eq. (11.49) may be subtracted from Eq. (11.48), to yield

$$Y_{n+1} - Y_{n+1}^{(0)} \approx \tfrac{270}{720}h^5 y^{\text{v}}(\xi) \tag{11.50}$$

where ξ lies in the above interval. Dividing Eq. (11.49) by Eq. (11.50) and rearranging,

$$e_c = y_{n+1} - Y_{n+1} = -\frac{19}{270}(Y_{n+1} - Y_{n+1}^{(0)}) \approx -\frac{1}{14}(Y_{n+1} - Y_{n+1}^{(0)}) \tag{11.51}$$

According to Eq. (11.51), the single-step truncation error in the corrector formula can be estimated from the difference between the corrected and predicted values. This fact can be utilized in a computer program to control the choice of the step size h. By estimating the error at every step, a decision can be made as to whether the step size needs to be made smaller for increased accuracy or may be expanded in order to save computer time.

11.8.2   Rate of Convergence

The analysis here is similar to that in Sec. 11.7.2, but the algebra is much more tedious and will be omitted here. Upon neglecting higher-order terms, the difference between successive approximations is

$$Y_{n+1}^{(r+1)} - Y_{n+1}^{(r)} = \frac{1}{9}\left(\frac{9h}{24}\right)^{r+1} [(28 - 55)y_n^{(r+1)} - (5 - 59)y_{n-1}^{(r+1)}$$
$$+ (1 - 37)y_{n-2}^{(r+1)} + 9y_{n-3}^{(r+1)}]   (11.52)$$

where the superscript $(r + 1)$ on the y terms on the right-hand side indicates the $(r + 1)$st derivative of y with respect to x. On the assumption that all the derivatives in Eq. (11.52) are bounded, the method is convergent. The actual rate depends on the above four derivatives as well as on h. If the derivatives are nearly the same, the bracketed term will be small and the convergence will be rapid.

11.8.3   Flow Charts

Flow charts for a program using the Adams's method are presented in Figs. 11.7 and 11.8. In the first, parameters are read, the necessary starting values are calculated by the modified Euler's method, and the integration is continued, a subroutine being used for Adams's method. Printing is also controlled by the main program. Because the step size is allowed to vary from one step to the next, the values of x used for the integration may happen not to be those at which printing is wanted. Consequently, an interpolating function $q(a,b,c,d,u)$ is used to find y at the desired points. The function is simply the Gregory-Newton backward-difference formula truncated after the third difference. Thus,

$$q(a,b,c,d,u) = d + (d - c)u + \frac{1}{2}(d - 2c + b)u(u + 1)$$
$$+ \frac{1}{6}(d - 3c + 3b - a)u(u + 1)(u + 2)$$

Input parameters are the same as those in the flow chart for the modified Euler's method (Fig. 11.5).

In the Adams's subroutine, the predicted value of y_{n+1}, called v_0, is first calculated. The terms in the corrector equation that do not depend on y_{n+1} are calculated and saved as the quantity s. The result of the iteration, called w, is used, together with v_0, to obtain an estimate of the error, as in Eq. (11.51). If the error is excessive, the integration is repeated, with h replaced by $h/2$, if $h > h_{\min}$. In order to do this, it is necessary to interpolate to find values of x, y, and y' at the points $i - \frac{1}{2}$, $i - 1$, and $i - \frac{3}{2}$, which is done by the same interpolating equation as was used in the main program. This is shown in Fig. 11.8b.

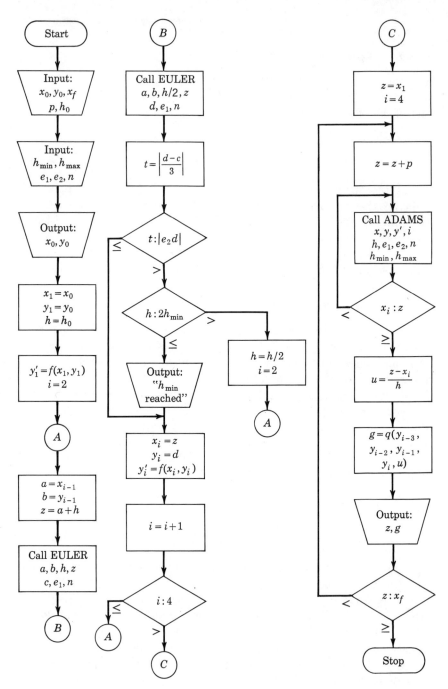

Fig. 11.7

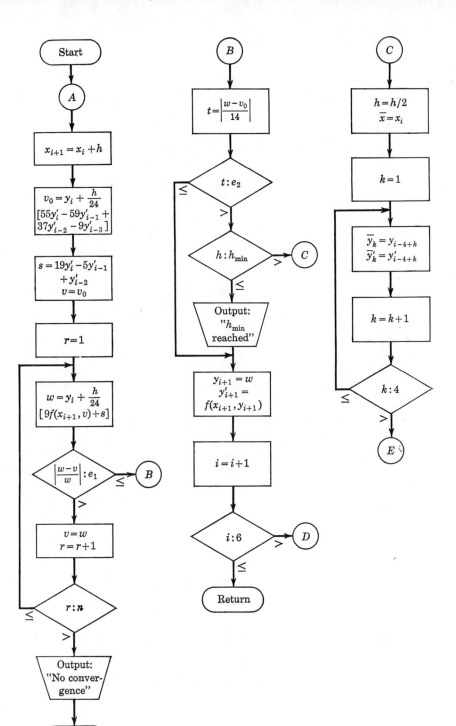

Fig. 11.8a

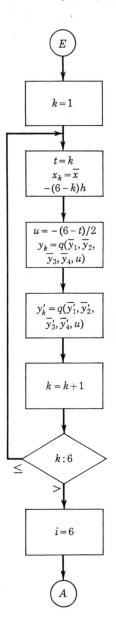
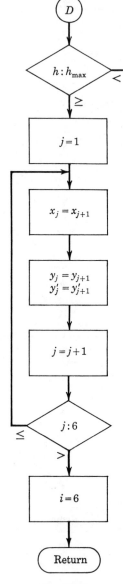
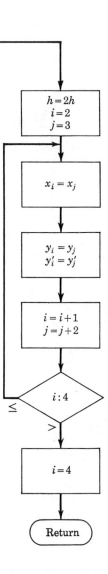

Fig. 11.8b

If the truncation error is small, an attempt is made to double h, in order to save computer time. In this case, values of x, y, and y' at $i - 2$, $i - 4$, and $i - 6$ are used to continue the integration.

Only the seven values of each of the variables most recently calculated are ever needed. The routine shown in Fig. 11.8b moves the new values into the locations occupied by the old, unneeded values after each step.

11.9   MILNE'S METHOD

In Euler's and Adams's methods, the function $y'(x)$ was integrated over the interval (x_n, x_{n+1}). A more general expression is

$$y_{n+1} = y_{n-k} + \int_{x_n - kh}^{x_n + h} y'(x)\, dx \tag{11.53}$$

where $k = 0, 1, 2, 3, \ldots$. Thus, the above methods are seen to be special cases of Eq. (11.53), for $k = 0$.

The substitution for $y'(x)$ in terms of the Gregory-Newton backward-difference formulas may again be made. Based on y'_n, the equation is

$$y_{n+1} = y_{n-k} + h \int_{-k}^{1} [y'_n + \frac{\nabla y'_n}{1!} u + \frac{\nabla^2 y'_n}{2!} u(u + 1)$$
$$+ \frac{\nabla^3 y'_n}{3!} u(u+1)(u+2) + \cdots] \, du \tag{11.54}$$

The results of the integration are

$k = 1$

$$y_{n+1} = y_{n-1} + h[2y'_n + (0)\nabla y'_n + \tfrac{1}{3}\nabla^2 y'_n + \tfrac{1}{3}\nabla^3 y'_n + \tfrac{29}{90}\nabla^4 y'_n$$
$$+ \tfrac{14}{45}\nabla^5 y'_n + \cdots] \tag{11.55}$$

$k = 2$

$$y_{n+1} = y_{n-2} + h(3y'_n - \tfrac{3}{2}\nabla y'_n + \tfrac{3}{4}\nabla^2 y'_n + \tfrac{3}{8}\nabla^3 y'_n - \tfrac{17}{80}\nabla^4 y'_n$$
$$+ \tfrac{469}{1440}\nabla^5 y'_n + \cdots) \tag{11.56}$$

$k = 3$

$$y_{n+1} = y_{n-3} + h[4y'_n - 4\nabla y'_n + \tfrac{8}{3}\nabla^2 y'_n + (0)\nabla^3 y'_n + \tfrac{14}{45}\nabla^4 y'_n$$
$$+ \tfrac{14}{45}\nabla^5 y'_n + \cdots] \tag{11.57}$$

$k = 5$

$$y_{n+1} = y_{n-5} + h[6y'_n - 12\nabla y'_n + 15\nabla^2 y'_n - 9\nabla^3 y'_n + \tfrac{33}{10}\nabla^4 y'_n$$
$$+ (0)\nabla^5 y'_n + \cdots] \tag{11.58}$$

An important point to note is that the coefficient of the kth difference is zero in the expressions for odd values of k. Consequently, these formulas retain the accuracy of k differences when truncated after only $k - 1$ differences.

The equations for iteration may be obtained as above by using backward differences from y'_{n+1}. The equation to be integrated is

$$y_{n+1} = y_{n-k} + h \int_{-k+1}^{0} \left[y'_{n+1} + \frac{\nabla y'_{n+1}}{1!} u + \frac{\nabla^2 y'_{n+1}}{2!} u(u+1) \right.$$
$$\left. + \frac{\nabla^3 y'_{n+1}}{3!} u(u+1)(u+2) + \cdots \right] du \quad (11.59)$$

The results for odd values of k are

$k = 1$

$$y_{n+1} = y_{n-1} + h[2y'_{n+1} - 2\nabla y'_{n+1} + \tfrac{1}{3}\nabla^2 y'_{n+1} + (0)\nabla^3 y'_{n+1} - \tfrac{1}{90}\nabla^4 y'_{n+1}$$
$$- \tfrac{1}{90}\nabla^5 y'_{n+1} + \cdots] \quad (11.60)$$

$k = 3$

$$y_{n+1} = y_{n-3} + h[4y'_{n+1} - 8\nabla y'_{n+1} + \tfrac{20}{3}\nabla^2 y'_{n+1} - \tfrac{8}{3}\nabla^3 y'_{n+1} + \tfrac{14}{45}\nabla^4 y'_{n+1}$$
$$+ (0)\nabla^5 y'_{n+1} + \cdots] \quad (11.61)$$

In Milne's *second-difference* method, Eq. (11.57) is used as the predictor and Eq. (11.60) as the corrector, with differences dropped after the second. Since the coefficient of the third difference is zero in both cases, this is equivalent to truncation after the third difference. Replacing the differences by their definitions produces

$$y_{n+1} = y_{n-3} + \tfrac{4}{3}h(2y'_n - y'_{n-1} + 2y'_{n-2}) \tag{11.62}$$

$$y_{n+1} = y_{n-1} + \frac{h}{3}(y'_{n+1} + 4y'_n + y'_{n-1}) \tag{11.63}$$

In Milne's *fourth-difference* method, Eqs. (11.58) and (11.61) are used in a similar fashion, differences after the fourth being dropped. In both cases, the coefficient of the fifth difference is zero. The final equations are

$$y_{n+1} = y_{n-5} + \frac{3h}{10}(11y'_n - 14y'_{n-1} + 26y'_{n-2} - 14y'_{n-3} + 11y'_{n-4}) \tag{11.64}$$

$$y_{n+1} = y_{n-3} + \frac{2h}{45}(7y'_{n+1} + 32y'_n + 12y'_{n-1} + 32y'_{n-2} + 7y'_{n-3}) \tag{11.65}$$

11.9.1   Error Analysis

In the second-difference method, differences after the second are dropped, but as shown in Eqs. (11.57) and (11.60), the coefficients of the third differences are zero. Hence, the error terms for the predictor and corrector formulas are given by the coefficients of the fourth differences. Thus, if y_{n+1} is the true value of $y(x)$ at x_{n+1}, the predictor equation corresponds to

$$y_{n+1} = y_{n-3} + \tfrac{4}{3}h(2y'_n - y'_{n-1} + 2y'_{n-2}) + \tfrac{14}{45}h^5 y^{\mathrm{v}}(\xi_1) \tag{11.66}$$

where $x_{n-3} < \xi_1 < x_{n+1}$. For the corrector formula, we have

$$y_{n+1} = y_{n-1} + \frac{h}{3}(y'_{n+1} + 4y'_n + y'_{n-1}) - \tfrac{1}{90}h^5 y^\mathrm{v}(\xi_2) \tag{11.67}$$

where $x_{n-1} < \xi_2 < x_{n+1}$. Letting $Y^{(0)}_{n+1}$ be the approximation to y_{n+1} from the predictor formula and Y_{n+1} that from the corrector, the errors are

$$e_p = y_{n+1} - Y^{(0)}_{n+1} = \tfrac{14}{45}h^5 y^\mathrm{v}(\xi_1) \tag{11.68}$$
$$e_c = y_{n+1} - Y_{n+1} = -\tfrac{1}{90}h^5 y^\mathrm{v}(\xi_2) \tag{11.69}$$

Assuming that $y^\mathrm{v}(\xi_1) \approx y^\mathrm{v}(\xi_2)$, we may derive the following relations:

$$Y_{n+1} - Y^{(0)}_{n+1} \approx \tfrac{29}{90}h^5 y^\mathrm{v}(\xi) \tag{11.70}$$

where $x_{n-3} < \xi < x_{n+1}$, and

$$y_{n+1} - Y_{n+1} \approx -\tfrac{1}{29}(Y_{n+1} - Y^{(0)}_{n+1}) \tag{11.71}$$

Equation (11.71) may be used to control the step size in the same way as Eq. (11.51) was used with Adams's method.

Similar derivations for the Milne four-point formulas yield

$$e_p = y_{n+1} - Y^{(0)}_{n+1} = \tfrac{41}{140}h^7 y^\mathrm{vii}(\xi_1) \tag{11.72}$$
$$e_c = y_{n+1} - Y_{n+1} = -\tfrac{8}{945}h^7 y^\mathrm{vii}(\xi_2) \tag{11.73}$$

where $x_{n-5} < \xi_1 < x_{n+1}$ and $x_{n-3} < \xi_2 < x_{n+1}$. Also,

$$y_{n+1} - Y_{n+1} \approx -\tfrac{1}{35}(Y_{n+1} - Y^{(0)}_{n+1}) \tag{11.74}$$

11.10   HAMMING'S METHOD

Starting with the very general linear equation

$$y_{n+1} = a_n y_n + a_{n-1}y_{n-1} + a_{n-2}y_{n-2} + h(b_{n+1}y'_{n+1} + b_n y'_n + b_{n-1}y'_{n-1} \\ + b_{n-2}y'_{n-2}) + O(h^5) \tag{11.75}$$

Hamming[1] has derived a class of corrector formulas that includes those used in the Milne second-difference formula and the Adams four-point formula as well as other possible forms. Of the seven constants in Eq. (11.75), five can be eliminated by expanding y_{n-1}, y_{n-2}, y'_{n+1}, y'_{n-1}, and y'_{n-2} in Taylor's series about the point n through terms in y_n^iv and then

[1] Richard W. Hamming, "Numerical Methods for Scientists and Engineers," chaps. 14 and 15, McGraw-Hill Book Company, New York, 1962.

equating the result to Taylor's series for y_{n+1}. Thus,

$$y_{n+1} = a_n y_n + a_{n-1}\left(y_n - hy_n' + \frac{h^2}{2!}y_n'' - \frac{h^3}{3!}y_n''' + \frac{h^4}{4!}y_n^{iv}\right)$$
$$+ a_{n-2}\left[y_n - (2h)y_n' + \frac{(2h)^2}{2!}y_n'' - \frac{(2h)^3}{3!}y_n''' + \frac{(2h)^4}{4!}y_n^{iv}\right]$$
$$+ h\left\{b_{n+1}\left(y_n' + hy_n'' + \frac{h^2}{2!}y_n''' + \frac{h^3}{3!}y_n^{iv}\right)\right.$$
$$+ b_n y_n' + b_{n-1}\left(y_n' - hy_n'' + \frac{h^2}{2!}y_n''' - \frac{h^3}{3!}y_n^{iv}\right)$$
$$\left. + b_{n-2}\left[y_n' - (2h)y_n'' + \frac{(2h)^2}{2!}y_n''' - \frac{(2h)^3}{3!}y_n^{iv}\right]\right\}$$
$$= y_n + hy_n' + \frac{h^2}{2!}y_n'' + \frac{h^3}{3!}y_n''' + \frac{h^4}{4!}y_n^{iv} \tag{11.76}$$

Equating coefficients of y_n and its derivatives gives

$$a_n + a_{n-1} + a_{n-2} = 1$$
$$-a_{n-1} - 2a_{n-2} + b_{n+1} + b_n + b_{n-1} + b_{n-2} = 1$$
$$\tfrac{1}{2}a_{n-1} + 2a_{n-2} + b_{n+1} - b_{n-1} - 2b_{n-2} = \tfrac{1}{2}$$
$$-\tfrac{1}{6}a_{n-1} - \tfrac{4}{3}a_{n-2} + \tfrac{1}{2}b_{n+1} + \tfrac{1}{2}b_{n-1} + 2b_{n-2} = \tfrac{1}{6}$$
$$\tfrac{1}{24}a_{n-1} + \tfrac{2}{3}a_{n-2} + \tfrac{1}{6}b_{n+1} - \tfrac{1}{6}b_{n-1} - \tfrac{4}{3}b_{n-2} = \tfrac{1}{24}$$

Two of the coefficients are arbitrary; in one solution Hamming has set $a_{n-1} = b_{n-2} = 0$. The remaining values are

$$b_{n+1} = \tfrac{3}{8}$$
$$a_n = \tfrac{9}{8} \qquad b_n = \tfrac{3}{4}$$
$$a_{n-2} = -\tfrac{1}{8} \qquad b_{n-1} = -\tfrac{3}{8}$$

Substituting into Eq. (11.75), we have, as a corrector equation,

$$y_{n+1} = \tfrac{1}{8}[9y_n - y_{n-2} + 3h(y_{n+1}' + 2y_n' - y_{n-1}')] \tag{11.77}$$

This equation may be used with a suitable predictor equation, such as that of Milne, Eq. (11.62). Hamming has shown that this method is more stable than the use of Eqs. (11.62) and (11.63), in the sense that it does not cause truncation and round-off errors to be amplified as the integration progresses.

Equations (11.62) and (11.77) may be used for prediction and iteration, but Hamming has suggested another procedure, in which iteration is eliminated, saving computer time. The predicted value is *modified* by an estimated error term, calculated from the previous step, giving a better value to be corrected. A new error estimate is used to give a final value.

The complete algorithm is

Predictor: $\quad y_{n+1}^{(0)} = y_{n-3} + \frac{4}{3}h(2y_n' - y_{n-1}' + 2y_{n-2}')$

Modifier: $\quad y_{n+1}^{(1)} = y_{n+1}^{(0)} - \frac{112}{121}(y_n^{(0)} - y_n^{(2)})$ $\qquad\qquad$ (11.78)

Corrector: $\quad y_{n+1}^{(2)} = \frac{1}{8}\{9y_n - y_{n-2} + 3h[f(x_{n+1},y_{n+1}^{(1)}) + 2y_n' - y_{n-1}']\}$

Final value: $\quad y_{n+1} = y_{n+1}^{(2)} + \frac{9}{121}(y_{n+1}^{(0)} - y_{n+1}^{(2)})$

In the modifier equation, the coefficient of the term being subtracted is obtained by noting that the errors in the predictor and corrector equations are, respectively,

$$e_p = \tfrac{14}{45}h^5 y^{\mathrm{v}}(\xi_1) \qquad \text{and} \qquad e_c = -\tfrac{1}{40}h^5 y^{\mathrm{v}}(\xi_2)$$

where ξ_1 and ξ_2 both lie within the interval (x_{n-3},x_{n+1}). Assuming that $\xi_1 \approx \xi_2$, we may form the ratio

$$\frac{e_p}{e_p - e_c} = \frac{\frac{14}{45}}{\frac{14}{45} - (-\frac{1}{40})} = \frac{112}{121}$$

Substitution of the definitions of e_p and e_c, together with the assumption that the error in the $(n+1)$st step is approximately the same as that in the nth step, gives the modifier equation. The equation for the final value may be derived in a similar manner from the expression

$$\frac{e_c}{e_p - e_c} = \frac{-\frac{1}{40}}{\frac{14}{45} - (-\frac{1}{40})} = -\frac{9}{121}$$

It should be noted that some other, self-starting method is needed to calculate values for the first four points in the integration. Either Euler's method or the Runge-Kutta method, discussed in the following section, may be used. In the first step in which Hamming's method is used, $y_n^{(0)} - y_n^{(2)}$ is taken to be zero.

11.11 RUNGE-KUTTA METHODS

In these methods, the interval (x_n,x_{n+1}) is broken into two or more subintervals, and the integral of the function $f(x,y)$ over the whole interval is calculated as the sum of the integrals over the subintervals. The function is taken to be constant over each subinterval, as in Euler's method, but by a judicious choice of the points at which the function is evaluated a much higher order of truncation error can be obtained than in Euler's method. In addition, the Runge-Kutta methods have the advantage of requiring only one initial point in order to get started. Consequently, they are useful not only for complete integrations but also for obtaining the starting values for Adams's, Milne's, and Hamming's methods. The disadvantage of the Runge-Kutta methods is that they require several evaluations of $f(x,y)$ for each point of the integration, which makes them somewhat slower than the other methods.

Actually, the modified Euler's method *without iteration* may be regarded as falling into the category of Runge-Kutta methods if it is written as the system of equations

$$k_0 = hf(x_n,y_n)$$
$$k_1 = hf(x_n + h, \ y_n + k_0) \tag{11.79}$$
$$y_{n+1} = y_n + \tfrac{1}{2}(k_0 + k_1)$$

In the integration over the first half of the interval, the slope at the left-hand side of the interval is used; over the second half, the slope at the right-hand side, as obtained by the first extrapolation, is used. It may be remembered, from Sec. 11.7.1, that the truncation error for this method is $O(h^3)$, compared with $O(h^2)$ for the simple Euler's method.

The most popular of the Runge-Kutta methods are the fourth-order methods, in which the interval h is divided into four subintervals. The integration is then given by the system of equations

$$k_0 = hf(x_n,y_n)$$
$$k_1 = hf(x_n + \alpha_1 h, \ y_n + \beta_{10}k_0)$$
$$k_2 = hf(x_n + \alpha_2 h, \ y_n + \beta_{20}k_0 + \beta_{21}k_1) \tag{11.80}$$
$$k_3 = hf(x_n + \alpha_3 h, \ y_n + \beta_{30}k_0 + \beta_{31}k_1 + \beta_{32}k_2)$$
$$y_{n+1} = y_n + (ak_0 + bk_1 + ck_2 + dk_3)$$

The large number of parameters in Eqs. (11.80) may be determined, at least in part, by equating the expression for y_{n+1} with the Taylor's-series expansion of y_{n+1} about (x_n,y_n). This can be done so as to achieve agreement through terms in h^4, giving solutions that have an error of $O(h^5)$. Two of the many possible solutions are presented here. The first, originally proposed by Runge, may be summarized as

$$k_0 = hf(x_n,y_n)$$
$$k_1 = hf\left(x_n + \frac{h}{2}, \ y_n + \frac{k_0}{2}\right)$$
$$k_2 = hf\left(x_n + \frac{h}{2}, \ y_n + \frac{k_1}{2}\right) \tag{11.81}$$
$$k_3 = hf(x_n + h, \ y_n + k_2)$$
$$y_{n+1} = y_n + \tfrac{1}{6}(k_0 + 2k_1 + 2k_2 + k_3)$$

where $y'(x) = f(x,y)$, with initial conditions (x_0,y_0). Here, the interval h is divided into the subintervals $h/6$, $h/3$, $h/3$, and $h/6$. The function is evaluated first at the left-hand side, then twice at extrapolated center points, and finally at the extrapolated right-hand side.

The equations for a second method, proposed by Kutta, are

$$k_0 = hf(x_n, y_n)$$
$$k_1 = hf\left(x_n + \frac{h}{3}, y_n + \frac{k_0}{3}\right)$$
$$k_2 = hf\left(x_n + \frac{2h}{3}, y_n - \frac{k_0}{3} + k_1\right) \tag{11.82}$$
$$k_3 = hf(x_n + h, y_n + k_0 - k_1 + k_2)$$
$$y_{n+1} = y_n + \tfrac{1}{8}(k_0 + 3k_1 + 3k_2 + k_3)$$

In this form, the interval h is divided into the subintervals $h/8$, $3h/8$, $3h/8$, and $h/8$, and the function is evaluated at interior points in the second and third subintervals.

For the special case that the right-hand side of the differential equation is a function of x alone, Eqs. (11.81) and (11.82) reduce to Simpson's one-third and three-eighths rules, respectively.

The method outlined in Eqs. (11.81) will now be applied to the integration of Eq. (11.27). Starting at (1,1) with $h = 0.1$, the calculations are

$$k_0 = 0.1(1.0^2 + 1.0) = 0.2$$
$$k_1 = 0.1[(1.05)^2 + 1.1] = 0.22025$$
$$k_2 = 0.1[(1.05)^2 + 1.110125] = 0.2212625$$
$$k_3 = 0.1[(1.1)^2 + 1.2212625] = 0.24312625$$
$$y(1.1) = 1.0 + \tfrac{1}{6}(1.32615125) = 1.22102521$$

The results of the complete integration, for step sizes of 0.1 and 0.01, are presented in Table 11.4. The second set of values agrees with the analytical results to at least 8 digits.

Table 11.4

x	Analytical	$h = 0.1$	$h = 0.01$
1.1	1.22102551	1.22102*521*	1.22102551
1.2	1.48841655	1.48841*586*	1.48841655
1.3	1.80915284	1.80915*168*	1.80915285
1.4	2.19094818	2.19094*641*	2.19094819
1.5	2.64232762	2.64232*512*	2.64232762
1.6	3.17271279	3.17270*940*	3.17271280
1.7	3.79251624	3.79251*177*	3.79251624
1.8	4.51324556	4.51323*981*	4.51324557
1.9	5.34761865	5.34761*137*	5.34761867
2.0	6.30969096	6.30968*187*	6.30969097

11.11.1   Error Analysis

As stated above, the coefficients in the Runge-Kutta formulas given in Eqs. (11.81) and (11.82) were developed so as to get agreement with Taylor's series through terms in h^4. The general derivation is lengthy and algebraically complex and will not be reproduced here. Instead, the discussion will be restricted to a comparison of Eqs. (11.81) with the Taylor's-series expansion of y_{n+1} about (x_n, y_n), which should illustrate the main ideas behind the derivation.

Writing the Taylor expansion first, we have

$$y_{n+1} = y_n + \left(hf + \frac{h^2}{2!} f' + \frac{h^3}{3!} f'' + \frac{h^4}{4!} f''' \right)_n + \frac{h^5}{5!} f^{iv}[\xi, y(\xi)] \qquad (11.83)$$

where $f_n \equiv f(x_n, y_n) \equiv y'(x_n)$, and $x_n < \xi < x_{n+1}$. In the next step, the derivatives in Eq. (11.83), which are functions of the two variables x and y, are expanded by the chain rule of differentiation. In order to simplify the notation, we may introduce the following operators:

$$D = \left(\frac{\partial}{\partial x} + f \frac{\partial}{\partial y} \right)$$

$$D^2 = \left(\frac{\partial^2}{\partial x^2} + 2f \frac{\partial^2}{\partial x\, \partial y} + f^2 \frac{\partial^2}{\partial y^2} \right)$$

$$D^3 = \left(\frac{\partial^3}{\partial x^3} + 3f \frac{\partial^3}{\partial x^2\, \partial y} + 3f^2 \frac{\partial^3}{\partial x\, \partial y^2} + f^3 \frac{\partial^3}{\partial y^3} \right)$$

$$D^4 = \left(\frac{\partial^4}{\partial x^4} + 4f \frac{\partial^4}{\partial x^3\, \partial y} + 6f^2 \frac{\partial^4}{\partial x^2\, \partial y^2} + 4f^3 \frac{\partial^4}{\partial x\, \partial y^3} + f^4 \frac{\partial^4}{\partial y^4} \right)$$

Then, using subscript notation for the partial derivatives,

$$f'(x,y) = f_x + ff_y = Df$$
$$f''(x,y) = f_{xx} + 2ff_{xy} + f^2 f_{yy} + f_y(f_x + ff_y) = D^2f + f_y\, Df$$

and, omitting intermediate steps,

$$f'''(x,y) = D^3f + 3\, Df_y\, Df + f_y\, D^2f + f_y{}^2\, Df$$
$$f^{iv}(x,y) = D^4f + f_y\, D^3f + (4\, Df_y + f_y{}^2)\, D^2f + (6\, D^2f_y + 7f_y\, Df_y$$
$$+ 3f_{yy}\, Df + f_y{}^3)\, Df \qquad (11.84)$$

Substituting into Eq. (11.83) produces

$$\begin{aligned}
y_{n+1} = y_n + &\left[hf + \frac{h^2}{2} Df + \frac{h^3}{6} (D^2f + f_y\, Df) \right. \\
&\left. + \frac{h^4}{24} (D^3f + 3\, Df_y\, Df + f_y\, D^2f + f_y{}^2\, Df) \right]_n \\
&+ \frac{h^5}{120} [D^4f + f_y\, D^3f + (4\, Df_y + f_y{}^2)\, D^2f \\
&\qquad + (6\, D^2f_y + 7f_y\, Df_y + 3f_{yy}\, Df + f_y{}^3)\, Df]_\xi \qquad (11.85)
\end{aligned}$$

Turning now to the Runge-Kutta formula of Eqs. (11.81), the functions k_0, k_1, k_2, and k_3 must be evaluated in terms of the derivatives at (x_n, y_n). Since some of these depend on the slopes at interior points of the interval (x_n, x_{n+1}), Taylor expansions are needed. Thus, starting with $k_0 = hf$, which requires no expansion, the details of the work for k_1 are

$$k_1 = hf\left(x_n + \frac{h}{2}, y_n + \frac{k_0}{2}\right)$$

$$= h\left[f + \left(\frac{h}{2}f_x + \frac{k_0}{2}f_y\right) + \frac{1}{2}\left(\frac{h^2}{4}f_{xx} + \frac{2hk_0}{4}f_{xy} + \frac{k_0^2}{4}f_{yy}\right)\right.$$

$$+ \frac{1}{6}\left(\frac{h^3}{8}f_{xxx} + \frac{3h^2k_0}{8}f_{xxy} + \frac{3hk_0^2}{8}f_{xyy} + \frac{k_0^3}{8}f_{yyy}\right)$$

$$+ \frac{1}{24}\left(\frac{h^4}{16}f_{xxxx} + \frac{4h^3k_0}{16}f_{xxxy} + \frac{6h^2k_0^2}{16}f_{xxyy} + \frac{4hk_0^3}{16}f_{xyyy}\right.$$

$$\left.\left. + \frac{k_0^4}{16}f_{yyyy}\right)\right]_n + O(h^6)$$

Substituting for k_0, collecting terms, and using the operator notation,

$$k_1 = h\left(f + \frac{h}{2}Df + \frac{h^2}{8}D^2f + \frac{h^3}{48}D^3f + \frac{h^4}{384}D^4f\right)_n + O(h^6)$$

The expressions for k_2 and k_3 are derived similarly. The results are

$$k_2 = h\left\{f + \frac{h}{2}Df + \frac{h^2}{8}(D^2f + 2f_y\,Df) + \frac{h^3}{48}(D^3f + 3f_y\,D^2f + 6\,Df_y\,Df)\right.$$

$$+ \frac{h^4}{384}[D^4f + 4f_y\,D^3f + 12\,Df_y\,D^2f + 12\,D^2f_y\,Df$$

$$\left. + 12f_{yy}(Df)^2]\right\}_n + O(h^6)$$

$$k_3 = h\left\{f + h\,Df + \frac{h^2}{2}(D^2f + f_y\,Df)\right.$$

$$+ \frac{h^3}{24}(4\,D^3f + 3f_y\,D^2f + 6f_y{}^2\,Df + 12\,Df_y\,Df)$$

$$+ \frac{h^4}{48}[2\,D^4f + f_y\,D^3f + 3f_y{}^2\,D^2f + 6\,Df_y\,D^2f + 16f_y\,Df_y\,Df$$

$$\left. + 12\,D^2f_y\,Df + 6f_{yy}(Df)^2]\right\}_n + O(h^6)$$

Substituting the expressions for k_0, k_1, k_2, and k_3 into Eqs. (11.81) gives

$$y_{n+1} = y_n + \left\{hf + \frac{h^2}{2}Df + \frac{h^3}{6}(D^2f + f_y\,Df) + \frac{h^4}{24}(D^3f + 3\,Df_y\,Df\right.$$

$$+ f_y\,D^2f + f_y{}^2\,Df) + \frac{h^5}{120}[\tfrac{15}{16}D^4f + \tfrac{5}{4}f_y\,D^3f + \tfrac{5}{4}f_y{}^2\,D^2f$$

$$+ \tfrac{15}{4}Df_y\,D^2f + \tfrac{25}{4}D^2f_y\,Df + \tfrac{15}{2}f_y\,Df_y\,Df$$

$$\left. + \tfrac{15}{4}f_{yy}(Df)^2]\right\}_n + O(h^6) \quad (11.86)$$

Equations (11.85) and (11.86) agree through terms in h^4; hence, the truncation error is $O(h^5)$.   An expression for the upper bound of the error has been derived by Lotkin,[1] but in practice it is more convenient to use the method suggested by Richardson, as given in Eq. (11.37).   For the Runge-Kutta method, the error estimate becomes

$$y - Y^{(2)} = \frac{Y^{(2)} - Y^{(1)}}{15} \tag{11.87}$$

where $Y^{(1)}$ and $Y^{(2)}$ are the results of integrations with step sizes of h and $h/2$.

11.12   SECOND-ORDER AND SIMULTANEOUS EQUATIONS

A differential equation of second or higher order can be reduced to a system of first-order equations by the introduction of auxiliary dependent variables.   For example, the second-order equation

$$y'' + 2yy' - 8y = x^2 \tag{11.88}$$

can be reduced to two first-order equations by setting $y' = u$.   Then, $y'' = du/dx = u'$, and the system is

$$y' = u \tag{11.89}$$
$$u' = f(x,y,u) = x^2 - 2yu + 8y \tag{11.90}$$

These equations can now be solved by any of the methods already presented.   Since both the dependent variables appear in each equation, the equations must be solved simultaneously.   If the modified Euler's method is used, for instance, the sequence of calculations is

Predictor: $\quad y_{n+1}^{(0)} = y_n + hu_n$
$\qquad\qquad\quad u_{n+1}^{(0)} = u_n + hu_n'$

Corrector: $\quad y_{n+1}^{(r+1)} = y_n + \dfrac{h}{2}(u_n + u_{n+1}^{(r)})$

$\qquad\qquad\quad u_{n+1}^{(r+1)} = u_n + \dfrac{h}{2}(u_n' + u_{n+1}'^{(r)})$

where u_n' is defined by Eq. (11.90).   Two conditions are needed, such as the initial conditions (x_0, y_0, y_0').   If the two conditions are not given at the same point, the system is classified as a *boundary-value* problem, which must be handled by more advanced methods.

[1] M. Lotkin, On the Accuracy of Runge-Kutta's Method, *Mathematical Tables and Other Aids to Computation*, vol. 5, p. 128, National Bureau of Standards, January, 1951.

If a pair of simultaneous equations of the form

$$y' = f_1(x,y,u) \qquad (11.91)$$
$$u' = f_2(x,y,u) \qquad (11.92)$$

are to be integrated by the Runge-Kutta method of Eqs. (11.81), the steps are

$$k_0 = hf_1(x_n, y_n, u_n)$$
$$l_0 = hf_2(x_n, y_n, u_n)$$
$$k_1 = hf_1\left(x_n + \frac{h}{2}, y_n + \frac{k_0}{2}, u_n + \frac{l_0}{2}\right)$$
$$l_1 = hf_2\left(x_n + \frac{h}{2}, y_n + \frac{k_0}{2}, u_n + \frac{l_0}{2}\right)$$
$$k_2 = hf_1\left(x_n + \frac{h}{2}, y_n + \frac{k_1}{2}, u_n + \frac{l_1}{2}\right) \qquad (11.93)$$
$$l_2 = hf_2\left(x_n + \frac{h}{2}, y_n + \frac{k_1}{2}, u_n + \frac{l_1}{2}\right)$$
$$k_3 = hf_1(x_n + h, y_n + k_2, u_n + l_2)$$
$$l_3 = hf_2(x_n + h, y_n + k_2, u_n + l_2)$$

and

$$y_{n+1} = y_n + \tfrac{1}{6}(k_0 + 2k_1 + 2k_2 + k_3)$$
$$u_{n+1} = u_n + \tfrac{1}{6}(l_0 + 2l_1 + 2l_2 + l_3)$$

As an example, consider the integration of the equation for the motion of a pendulum, given in Sec. 10.6.3 as

$$\frac{d^2\theta}{dt^2} + \frac{g}{L}\sin\theta = 0 \qquad (11.94)$$

The pendulum is released from rest at $t = 0$, with $\theta = 0.8$ rad, $L = 2$ ft. Equation (11.94) can be rewritten as the system

$$\frac{d\theta}{dt} = u$$
$$\frac{du}{dt} = -\frac{g}{L}\sin\theta \qquad (11.95)$$

with initial conditions of $t_0 = 0$, $\theta_0 = 0.8$, and $u_0 = 0$. Taking g as 32.2 ft/sec² , and letting $h = 0.1$ sec, integration by the Runge-Kutta method for the first interval proceeds as follows:

$k_0 = 0.1(0) = 0$

$l_0 = 0.1(-16.1) \sin (0.8) = -1.155$

$k_1 = 0.1\left(-\dfrac{1.155}{2}\right) = -0.0578$

$l_1 = 0.1(-16.1) \sin (0.8) = -1.155$

$k_2 = 0.1\left(-\dfrac{1.155}{2}\right) = -0.0578$

$l_2 = 0.1(-16.1) \sin \left(0.8 - \dfrac{0.0578}{2}\right) = -1.122$

$k_3 = 0.1(-1.122) = -0.1122$

$l_3 = 0.1(-16.1) \sin (0.8 - 0.0578) = -1.088$

$\theta_1 = 0.8 + \frac{1}{6}(-0.3432) = 0.7428 \text{ rad}$

$u_1 = 0 + \frac{1}{6}(-6.797) = -1.133 \text{ rad/sec}$

$\text{Velocity} = Lu = -2.266 \text{ ft/sec}$

11.13   APPLICATIONS

Several illustrations of the use of ordinary differential equations in engineering problems have already been given in the introduction to this chapter.   Three others are considered here: the equations for the motion of a rocket, for the motion of a body attached to a spring, and for the flux in a certain electrical circuit.   The second and third physical systems are shown to lead to the same differential equation.

11.13.1   Motion of a Rocket

Figure 3.27 shows a rocket in vertical flight, with fuel being consumed and expelled from the rear to produce a forward thrust.   According to Newton's second law of motion, the sum of the forces acting on the rocket must equal the product of its mass m and acceleration a.   The forces include (1) the forward thrust T, (2) the weight of the rocket W, directed downward, and (3) a drag force D, due to friction with the surrounding atmosphere, directed toward the rear of the rocket.   In equation form,

$$T - W - D = ma \tag{11.96}$$

If the unit of force in Eq. (11.96) is the pound, the unit of mass is the slug, equal to 32.16 lb mass.   The weight W may be replaced by mg, where g is the acceleration of gravity.   On the assumption that the fuel is being burned at a constant rate q, the mass at any time t is

$$m = m_0 - qt$$

where m_0 is the initial mass.   Substituting into Eq. (11.96), replacing a by dV/dt, where V is the velocity of the rocket, and rearranging give

$$\frac{dV}{dt} = \frac{T - D}{m_0 - qt} - g \tag{11.97}$$

The term that causes difficulty in Eq. (11.97) is the drag force.   Experiment has shown that it is proportional to V^2, to the density ρ of the air through which the rocket is traveling, and to the cross-sectional area A of the rocket.   Hence,

$$\frac{dV}{dt} = \frac{T - CV^2 \rho A}{m_0 - qt} - g \tag{11.98}$$

where C is a constant that depends on the shape of the rocket.

Equation (11.98) may be combined with

$$\frac{ds}{dt} = V \tag{11.99}$$

to find the distance s traveled by the rocket.   Unless ρ is assumed to be constant, the equations must be integrated simultaneously, since ρ will be a function of s.   Also, for large values of s, the variation of g with height above the earth must be taken into account.

11.13.2   Motion of a Body Attached to a Spring

In Fig. 11.9a a body of mass m rests on a horizontal surface and is attached to a spring, as shown.   If the position of the body as measured from the left end of the spring is denoted as x, the force exerted by the spring may be expressed as $f(x)$.   Then, by Newton's second law of motion

$$m\frac{d^2x}{dt^2} + f(x) = 0 \tag{11.100}$$

Some form of the function for the force of the spring must be assumed for (11.100) to be integrated.   Three possibilities are sketched in Fig. 11.9b, for a function $f(x) = ax + bx^3$.   If $b = 0$, the spring is linear; otherwise it is a hard spring, for $b > 0$, or a soft spring, for $b < 0$.   Substituting into Eq. (11.100) produces

$$m\frac{d^2x}{dt^2} + (ax + bx^3) = 0 \tag{11.101}$$

The problem becomes more complex (and realistic) if a damping force is assumed, such as a frictional resistance to the motion of the body on the horizontal surface.   Letting this force be $\phi(V)$, where V is the velocity of the body, Eq. (11.100) becomes

$$m\frac{d^2x}{dt^2} + \phi(V) + f(x) = 0 \tag{11.102}$$

Often $\phi(V) = CV|V|$, producing again a nonlinear equation.

It is interesting to find that the same basic differential equation may be used to describe the flux in an electrical circuit containing a capacitance and an inductance, as shown in Fig. 11.10. In the terminology of this system

$$\frac{d\phi}{dt} + \frac{q}{c} = 0 \tag{11.103}$$

and

$$\frac{d^2\phi}{dt^2} + \frac{i}{c} = 0 \tag{11.104}$$

where

$\phi = $ flux
$q = $ charge on capacitor
$c = $ capacitance
$t = $ time
$i = $ current $ = dq/dt$

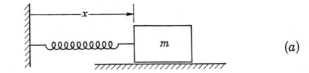

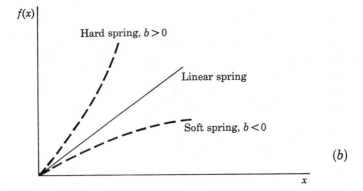

Fig. 11.9

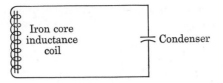

Fig. 11.10

For a linear system, $i = \phi/L$, where $L =$ inductance. For a nonlinear system, a possible expression is $i = a\phi - b\phi^3$, where $a > 0$, $b > 0$. Substituting the latter into Eq. (11.104), we have

$$\frac{d^2\phi}{dt^2} + \frac{1}{c}\left(a\phi - b\phi^3\right) = 0 \tag{11.105}$$

which is the same form as Eq. (11.102) without the damping term.

PROBLEMS

♦**11.1** Using Picard's method, obtain an approximate solution to the following differential equations, through terms in x^5. Find the value of y at $x = 0.2$.

(a) $y' = x + y$, $y = 1$ at $x = 0$
(b) $y' = x + y^2$, $y = 1$ at $x = 0$

♦**11.2** Use Taylor's series to obtain approximate solutions to the equations of Prob. 11.1, through terms in x^5. Find the value of y at $x = 0.2$.

11.3 A certain reversible chemical reaction has the stoichiometry

$$2A \underset{k_2}{\overset{k_1}{\rightleftharpoons}} B$$

The reaction is second-order in the forward direction and linear in the reverse direction. Hence, the rate of disappearance of compound A is given by

$$-\frac{dA}{dt} = k_1 A^2 - k_2 B$$

where k_1 is the rate constant for the forward reaction and k_2 is that for the reverse reaction. Using initial conditions of $A_0 = 1.0$ g mole/liter, $B_0 = 0$, integrate the equation by Adams's method, printing concentrations of A and B at 60-sec intervals for a reaction time of 1 hr. Use a time increment of 1 sec. The rate constants have values of 0.01 (min^{-1})(g moles/liter)$^{-1}$ and 0.02 min^{-1}, respectively.

♦**11.4** A molecule of compound A splits in half into two molecules of compound B, which recombine to form compound C, a chemical isomer of A. The reaction scheme, with rate constants, is

$$A \xrightarrow{k_1} 2B \xrightarrow{k_2} C$$

Assuming that B is the most valuable of the three compounds, it is desirable to stop the reaction when the concentration of B reaches a maximum. The rate of formation of B is given by

$$\frac{dB}{dt} = 2k_1 A - k_2 B^2$$

(a) Write the differential equation for the rate of disappearance of A, integrate the equation, and substitute the resulting expression for A into the above equation.

(b) Using initial conditions of $A_0 = 1.0$ g mole/liter, $B_0 = C_0 = 0$, integrate the differential equation for B by the Runge-Kutta method. Let $k_1 = 0.03$ min^{-1} and $k_2 = 0.02$ (min^{-1})(g moles/liter)$^{-1}$. Use $h = 1.0$ sec, and obtain the concentrations of A and B at 1-min intervals for 1 hr.

11.5 Write a program to integrate Eqs. (11.98) and (11.99) numerically to find V and s as functions of time. Use (a) the modified Euler's method and (b) Hamming's method. As initial conditions, use $V = 0$ and $s = 0$. Use $h = 0.1$ sec, and carry the integration to $t = 60$ sec, printing the results at 1-sec intervals. Other values include $m_0 = 900,000$ lb mass, $q = 2700$ lb mass/sec, $T = 10^6$ lb force, $C = 0.01$ (lb force)/(lb mass)(ft/sec^2), and $A = 100$ ft^2. Compare the results obtained by assuming ρ constant at the sea-level value ρ_0 with those obtained by neglecting the drag force completely.

11.6 Equation (11.37) may be used to estimate the single-step truncation error in an integration if two step sizes h and $h/2$ are used. Derive a similar equation for estimating the error if the step sizes are h and $h/10$.

♦**11.7** Meshaka[1] has published the following predictor-corrector method for integrating a first-order differential equation $y' = f(x,y)$:

Predictor: $y_{n+1}^{(0)} = 5y_n - 10y_{n-1} + 10y_{n-2} - 5y_{n-3} + y_{n-4}$

Corrector: $y_{n+1}^{(r+1)} = \frac{60}{137}hf(x_{n+1}, y_{n+1}^{(r)}) + \frac{1}{137}(300y_n - 300y_{n-1} + 200y_{n-2}$
$$- 75y_{n-3} + 12y_{n-4})$$

What are the single-step truncation errors e_p and e_c, defined by

$$e_p = y_{n+1} - y_{n+1}^{(0)}$$
$$e_c = y_{n+1} - y_{n+1}^{(c)}$$

where y_{n+1} is the exact value, $y_{n+1}^{(0)}$ is the predicted value, and $y_{n+1}^{(c)}$ is the final corrected value?

11.8 Integrate one of the following differential equations by one of the numerical methods described in this chapter. Use an interval $0 \leq x \leq 1$, with an increment of 0.1 and an initial condition of $y = 1$ at $x = 0$. Compare your results with those obtained from an analytical integration wherever possible.

(a) $y' = x$ (e) $y' = x^2 + y^2$
(b) $y' = x^2$ (f) $y' = (x^2 + y^2)^{\frac{1}{2}}$
(c) $y' = 1 + y$ (g) $y' = e^x$
(d) $y' = 1 + y^2$ (h) $y' = y^2 + e^x$

[1] P. Meshaka, Two Methods of Numerical Integration for Differential Systems, *Rev. Franc. Trait. Inf.*, **7:** 135–148 (1964). Abstracted in *Comp. Rev.*, **6:** 123 (1965).

11.9 Repeat Prob. 11.8, using one of the following differential equations, with an interval $1 \leq x \leq 2$ and an initial condition of $y = 1$ at $x = 1$.

(a) $y' = 1/x$
(b) $y' = x \ln y$
(c) $y' = x^2 + \ln y$

11.10 Repeat Prob. 11.8, using one of the following differential equations with an interval $0 \leq x \leq \pi/2$ and an initial condition of $y = 1$ at $x = 0$.

(a) $y' = \sin x$
(b) $y' = \sin y$
(c) $y' = x^2 + \sin y$

11.11 Repeat Prob. 11.8, using one of the following second-order differential equations, with an interval $0 \leq x \leq 1$ and initial conditions of $y = 1$ and $y' = 0$ at $x = 0$.

(a) $y'' + y' + y = 0$
(b) $y'' + y'/x + y/x^2 = 0$

11.12 A pendulum is released from rest at $\theta = 0.8$ rad. Determine $\theta(t)$ through one complete cycle. The differential equation is given in Eq. (11.94).

11.13 Solve Prob. 11.12 with initial conditions of $\theta = 0.8$ rad and $d\theta/dt = 2$ rad/sec at $t = 3$ sec.

11.14 In Prob. 10.9 assume that the mass m is released from rest at $x = 2$ ft. Determine $x(t)$ through one complete cycle of the motion. The spring constant $k = 5.0$ lb/ft and $m = 0.25$ slug.

11.15 In Prob. 11.14 the relationship between the force f and the distance x is

$$f = 5.0x + 0.5x^3$$

Determine $x(t)$ through one complete cycle of the motion.

11.16 Repeat Prob. 11.15, including the effect of air resistance, given by

$$\phi(V) = CV|V|$$

Assume that $C = 0.02$ lb/(ft/sec)^2.

◆**11.17** Water is flowing into a tank at the constant rate of 10 ft^3/sec and flowing out at the rate given by

$$F_2 = 2.5 \sqrt{h} \qquad \text{ft}^3/\text{sec}$$

where h is the depth of the water in feet. The cross-sectional area of the tank is 10 ft^2. Determine the height of the water after 60 sec, assuming that $h_0 = 10$ ft.

11.18 In Fig. 11.10, $c = 5 \times 10^{-6}$ farad and $L = 0.10$ henry. Determine $\phi(t)$ through one complete cycle. At $t = 0$, $\phi = 0.5$ and $d\phi/dt = 0$.

11.19 Repeat Prob. 11.18, assuming that the current i is related to the flux ϕ by

$$i = 10.0\phi - 3.0\phi^3$$

♦**11.20** The transient behavior of the neutron flux in a pressurized-water nuclear reactor in which there has been a sudden change in the water temperature may be approximated by the following differential equations:

$$\frac{d\phi}{dt} = \frac{\rho - \beta}{l} \phi + \lambda vc$$

$$\frac{dc}{dt} = \frac{\beta}{vl} \phi - \lambda c$$

where

ϕ = neutron flux, neutrons/(cm²)(sec)
t = time, sec
c = precursor concentration, atoms/cm³
β = fraction of delayed neutrons = 7.55×10^{-3}
l = mean thermal-neutron lifetime = 5×10^{-4} sec
v = neutron velocity = 3×10^5 cm/sec
λ = average decay constant for precursors = 8.2×10^{-2} sec⁻¹
ρ = reactivity, a function of time and also a function of whether or not the control rods of the reactor have been inserted. As soon as ϕ/ϕ_0 reaches a value of 2, the control rods are started in. The various expressions for ρ are:

$\rho = 0.03t$ for $t < 0.7$ and the control rods not inserted
$\rho = 0.021$ for $t > 0.7$ and the control rods not inserted
$$\rho = \left(1 - \frac{t - t_2}{0.7}\right) \rho_2 \text{ if the control rods have been inserted, until } \rho = -0.01$$

where t_2 = time at which $\phi/\phi_0 = 2$
        $\rho_2 = \rho$ at time t_2

$\rho = -0.01$ if the control rods are fully inserted

The initial values are:

$\phi_0 = 5 \times 10^{13}$
$c_0 = 3.08 \times 10^{10}$
$\rho_0 = 0$

Using Euler's method of numerical integration, determine the neutron flux ϕ as a function of time. Continue the calculation until ϕ returns to its initial value ϕ_0. Use time increments of 0.01 and 0.001 sec, printing results only at intervals of 0.01 sec in each case.

11.21 Repeat Prob. 11.20, using the Runge-Kutta method of integration.

11.22 Repeat Prob. 11.20, using Hamming's method of integration. Obtain the necessary starting values by using the Runge-Kutta method.

12   EMPIRICAL FORMULAS AND APPROXIMATION

12.1   INTRODUCTION

Engineers and scientists frequently conduct experiments of a quantitative nature, such as making measurements of a physical property of a material at several temperatures. The results of the experiment are usually obtained as a set of values of the independent variable with corresponding values of the dependent variable. For example, an experiment to determine the electrical resistance of aluminum as a function of temperature might produce the data shown in Table 12.1.

Table 12.1

Temperature t, °C	Specific resistance ρ, ohm–cir mils/ft
0	17.8
10	18.6
20	19.3
30	20.2
40	20.8
50	21.6
60	22.2
70	23.1
80	23.8
90	24.6

These data must now be analyzed, possibly for the purpose of seeing whether they obey a known physical law or else to see whether they can be reasonably well represented by some empirical function. Also, reducing the data from a tabular form to that of an analytical function has obvious advantages in the ease of subsequent interpolation, differentiation, and integration.

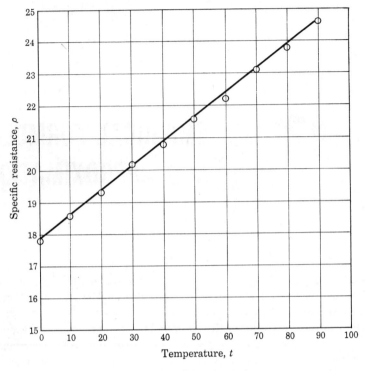

Fig. 12.1

If the form of the function is not known in advance, a plot of the data points will often be of great assistance in choosing the function. Thus, the data of Table 12.1 are plotted on rectangular coordinates in Fig. 12.1, and it is evident that a straight line will pass through or near all the points. (Admittedly, this is a very simple example.) Hence, in this case, the data can be reduced from the tabular form to a function that contains only two parameters. In doing so, we are assuming either that the linear function is theoretically sound and that deviations of the data points from this function are caused by random errors in their measurement or that the linear function is an empirical representation of the data that is not exact but is good enough to be useful. In either case, we are replacing a table of discrete points by a smooth function that covers all values of the independent variable in the given interval.

There is still a problem to be resolved, namely: *Which* straight line should be used? A very large number of lines can be drawn, and we need a criterion for choosing the one that best fits the data. In other words, we need a definition of *best fit* of an empirical function to a set of data points.

There is no such unique definition, although as we shall see, one particular criterion is most often used.

12.2   MINIMIZING SUM OF DEVIATIONS

Perhaps the simplest method would be to draw the line in such a way that the magnitude of the sum of the deviations of the data points lying above the line equaled that of those lying below the line.   Letting the data points be Y_k and the functional values $y_k = a + bx_k$, we have for the deviations

$$d_k = Y_k - y_k = Y_k - a - bx_k \qquad (12.1)$$

An approach called the *method of averages* can be used to solve for the parameters in this case.   The data are first divided into n equal groups, where n is the number of parameters, and then the expressions for the deviations of the individual data points are written down.   The deviations within each group are summed and set equal to zero, giving n simultaneous equations to be solved for the n parameters.   Using the data of Table 12.1, the procedure is as follows:

Group 1	Group 2
$17.8 - a - (0)b$	$21.6 - a - 50b$
$18.6 - a - 10b$	$22.2 - a - 60b$
$19.3 - a - 20b$	$23.1 - a - 70b$
$20.2 - a - 30b$	$23.8 - a - 80b$
$20.8 - a - 40b$	$24.6 - a - 90b$
$96.7 - 5a - 100b = 0$	$115.3 - 5a - 350b = 0$

Solving the equations of the sums gives $a = 17.9$, $b = 0.0744$.   It is apparent that this method does not lead to a unique solution for the parameters but is dependent on how the groups are selected.   Consider, for example, the case of three data points to be fitted by a straight line.

If, instead of minimizing the sum of the algebraic deviations, we try to work with absolute values, we have the function

$$\sum_{k=1}^{N} |d_k| = \sum_{k=1}^{N} |Y_k - a - bx_k| \qquad (12.2)$$

According to this criterion of best fit, the sum in Eq. (12.2) should be minimized.   Since the sum is a function of a and b, we would like to use the standard method of calculus of finding the minimum by taking the partial derivatives of the right-hand side of Eq. (12.2) with respect to a and b, setting them equal to zero, and solving the resulting equations for

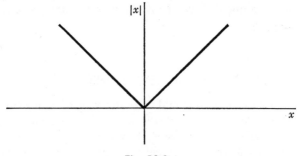

Fig. 12.2

a and *b*.   This is not possible, however, because the absolute value function is not differentiable at its minimum, as shown in Fig. 12.2.   Furthermore, a short consideration of the proposed criterion shows that it would not lead to a unique solution for the parameters.

12.3   METHOD OF LEAST SQUARES

The criterion of best fit that is most widely used is that the sum of the squares of the deviations shall be a minimum.   Without going into the statistical reasons for using this criterion, we can see qualitatively that it assumes that positive and negative deviations (errors) are equally likely and that large deviations are less likely to occur than small deviations. In the latter respect it is different from the criteria assumed in the previous section.   There is also the assumption that the magnitude of any deviation is independent of the value of the variable.   If, for example, a scale were graduated logarithmically, this would not be true, since an error in reading at one end of the scale would correspond to a larger *absolute* deviation than at the other end.   A final assumption is that the errors in measurement of the independent variable are insignificant compared with those in the dependent variable.

The function to be minimized is

$$\sum_{k=1}^{N} d_k{}^2 = \sum_{k=1}^{N} (Y_k - a - bx_k)^2 = g(a,b) \tag{12.3}$$

This function *is* differentiable at its minimum, and taking partial derivatives with respect to *a* and *b*, we have

$$\frac{\partial g}{\partial a} = -2 \sum_{k=1}^{N} (Y_k - a - bx_k)(1)$$

$$\frac{\partial g}{\partial b} = -2 \sum_{k=1}^{N} (Y_k - a - bx_k)x_k$$

Setting the derivatives equal to zero and rearranging,

$$aN \quad + b\Sigma x_k = \Sigma Y_k$$
$$a\Sigma x_k + b\Sigma x_k{}^2 = \Sigma x_k Y_k \tag{12.4}$$

where the summations all extend from $k = 1$ to $k = N$. These equations are called the *normal equations* for the method of least squares. They are linear in the unknowns a and b and can be solved by the methods of Chap. 7, such as Gaussian elimination, except under conditions where the coefficient matrix approaches zero. This point is discussed in Sec. 12.5.

As an example, let us apply the method of least squares to the data of Table 12.1. The required sums, for $N = 10$, are

$$\Sigma t_k = 450 \qquad \Sigma \rho_k = 212.0$$
$$\Sigma t_k{}^2 = 28,500 \qquad \Sigma t_k \rho_k = 10,157$$

and the normal equations are

$$10a + \quad 450b = 212.0$$
$$450a + 28,500b = 10,157$$

Solving for a and b gives $a = 17.8$, $b = 0.0748$.

The derivation of the normal equations for higher-order polynomials is accomplished in the same way as Eq. (12.4) was derived. For a second-order polynomial $y = a + bx + cx^2$, partial differentiation with respect to the parameters a, b, and c leads to the system

$$aN \quad + b\Sigma x_k + c\Sigma x_k{}^2 = \Sigma Y_k$$
$$a\Sigma x_k + b\Sigma x_k{}^2 + c\Sigma x_k{}^3 = \Sigma x_k Y_k \tag{12.5}$$
$$a\Sigma x_k{}^2 + b\Sigma x_k{}^3 + c\Sigma x_k{}^4 = \Sigma x_k{}^2 Y_k$$

where the summations all extend from $k = 1$ to $k = N$. In general, for an nth-order polynomial, the elements of the coefficient matrix $\mathbf{A}$ may be represented as

$$a_{ij} = \sum_{k=1}^{N} x_k{}^{i+j-2} \qquad i = 1, n+1; j = 1, n+1 \tag{12.6}$$

and those of the right-hand column vector $\mathbf{C}$ as

$$c_i = \sum_{k=1}^{N} x_k{}^{i-1} Y_k \qquad i = 1, n+1 \tag{12.7}$$

12.4 EQUATIONS NONLINEAR IN PARAMETERS

The polynomial equations discussed in Sec. 12.3 are linear with respect to the parameters, which leads to a system of simultaneous linear equations in the least-squares derivation. If the same method of derivation is applied to equations that are not linear in the parameters, the resulting system of simultaneous equations will also not be linear and cannot be readily solved. Consider, for example, the equation

$$y = ae^{bx} \tag{12.8}$$

where it is desired to determine a and b in a least-squares sense. Following the regular procedure, the function to be minimized is

$$\sum_{k=1}^{N} d_k{}^2 = \sum_{k=1}^{N} (Y_k - ae^{bx_k})^2 = g(a,b)$$

Taking the partial derivatives of $g(a,b)$, we have

$$\frac{\partial g}{\partial a} = -2 \sum_{k=1}^{N} (Y_k - ae^{bx_k})e^{bx_k}$$

$$\frac{\partial g}{\partial b} = -2a \sum_{k=1}^{N} (Y_k - ae^{bx_k})x_k e^{bx_k}$$

and the equations to be solved for a and b are

$$\Sigma Y_k e^{bx_k} = a\Sigma e^{2bx_k}$$
$$\Sigma Y_k x_k e^{bx_k} = a\Sigma x_k e^{2bx_k}$$

which are clearly not in a useful form.

On the other hand, Eq. (12.8) can easily be converted to a linear form by taking the logarithm of both sides and defining new variables. Thus,

$$\ln y = \ln a + bx$$

or

$$\bar{y} = \bar{a} + bx \tag{12.9}$$

The least-square values of the parameters $\bar{a}$ and b in Eq. (12.9) may now be obtained in the usual way. However, in general, they will not be the correct values for Eq. (12.8), because of the changes of variable. By using Eq. (12.9), the sums of the squares of the deviations of $\ln y$ are being minimized, rather than those of the true variable y. This will tend to overemphasize the deviations at the low end of the curve and underemphasize those at the high end.

An iterative correction procedure may be applied to the parameters in the following manner: If we let the values obtained from the use of Eq. (12.9) be a_0 (calculated from $e^{\bar{a}}$) and b_0, then the true values of a and b may be written as

$$a = a_0 + \Delta a$$
$$b = b_0 + \Delta b \tag{12.10}$$

The problem is now that of finding Δa and Δb. Regarding y as a function of a and b,

$$y = f(a,b) = f(a_0 + \Delta a, b_0 + \Delta b) \tag{12.11}$$

and expanding in Taylor's series about the point (a_0, b_0) through the first derivatives,

$$y = f(a_0, b_0) + \Delta a \left(\frac{\partial f}{\partial a}\right)_0 + \Delta b \left(\frac{\partial f}{\partial b}\right)_0 \tag{12.12}$$

where the partial derivatives are to be evaluated at (a_0, b_0). From Eq. (12.8),

$$\frac{\partial f}{\partial a} = e^{bx} \qquad \frac{\partial f}{\partial b} = axe^{bx}$$

Substituting into Eq. (12.12),

$$y = f(a_0, b_0) + \Delta a\, e^{b_0 x} + \Delta b\, a_0 x e^{b_0 x} \tag{12.13}$$

Equation (12.13) is linear in Δa and Δb, and the method of least squares is directly applicable to the calculation of Δa and Δb. The function to be minimized is

$$\sum_{k=1}^{N} d_k^2 = \sum_{k=1}^{N} [Y_k - f_k(a_0, b_0) - \Delta a\, e^{b_0 x_k} - \Delta b\, a_0 x_k e^{b_0 x_k}]^2 \tag{12.14}$$

Differentiating with respect to Δa and Δb and rearranging give the normal equations,

$$\begin{aligned}
\Sigma e^{b_0 x_k} d_k^* &= \Delta a\, \Sigma e^{2b_0 x_k} + \Delta b\, a_0 \Sigma x_k e^{2b_0 x_k} \\
\Sigma x_k e^{b_0 x_k} d_k^* &= \Delta a\, \Sigma x_k e^{2b_0 x_k} + \Delta b\, a_0 \Sigma x_k^2 e^{2b_0 x_k}
\end{aligned} \tag{12.15}$$

where the summations extend from $k = 1$ to N and

$$d_k^* \equiv Y_k - f_k(a_0, b_0) = Y_k - a_0 e^{b_0 x_k}$$

which is the deviation between the observed data and the functional values calculated from a_0 and b_0.

    The improved values of a and b may be used in place of a_0 and b_0 in Eq. (12.15) to calculate another Δa and Δb, and so on, iteratively. Usually the process converges rapidly, depending on the functions being used and the error in the initial values.

    Although there is ordinarily no reason to apply the above linearizing technique to an equation that is already linear in the parameters, nevertheless it will work, as shown in the following example: Consider again the data of Table 12.1, which were fitted by a straight line in Sec. 12.3. This time, instead of calculating the least-square values of the parameters directly, assume that an initial estimate of $a_0 = 17.0$ and $b_0 = 0.08$ has been made and is to be improved. The functional equation is

$$\rho = a + bt = f(a,b) = f(a_0 + \Delta a,\, b_0 + \Delta b)$$

and, by Taylor's series,

$$f(a,b) = f(a_0,b_0) + \Delta a \left(\frac{\partial f}{\partial a}\right)_0 + \Delta b \left(\frac{\partial f}{\partial b}\right)_0 = f(a_0,b_0) + \Delta a + t\,\Delta b$$

The function to be minimized is

$$\sum_{k=1}^{N} d_k{}^2 = \sum_{k=1}^{N} [\rho_k - f_k(a_0,b_0) - \Delta a - t_k\,\Delta b]^2$$

Taking derivatives with respect to Δa and Δb and rearranging give the normal equations,

$$\begin{aligned}
\Sigma d_k^* &= \Delta a\, N + \Delta b\, \Sigma t_k \\
\Sigma t_k d_k^* &= \Delta a\, \Sigma t_k + \Delta b\, \Sigma t_k{}^2
\end{aligned} \tag{12.16}$$

where $d_k^* = \rho_k - a_0 - b_0 t_k$ and the summations extend from $k = 1$ to N.
The values needed in the calculations are shown in Table 12.2.

$$\Sigma t_k = 450 \qquad \Sigma t_k{}^2 = 28{,}500 \qquad \Sigma d_k^* = 6.0 \qquad \Sigma t_k d_k^* = 227.0$$

Substituting into Eq. (12.16) and solving yield $\Delta a = 0.835$ and $\Delta b = -0.0052$. Hence, $a = 17.8$, and $b = 0.0748$, in agreement with the values obtained previously.

Table 12.2

t	ρ	$f(a_0,b_0)$	d^*	td^*
0	17.8	17.0	0.8	0.0
10	18.6	17.8	0.8	8.0
20	19.3	18.6	0.7	14.0
30	20.2	19.4	0.8	24.0
40	20.8	20.2	0.6	24.0
50	21.6	21.0	0.6	30.0
60	22.2	21.8	0.4	24.0
70	23.1	22.6	0.5	35.0
80	23.8	23.4	0.4	32.0
90	24.6	24.2	0.4	36.0

12.5   USE OF ORTHOGONAL POLYNOMIALS

As presented in the preceding sections, the method of least squares is practical for determining only a relatively small number of constants.

Inspection of the system of Eqs. (12.5) shows that even for as few as three constants the range of values in the coefficient matrix is from N to the sum of $x_k{}^4$. As the number of constants and equations increases, the range becomes greater and the matrix becomes increasingly difficult to handle computationally. More and more figures must be retained at each step in order to keep the round-off error under control, and double precision may be needed for as few as six equations.

The basis of the difficulty is that the coefficient matrix has a determinant value that approaches zero as the number of terms increases; i.e., it becomes *near-singular*. It may also be characterized as an *ill-conditioned* matrix, where a measure of ill conditioning is the relative magnitude of the elements of the matrix and its determinant value.

Assuming that the x_k values are equally spaced in some interval, say $(0,1)$, the summation form of Eq. (12.6) may be approximated by an integral as

$$a_{ij} = \sum_{k=1}^{N} x_k{}^{i+j-2} \approx N \int_0^1 x_k{}^{i+j-2}\, dx = \frac{N}{i+j-1}$$

The coefficient matrix for n equations is, then, approximately

$$\begin{bmatrix} \dfrac{N}{1} & \dfrac{N}{2} & \cdots & \dfrac{N}{n} \\[2mm] \dfrac{N}{2} & \dfrac{N}{3} & \cdots & \dfrac{N}{n+1} \\[2mm] \cdots & \cdots & \cdots & \cdots \\[2mm] \dfrac{N}{n} & \dfrac{N}{n+1} & \cdots & \dfrac{N}{2n-1} \end{bmatrix} = N^n \begin{bmatrix} \dfrac{1}{1} & \dfrac{1}{2} & \dfrac{1}{3} & \cdots & \dfrac{1}{n} \\[2mm] \dfrac{1}{2} & \dfrac{1}{3} & \dfrac{1}{4} & \cdots & \dfrac{1}{n+1} \\[2mm] \cdots & \cdots & \cdots & \cdots \\[2mm] \dfrac{1}{n} & \dfrac{1}{n+1} & \dfrac{1}{n+2} & \cdots & \dfrac{1}{2n-1} \end{bmatrix}$$

$$(12.17)$$

The matrix on the right-hand side of Eq. (12.17) is known as the *Hilbert matrix*. Values of its determinant, H_n for n up to nine, are shown in Table 12.3.

Table 12.3

n	H_n
1	1.0
2	8.3×10^{-2}
3	4.6×10^{-4}
4	1.7×10^{-7}
5	3.7×10^{-12}
6	5.4×10^{-18}
7	4.8×10^{-25}
8	2.7×10^{-33}
9	9.7×10^{-43}

The computational problems inherent in handling the coefficient matrix of the least-squares method can be avoided if the matrix can be initially developed in diagonal form. Then the simultaneous equations will be independent and will, in effect, be already solved. Restricting the discussion to polynomials in x, this can be accomplished by representing the function for y as a sum of *orthogonal functions* of x rather than as a sum of powers of x. Accordingly, instead of writing

$$y = a_0 x^0 + a_1 x + a_2 x^2 + \cdots + a_n x^n$$

we shall write

$$y = A_0 \phi_0(x) + A_1 \phi_1(x) + A_2 \phi_2(x) + \cdots + A_n \phi_n(x) \tag{12.18}$$

where the A_i are the parameters to be determined and the $\phi_i(x)$ are suitable functions of x.

It may be remembered that an orthogonal set of vectors is a set of mutually perpendicular vectors and that the inner product of two perpendicular vectors $\mathbf{A}$ and $\mathbf{B}$, with components a_k and b_k, is zero. In equation form,

$$\sum_{k=1}^{N} a_k b_k = 0$$

This idea may be extended to functions by defining an orthogonal set as one for which the following holds true,

$$\sum_{k=1}^{N} \phi_i(x_k) \phi_j(x_k) = \begin{cases} 0 & i \neq j \\ \lambda_i > 0 & i = j \end{cases} \tag{12.19}$$

where the x_k are at specific points. If the set of functions is normalized, i.e., *orthonormal*,

$$\sum_{k=1}^{N} \phi_i(x_k) \phi_j(x_k) = \begin{cases} 0 & i \neq j \\ 1 & i = j \end{cases} \tag{12.20}$$

On the assumption, for the sake of simplicity, that the $\phi_i(x)$ are orthonormal, then the coefficients A_i may be found from

$$A_i = \sum_{k=1}^{N} y_k \phi_i(x_k) \tag{12.21}$$

since, by substituting Eq. (12.18) into Eq. (12.21), we have

$$A_i = \sum_{k=1}^{N} [A_0 \phi_0(x_k) + A_1 \phi_1(x_k) + \cdots + A_i \phi_i(x_k)$$
$$+ \cdots + A_n \phi_n(x_k)] \phi_i(x_k) \tag{12.22}$$

and all the terms on the right-hand side vanish except for

$$A_i \sum_{k=1}^{N} \phi_i(x_k)\phi_i(x_k) = A_i$$

Before specifying the functions $\phi_i(x)$ that we shall use, let us next examine the application of the least-squares method to Eq. (12.18). Using an equation of only two terms, we have for the function to be minimized:

$$\sum_{k=1}^{N} d_k^2 = \sum_{k=1}^{N} [Y_k - A_0\phi_0(x_k) - A_1\phi_1(x_k)]^2 = g(A_0,A_1)$$

The partial derivatives with respect to A_0 and A_1 are

$$\frac{\partial g}{\partial A_0} = -2 \sum_{k=1}^{N} [Y_k - A_0\phi_0(x_k) - A_1\phi_1(x_k)]\phi_0(x_k)$$

$$\frac{\partial g}{\partial A_1} = -2 \sum_{k=1}^{N} [Y_k - A_0\phi_0(x_k) - A_1\phi_1(x_k)]\phi_1(x_k)$$

Because of the orthogonality of the functions, the equations reduce to the diagonal form

$$\sum_{k=1}^{N} Y_k\phi_0(x_k) = A_0$$

$$\sum_{k=1}^{N} Y_k\phi_1(x_k) = A_1$$

and the problems associated with the usual coefficient matrix have been avoided.

12.6   CHEBYSHEV POLYNOMIALS[1]

There are many sets of orthogonal polynomials, but we shall consider here only the Chebyshev polynomials. They are named after the mathematician who discovered many of their properties and are defined by

$$T_n(x) = \cos (n \arccos x) \tag{12.23}$$

For $n = 0$, $T_0(x) = \cos (0) = 1$; for $n = 1$, $T_1(x) = \cos (\arccos x) = x$; for $n = 2$, $T_2(x) = \cos (2 \arccos x) = 2x^2 - 1$; etc. The polynomials for

[1] An excellent survey of the theory and uses of Chebyshev polynomials is given by C. Lanczos, Tables of Chebyshev Polynomials $S_n(x)$ and $C_n(x)$, *Natl. Bur. Std.* (*U.S.*), *Appl. Math. Ser.*, vol. 9, December, 1952.

$n = 0$ to 8 are shown in Table 12.4.   A few points of interest are that the

Table 12.4
$T_0(x) = 1$
$T_1(x) = x$
$T_2(x) = 2x^2 - 1$
$T_3(x) = 4x^3 - 3x$
$T_4(x) = 8x^4 - 8x^2 + 1$
$T_5(x) = 16x^5 - 20x^3 + 5x$
$T_6(x) = 32x^6 - 48x^4 + 18x^2 - 1$
$T_7(x) = 64x^7 - 112x^5 + 56x^3 - 7x$
$T_8(x) = 128x^8 - 256x^6 + 160x^4 - 32x^2 + 1$

coefficient of the first term is 2^{n-1}, that the signs alternate, and that the polynomials contain only either odd powers of x or even powers, but not both.   A simple recurrence relation may be derived from the trigonometric identity

$$\cos (n + 1)\theta + \cos (n - 1)\theta = 2 \cos \theta \cos n\theta \qquad (12.24)$$

By letting $\cos \theta = x$ and applying Eq. (12.23), we find

$$T_{n+1}(x) + T_{n-1}(x) = 2xT_n(x)$$

or

$$T_{n+1}(x) = 2xT_n(x) - T_{n-1}(x) \qquad (12.25)$$

Again by letting $\cos \theta = x$, $\theta = \arccos x$, and another form for $T_n(x)$ is

$$T_n(x) = \cos n\theta \qquad (12.26)$$

The set of functions $\cos \theta$, $\cos 2\theta$, . . . , $\cos n\theta$ is that of the Fourier series; hence the Chebyshev polynomials are a somewhat disguised form of the Fourier series.   The orthogonality property of these functions is

$$\sum_{k=0}^{N-1} \cos m\theta_k \cos n\theta_k = \begin{cases} 0 & m \neq n \\ \dfrac{N}{2} & m = n \neq 0 \\ N & m = n = 0 \end{cases}$$

where $m, n < N$ and the θ_k are taken at the N equidistant points in the interval $0 < \theta_k < 2\pi$,

$$\theta_k = \frac{\pi(2k + 1)}{2N}$$

On the assumption, for example, that $N = 4$, $m = 1$, and $n = 2$, the sum is given by

$$\cos\frac{\pi}{8}\cos\frac{\pi}{4} + \cos\frac{3\pi}{8}\cos\frac{3\pi}{4} + \cos\frac{5\pi}{8}\cos\frac{5\pi}{4} + \cos\frac{7\pi}{8}\cos\frac{7\pi}{4}$$

$$= (0.9205)(0.7071) + (0.3827)(-0.7071)$$
$$+ (-0.3827)(-0.7071) + (0.9205)(0.7071)$$
$$= 0$$

Upon translation back to the polynomial form, the orthogonality relationship is

$$\sum_{k=0}^{N-1} T_m(x_k)T_n(x_k) = \begin{cases} 0 & m \neq n \\ \dfrac{N}{2} & m = n \neq 0 \\ N & m = n = 0 \end{cases} \tag{12.27}$$

where m, $n < N$ and $x_k = \cos[\pi(2k+1)/2N]$.

We are now in the position to calculate the least-square coefficients A_i in the equation

$$y = A_0 T_0(x) + A_1 T_1(x) + A_2 T_2(x) + \cdots + A_n T_n(x)$$

From the orthogonality relationships of Eq. (12.27), they are given by

$$A_0 = \frac{1}{N}\sum_{k=0}^{N-1} Y(x_k)T_0(x_k) = \frac{1}{N}\sum_{k=0}^{N-1} Y(x_k)$$

$$A_i = \frac{2}{N}\sum_{k=0}^{N-1} Y(x_k)T_i(x_k) = \frac{2}{N}\sum_{k=0}^{N-1} Y(x_k)\cos\frac{i\pi(2k+1)}{2N} \tag{12.28}$$

for $i = 1, 2, 3, \ldots$, and the x_k are the same as those given in Eq. (12.27). By the nature of its definition x_k lies in the range from -1 to $+1$. To convert from a different range of the independent variable z, let

$$x = \frac{2z - (a+b)}{b-a}$$

or

$$z = \frac{1}{2}(a+b) + \frac{x}{2}(b-a) \tag{12.29}$$

where a and b are the limits of z.

As an example in the application of Eqs. (12.28), the data of Table 12.1 will again be used, this time in fitting the function

$$\rho = A_0 T_0(x) + A_1 T_1(x) \tag{12.30}$$

Since $N = 10$,

$$x_k = \cos \frac{i\pi(2k + 1)}{20} \tag{12.31}$$

and, by converting to the range of t,

$$t_k = \tfrac{1}{2}(0 + 90) + 45x_k = 45(1 + x_k)$$

It is immediately apparent that the values of t_k calculated from Eq. (12.31) will not, in general, coincide with those at which $\rho(t)$ was measured. The data were taken at equal intervals of t, but the use of the Chebyshev polynomials requires equal intervals of the angle in Eq. (12.31). It is necessary either to interpolate the measured values or to use alternative formulas that involve numerical integration.

In this example, linear interpolation was used to find $\rho(t_k)$. The calculations are summarized in Table 12.5. The sum of the ρ column is 211.99

Table 12.5

k	θ	$\cos \theta$	t	ρ	$\rho \cos \theta$
0	$\pi/20$	0.9877	89.45	24.56	24.26
1	$3\pi/20$	0.8910	85.10	24.21	21.57
2	$5\pi/20$	0.7071	76.82	23.58	16.67
3	$7\pi/20$	0.4540	65.43	22.69	10.30
4	$9\pi/20$	0.1564	52.04	21.72	3.40
5	$11\pi/20$	−0.1564	37.96	20.68	− 3.23
6	$13\pi/20$	−0.4540	24.57	19.71	− 8.95
7	$15\pi/20$	−0.7071	13.18	18.82	−13.31
8	$17\pi/20$	−0.8910	4.90	18.18	−16.20
9	$19\pi/20$	−0.9877	0.55	17.84	−17.62

and of the $\rho \cos \theta$ column is 16.89. Then,

$$A_0 = \frac{211.99}{10} = 21.20 \qquad A_1 = \frac{16.89}{5} = 3.378$$

Putting the equation in terms of t,

$$\rho = 21.20 + 3.378 \left(\frac{t}{45} - 1\right) = 17.82 + 0.075t$$

This agrees with the previous calculations.

For a continuous function, specified at all points in the interval $(0,\pi)$, an integral replaces the summation in the orthogonality relationships. For the cosine functions, the relationship becomes

$$\int_0^\pi \cos m\theta \cos n\theta \, d\theta = \begin{cases} 0 & m \neq n \\ \dfrac{\pi}{2} & m = n \neq 0 \\ \pi & m = n = 0 \end{cases} \tag{12.32}$$

Upon making the usual transformation by letting $x = \cos \theta$, the relationship for the Chebyshev polynomials is

$$\int_{-1}^1 \frac{T_m(x)\, T_n(x)}{1 - x^2} \, dx = \begin{cases} 0 & m \neq n \\ \dfrac{\pi}{2} & m = n \neq 0 \\ \pi & m = n = 0 \end{cases} \tag{12.33}$$

The coefficients of the expansion of Eq. (12.18) are now given by

$$\begin{aligned} A_0 &= \frac{1}{\pi} \int_{-1}^1 f(x) \, dx \\ A_i &= \frac{2}{\pi} \int_{-1}^1 f(x)\, T_i(x) \, dx \end{aligned} \tag{12.34}$$

for $i = 1, 2, 3, \ldots$. Hence, an alternative to evaluating the sums of Eqs. (12.28) is a numerical integration of Eq. (12.34). Because of the change of variable necessary to fall within the range $(-1,1)$, the ordinates used in the integration will not be calculated at equally spaced values of x.

12.7 GENERALIZED ORTHOGONAL POLYNOMIALS

Forsythe[1] has developed a set of orthogonal polynomials for which the data points may be spaced arbitrarily, making the polynomials more convenient to use than the Chebyshev polynomials. They are defined by

$$\begin{aligned} P_{-1}(x) &= 0 \\ P_0(x) &= 1 \\ P_1(x) &= (x - \alpha_1)P_0(x) - \beta_0 P_{-1}(x) \\ P_2(x) &= (x - \alpha_2)P_1(x) - \beta_1 P_0(x) \\ &\cdots\cdots\cdots\cdots\cdots\cdots\cdots \\ P_{j+1}(x) &= (x - \alpha_{j+1})P_j(x) - \beta_j P_{j-1}(x) \end{aligned} \tag{12.35}$$

[1] G. E. Forsythe, Generation and Use of Orthogonal Polynomials for Data-fitting with a Digital Computer, *J. Soc. Ind. Appl. Math.*, vol. 5, no. 2, June, 1957.

where $\beta_0 = 0$ and α_j and β_j depend on the spacing of the data points and are chosen in such a way that the set of polynomials is orthogonal under summation, that is,

$$\sum_{k=1}^{N} P_m(x_k)P_n(x_k) = 0 \qquad m \neq n \tag{12.36}$$

In order to find expressions for α_j and β_j, we may multiply the recursion relationship of Eqs. (12.35) by $P_i(x_k)$ and form the sums

$$\sum_{k=1}^{N} P_i(x_k)P_{j+1}(x_k) = \sum_{k=1}^{N} (x_k - \alpha_{j+1})P_i(x_k)P_j(x_k) - \beta_j \sum_{k=1}^{N} P_i(x_k)P_{j-1}(x_k) \tag{12.37}$$

If we assume that the functions *are* orthogonal, the coefficient of β_j equals zero for $i = j$. If we now set the left-hand side of Eq. (12.37) equal to zero and solve for α_{j+1}, we get

$$\alpha_{j+1} = \frac{\displaystyle\sum_{k=1}^{N} x_k[P_j(x_k)]^2}{\displaystyle\sum_{k=1}^{N} [P_j(x_k)]^2} \tag{12.38}$$

Similarly, for $i = j - 1$, the coefficient of α_{j+1} in Eq. (12.37) equals zero. Setting the left-hand side of the equation equal to zero, we obtain

$$\beta_j = \frac{\displaystyle\sum_{k=1}^{N} x_k P_{j-1}(x_k)P_j(x_k)}{\displaystyle\sum_{k=1}^{N} [P_{j-1}(x_k)]^2} \tag{12.39}$$

The choice of α_{j+1} by Eq. (12.38) is such that $\Sigma P_j(x_k)P_{j+1}(x_k)$ and $\Sigma P_{j-1}(x_k)P_j(x_k)$ are zero. Thus, *adjacent* polynomials in the set are orthogonal. The choice of β_j in Eq. (12.39) makes $\Sigma P_{j-1}(x_k)P_{j+1}(x_k)$ zero, or polynomials that are *once removed* are orthogonal. The proof of mutual orthogonality of the complete set may be derived from these relationships and the definitions of Eqs. (12.35).

The results of an application of the generalized polynomials are presented in Table 12.6. The data of Table 12.1 have been used in fitting the function

$$\rho = b_0 P_0(x) + b_1 P_1(x)$$

The necessary relationships, together with the results from the table, are

$$b_0 = \frac{\sum_{k=1}^{N} \rho_k P_0(t_k)}{\sum_{k=1}^{N} [P_0(t_k)]^2} = \frac{\sum_{k=1}^{N} \rho_k}{N} = \frac{212.0}{10} = 21.2$$

$$\alpha_1 = \frac{\sum_{k=1}^{N} t_k [P_0(t_k)]^2}{\sum_{k=1}^{N} [P_0(t_k)]^2} = \frac{\sum_{k=1}^{N} t_k}{N} = \frac{450}{10} = 45$$

$$P_1(t_k) = (t_k - \alpha_1)P_0(t_k) = t_k - \alpha_1$$

$$b_1 = \frac{\sum_{k=1}^{N} \rho_k P_1(t_k)}{\sum_{k=1}^{N} [P_1(t_k)]^2} = \frac{617.0}{8250.0} = 0.0748$$

Thus,

$$\rho = 21.2 + 0.0748(t - 45) = 17.83 + 0.0748t$$

Table 12.6

k	t	ρ	$P_1(t)$	$\rho[P_1(t)]$	$[P_1(t)]^2$
1	0	17.8	-45	-801.0	2025
2	10	18.6	-35	-651.0	1225
3	20	19.3	-25	-482.5	625
4	30	20.2	-15	-303.0	225
5	40	20.8	-5	-104.0	25
6	50	21.6	5	108.0	25
7	60	22.2	15	333.0	225
8	70	23.1	25	577.5	625
9	80	23.8	35	833.0	1225
10	90	24.6	45	1107.0	2025

12.8   APPROXIMATING A POWER SERIES

In Chap. 10, methods were presented for approximating a continuous function by Taylor's series.   For example, if $f(x) = \ln(1 + x)$, the first five terms of the expansion in the interval $0 \le x \le 1$ about $x = 0$ are

$$\ln(1 + x) = x - \frac{x^2}{2} + \frac{x^3}{3} - \frac{x^4}{4} + \frac{x^5}{5}$$

This series fits the function exactly at one point, namely, $x = 0$, and has the greatest error at the end of the interval.   That this will be so can be seen from the fact that all the functions x, x^2, x^3, x^4, and x^5 have their maximum values at the end of the interval.   The series is just a weighted sum of these functions.

It would seem preferable to be able to approximate a function in such a way that the error changes sign in the interval and has a lower maximum than in the above case.   In order to do this, the function must be expanded in terms of other functions that behave in a certain way.   In particular, they should not all take on their maximum values at the same point in the interval if the error is to be uniformly distributed.   Functions that satisfy this criterion are the sine and cosine functions in the interval $(0,\pi)$, as shown in Fig. 12.3 for the functions 1, $\cos \theta$, $\cos 2\theta$, and $\cos 3\theta$.   However, rather than use the cosine functions directly, it is more convenient to transform them into the Chebyshev polynomials.

The most remarkable property of the Chebyshev polynomials is concerned with this matter of error distribution.   Chebyshev proved that of

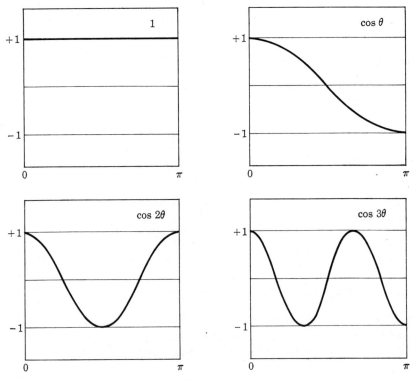

Fig. 12.3

all polynomials $p(x)$ of degree n, with the coefficient of the high-order term equal to 1, the corresponding Chebyshev polynomial

$$\frac{T_n(x)}{2^{n-1}}$$

has the smallest maximum magnitude in the interval $(-1,1)$. Since the maximum value of $T_n(x)$ is 1, this upper bound is $1/2^{n-1}$. The proof is given indirectly by assuming that it is not the case, i.e., that there *is* an nth-degree polynomial $p_n(x)$ that has a smaller maximum magnitude, and then showing that this leads to a logical contradiction. Letting $q_{n-1}(x)$ be the difference

$$q_{n-1}(x) = \frac{T_n(x)}{2^{n-1}} - p(x)$$

we may note that $q_{n-1}(x)$ is of degree $n - 1$, since the coefficients of x^n in the two polynomials are both 1. Because $T_n(x)$ may be written as

$$T_n(x) = \cos n\theta$$

it can be seen that it takes its maximum magnitude $n + 1$ times in the interval $-1 < x < 1$ or $0 < \theta < \pi$. If, now, $p(x)$ has a smaller maximum magnitude than $T_n(x)/2^{n-1}$, the difference $q_{n-1}(x)$ will be alternately positive and negative $n + 1$ times and must therefore have n real zeros. Since it is of degree $n - 1$, this is impossible and the polynomial $p(x)$, as postulated, cannot exist.

One of the consequences of this property of the Chebyshev polynomials is that the expansion of a function in terms of $T_n(x)$ achieves a very high rate of convergence. Practical use is made of this fact in the economization or telescoping of a polynomial.[1] In order to demonstrate this technique, we shall need to replace powers of x by combinations of $T_n(x)$. These may be derived from the functions in Table 12.4 and are presented in Table 12.7.

Table 12.7

x_n	Expansion into $T_n(x)$
1	T_0
x	T_1
x^2	$(T_0 + T_2)/2$
x^3	$(3T_1 + T_3)/4$
x^4	$(3T_0 + 4T_2 + T_4)/8$
x^5	$(10T_1 + 5T_3 + T_5)/16$
x^6	$(10T_0 + 15T_2 + 6T_4 + T_6)/32$
x^7	$(35T_1 + 21T_3 + 7T_5 + T_7)/64$
x^8	$(35T_0 + 56T_2 + 28T_4 + 8T_6 + T_8)/128$

[1] Lanczos, *op. cit.*

By use of the power series for $\ln (1 + x)$, the replacement proceeds as follows:

$$\ln (1 + x) = T_1 - \tfrac{1}{4}(T_0 + T_2) + \tfrac{1}{12}(3T_1 + T_3)$$
$$- \tfrac{1}{32}(3T_0 + 4T_2 + T_4) + \tfrac{1}{80}(10T_1 + 5T_3 + T_5)$$

This reduces to

$$\ln (1 + x) = -\tfrac{11}{32}T_0(x) + \tfrac{11}{8}T_1(x) - \tfrac{3}{8}T_2(x) + \tfrac{7}{48}T_3(x) - \tfrac{1}{32}T_4(x)$$
$$+ \tfrac{1}{80}T_5(x)$$

The maximum error in the original series in the interval $(0,1)$ occurs at $x = 1$ and is $\tfrac{1}{5}$. In the Chebyshev form, however, the coefficient of the high-order term is only $\tfrac{1}{80}$. Since the maximum magnitude of $T_5(x)$ is 1 in this interval, the error caused by dropping the last term is only $\tfrac{1}{80}$. In other words, the original fifth-degree polynomial can be *almost* as well represented by a fourth-degree polynomial, or, more important, the fourth-degree polynomial resulting from dropping $T_5(x)$ will have a smaller error than the simple fourth-degree polynomial of Taylor's series. In fact, both $T_4(x)$ and $T_5(x)$ can be dropped and a better fit still retained than that of the fourth-degree Taylor's series.

After the terms have been dropped, the polynomials for $T_0(x)$, $T_1(x)$, $T_2(x)$, and $T_3(x)$ may be resubstituted into the equation. This gives

$$\ln (1 + x) = \tfrac{1}{32} + \tfrac{15}{16}x - \tfrac{3}{4}x^2 + \tfrac{7}{12}x^3$$

The polynomial no longer fits exactly at the point $x = 0$. Instead, the exact fit at a point has been replaced by a fit of lower maximum error over the interval.

It is not actually necessary to go through the two substitutions that were carried out above. The high-order term, $x^5/5$, can be dropped by simply subtracting

$$\frac{1}{5}\frac{T_5(x)}{2^4} = \frac{1}{80}(16x^5 - 20x^3 + 5x)$$

Similarly, the next higher term can be dropped by subtracting

$$\frac{1}{4}\frac{T_4(x)}{2^3} = \frac{1}{32}(8x^4 - 8x^2 + 1)$$

PROBLEMS

♦**12.1** The following data for the heat capacity of methane gas at constant pressure ($p = 0$ atm absolute) were reported by E. Justi and H. Lüder.[1]

[1] E. Justi and H. Lüder, *Forsch. Gebiete Ingenieurw.*, **6:** 210–211 (1935).

Temperature t, °C	Heat capacity C_p, cal/(g mole)(°C)
0	8.24
100	9.40
200	10.70
300	12.15
400	13.40
500	14.60
600	15.65
700	16.60
800	17.40
900	18.23
1000	18.93

Using the method of least squares, determine the constants in the empirical equation

$$C_p = a + bt + ct^2$$

Find the maximum deviation of the data from this function and also the standard deviation, defined by

$$\sigma = \sqrt{\frac{\sum\limits_{k=1}^{N} d_k^2}{N-1}}$$

where N is the number of data points.

12.2   Using the data of Prob. 12.1, find the least-squares values of the constants in the empirical equation

$$C_p = a + bT + \frac{c}{T^2}$$

where T is the absolute temperature in degrees Kelvin.   Also, find the maximum deviation of the data from this function and the standard deviation.

12.3   Using the data of Prob. 12.1 and the initial estimates $a_0 = 8.24$, $b_0 = 0.0116$, and $c_0 = 3.5 \times 10^{-6}$, find improved values of a, b, and c in the function

$$C_p = a + bt + ct^2$$

◆**12.4**   In the following table are listed values of the vapor pressure of water in the range of 0 to 100°C.†

Using the method of least squares, find the constants a and b in the empirical equation

$$p = a10^{-b/T}$$

where T is in degrees Kelvin.   Also, calculate the maximum deviation and the standard deviation.

† From J. H. Keenan and F. G. Keyes, "Thermodynamic Properties of Steam," John Wiley & Sons, Inc., New York, 1937.   Reprinted with permission of the copyright holders.

Temperature t, °C   Vapor pressure p, in. Hg

0	0.1803
10	0.3626
20	0.6903
30	1.2527
40	2.1775
50	3.6420
60	5.8812
70	9.200
80	13.983
90	20.703
100	29.922

12.5 Using the data of Prob. 12.4, apply the method of least squares to find the constants a, b, and c in the empirical equation

$$p = a10^{-b/(T+c)}$$

Calculate the maximum deviation and the standard deviation, and compare these results with those of Prob. 12.4 if they are available.

◆**12.6** Write the Taylor's-series expansion for e^x in the interval $(0,1)$, including enough terms so that the error made in dropping the last term is less than 10^{-3}. If two more terms are added, how many terms can be dropped by Chebyshev economization without causing the maximum error in the interval to be greater than 10^{-3}?

12.7 Taking five terms in the Taylor's-series expansion for $\sin x$ in the interval $(0, \pi/2)$, find the third-degree polynomial that best approximates the series in this interval.

12.8 Determine a parabolic relationship for the specific weight of air as a function of altitude, using the data given in Prob. 3.113.

12.9 Determine an exponential relationship for the specific weight of air as a function of altitude, using the data given in Prob. 3.113.

12.10 Construct a flow chart and write a FORTRAN program to convert the coefficients of a sum of Chebyshev polynomials to those of the normal power series.

12.11 The following stress-strain data were taken from a tensile test on a steel specimen:

σ, psi	ϵ, in./in.
0	0
5,000	150×10^{-6}
10,000	350×10^{-6}
15,000	500×10^{-6}
20,000	650×10^{-6}
25,000	850×10^{-6}
30,000	1000×10^{-6}

Using the method of least squares, calculate the modulus of elasticity of the specimen.

BIBLIOGRAPHY

References on specific topics have been indicated by footnotes throughout the text. Other, more general references that have been used by the authors are listed below:

Arden, B. W.: "An Introduction to Digital Computing," Addison-Wesley Publishing Company, Inc., Reading, Mass., 1963.

Bennett, A. A., W. E. Milne, and H. Bateman: "Numerical Integration of Differential Equations," Dover Publications, Inc., New York, 1956.

Colman, H. L., and C. P. Smallwood: "Computer Language," McGraw-Hill Book Company, New York, 1962.

Curtis, C. W.: "Linear Algebra," Allyn and Bacon, Inc., Boston, 1963.

Faddeeva, V. N.: "Computational Methods of Linear Algebra" (translated by C. D. Benster), Dover Publications, Inc., New York, 1959.

Fox, A. H.: "Fundamentals of Numerical Analysis," The Ronald Press Company, New York, 1963.

Fuller, L. E.: "Basic Matrix Theory," Prentice-Hall, Inc., Englewood Cliffs, N.J., 1962.

Hamming, R. W.: "Numerical Methods for Scientists and Engineers," McGraw-Hill Book Company, New York, 1962.

Harris, L. D.: "Numerical Methods Using Fortran," Charles E. Merrill Books, Inc., Columbus, Ohio, 1964.

Henrici, P.: "Elements of Numerical Analysis," John Wiley & Sons, Inc., New York, 1964.

Hildebrand, F. B.: "Introduction to Numerical Analysis," McGraw-Hill Book Company, New York, 1956.

Householder, A. S.: "Principles of Numerical Analysis," McGraw-Hill Book Company, New York, 1953.

International Business Machines Corporation: 7090/7094 Fortran IV Language, *Form* C28-6274-1, 1963.

————: 7090/7094 Macro Assembly Program (MAP) Language, *Form* C28-6311-3, 1964.

James, M. L., G. M. Smith, and J. C. Wolford: "Analog and Digital Computer Methods in Engineering Analysis," International Textbook Company, Scranton, Pa., 1964.

Kunz, K. S.: "Numerical Analysis," McGraw-Hill Book Company, New York, 1957.

Lapidus, L.: "Digital Computation for Chemical Engineers," McGraw-Hill Book Company, New York, 1962.

McCormick, J. M., and M. G. Salvadori: "Numerical Methods in Fortran," Prentice-Hall, Inc., Englewood Cliffs, N.J., 1964.

McCracken, D. D., and W. S. Dorn: "Numerical Methods and Fortran Programming," John Wiley & Sons, Inc., New York, 1964.

Nielsen, K. L.: "Methods in Numerical Analysis," The Macmillan Company, New York, 1956.

Organick, E. I.: "A Fortran Primer," Addison-Wesley Publishing Company, Inc., Reading, Mass., 1963.

Ralston, A., and H. S. Wilf: "Mathematical Methods for Digital Computers," John Wiley & Sons, Inc., New York, 1960.

Scarborough, J. B.: "Numerical Mathematical Analysis," 5th ed., The Johns Hopkins Press, Baltimore, 1962.

Stanton, R. G.: "Numerical Methods for Science and Engineering," Prentice-Hall, Inc., Englewood Cliffs, N.J., 1962.

Todd, J. (ed.): "Survey of Numerical Analysis," McGraw-Hill Book Company, New York, 1962.

SOLUTIONS TO SELECTED PROBLEMS

Among the following solutions are several that involve flow charts and FORTRAN programs. The solutions that are presented here should not be regarded as the only solutions or necessarily the best solutions. Almost any problem can be solved in terms of many different programs. It will help the student in learning to program if he will try to find more than one answer to the programming problems.

2.2

2.9

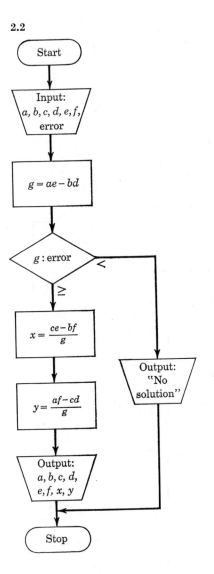

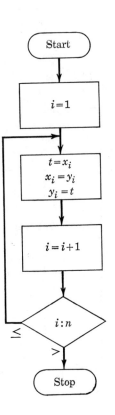

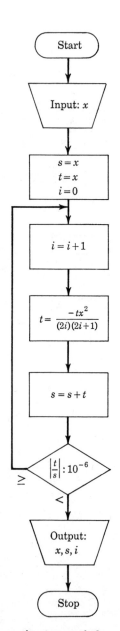

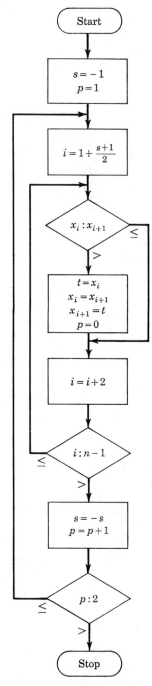

Note: Successive terms of the series are calculated recursively so as to save computer time. The ratio of successive terms is indicated as t.

Note: A switch s causes the alternation of comparisons in successive passes. A counter p is used to ensure that two successive passes occur with no exchanges.

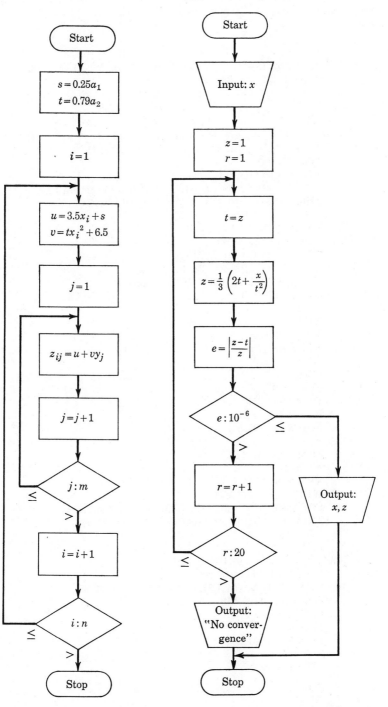

2.21

3.1   TEMPERATURE contains too many characters.   Also, it is probably not an integer variable; if so, the I7 specification is incorrect.

3.8

1		1		2		3	3
1	5		2		4	1	4
		39.4E-17		157.38			3129

3.13   READ (5,13) I1, I2, I3, I4, I5, I6, I7, I8, I9, A1, A2, A3,
$B1, B2, B3, B4, B5, B6
        13   FORMAT (8I10/I10, 3FI2.7/6E12.7)

3.18 No errors.   **3.26** $\phi = \dfrac{A_{mm}}{L_1/h_1 a_1 + L_2/u_2 a_2}$

3.32 RHO = DP/(V*V*A + DX/(A*X))   **3.40** N = 4   **3.41** N = 5

3.43   REAL P1, P2, V1, V2, T1, T2
        WRITE (6,1)
     1 FORMAT (1H1, 9X,
        $32HPRESSURE-VOLUME-TEMPERATURE DATA/
        $1H0, 5X, 8HPRESSURE, 13X, 6HVOLUME,
        $12X, 11HTEMPERATURE)
    10 READ (5,2) P1, V1, T1, V2, T2
     2 FORMAT (5F10.5)
        P2 = P1*(V1/V2)*(T2/T1)
        WRITE (6,3) P1, V1, T1, P2, V2, T2
     3 FORMAT (5X, F10.3, 2(10X, F10.3))
        GO TO 10
        END

3.55   REAL K
    10 READ (5,1) N
     1 FORMAT (I5)
        IF (N) 2, 2, 3
     2 STOP
     3 SUM = 0.
        DO 4 I = 1, N
        READ (5,5) P, E
     5 FORMAT (2E15.6)
        K = P/E
     4 SUM = SUM + K
        AN = N
        AVE = SUM/AN
        WRITE (6,6) N, AVE
     6 FORMAT (1H0, 15HNUMBER OF TESTS, 2X, I5/1X,
        $24HAVERAGE ELASTIC CONSTANT, 2X, E15.5)
        GO TO 10
        END

3.60 .FALSE.   **3.70** .TRUE.

3.82   DIMENSION A(100,100)
    10 READ (5,1) N, ((A(I,J), I = 1, N), J = 1, N)
     1 FORMAT (I10, 7F10.5/(8F10.5))
        IF (N.LE.0) STOP
        M = N − 1
        DO 2 I = 1, M
        K = I + 1
        DO 2 J = K, N

```
         IF (A(I,J) − A(J,I)) 3,2,3
      3  WRITE (6,4)
      4  FORMAT (1H0, 15HNOT SYMMETRICAL)
         GO TO 10
      2  CONTINUE
         WRITE (6,5)
      5  FORMAT (1H0, 21HMATRIX IS SYMMETRICAL)
         GO TO 10
         END
```

3.88   T = SQRT(2.*G*S)   **3.99**   Y = DIM(X2,X1)

3.104   FS(D,AM,S) = SIN(3.14159*(D + AM/60. + S/3600.)/180.)

3.111
```
         REAL FUNCTION I(N,V,R)
         DIMENSION R(N)
         I = 0.
         DO 1 J = 1, N
      1  I = I + V/R(J)
         RETURN
         END
      C  MAIN PROGRAM
         DIMENSION R(1000)
         READ (5,1) N,V,(R(K), K = 1, N)
      1  FORMAT (I10, 7F10.5/(8F10.5))
         EQ = V/I(N,V,R)
         WRITE (6,2) EQ
      2  FORMAT (1H0, 21HEQUIVALENT RESISTANCE, 2X, E15.8)
         STOP
         END
```

3.114
```
         SUBROUTINE ANGLE(DEG, MIN, SEC, TANG)
         INTEGER DEG, SEC
         RAD = ATAN(TANG)
         A = 180.*RAD/3.14159
         DEG = A
         B = ABS(A − FLOAT(DEG))
         C = B*60.
         MIN = C
         D = ABS(C − FLOAT(MIN))
         SEC = D*60.
         RETURN
         END
      C  MAIN PROGRAM
     10  READ (5,1) A, B
      1  FORMAT (2F10.5)
         IF (A.LT.0.) STOP
         C = SQRT(A*A + B*B)
         CALL ANGLE(DEG1, MIN1, SEC1, A/B)
         CALL ANGLE(DEG2, MIN2, SEC2, B/A)
         WRITE (6,2) A, B, C, DEG1, MIN1, SEC1, DEG2, MIN2, SEC2
      2  FORMAT (5X, 3E20.7, 6I5)
         GO TO 10
         END
```

4.3 $627_8 = 6_8 \times 8^2 + 2_8 \times 8 + 7$

4.6

Decimal	Binary	Octal	Hexadecimal
78	0100 1110	116	4e
279	100 010 111	427	117
52	110 100	64	34
933	11 1010 0101	1645	3a5

4.14

Decimal	Binary	Octal	Hexadecimal
0.99	0.1111 1101 0111	0.7727	0.fd7
0.114453125	0.100 100 001	0.441	0.908
12.171875	1100.001 011	14.13	c.2c
0.999755859375	0.1111 1111 1111	0.7777	0.fff

4.16 $1\ 101\ 110\ 111_2$ **4.24** 630_8 **4.32** $d3593_{16}$
4.39 25_{16} **4.43** $001\ 100_2$ **4.57** -124
4.64 8.38   **4.69** 56631_8
4.80 All locations in octal:

Location	Data	Location	Instruction	
101	a	000	LQ	102
102	b	001	FM	102
103	c	002	STQ	104
104	x	003	LQ	105
105	4.0	004	FM	101
		005	FM	103
		006	STQ	101
		007	L	104
		010	FS	101
		011	ST	104
		012	STOP	

4.84 All locations in octal:

Location	Data	Location	Instruction	
101	sum	000	L	104
102	n	001	LX	103
103	2	002	LXL	102
104	0.	003	FA	200,X
.		004	AX	103
.		005	BXLE	003
201	a_1	006	ST	101
202	a_2	007	STOP	

5.2 (a) 600.05, 0.00807   (b) 1000.05, 2.04   (c) 5.766×10^{12}, 0.0242
(d) 0.0015, 0.015   (e) 0.0578, 0.02295   (f) 0.00454, 0.00365
(g) 0.0335, 0.0888
5.3 $V = 22.1$ ft/sec, $\Delta V = 0.79$ ft/sec
5.6 $x = (2.25, 2.413), (1.637, 1.800)$
5.9 $h = 316.47$ ft, $\Delta h = 0.21$ ft
6.2 3.6, 1.2   **6.4** 21.3, 0.63, -1.31 **6.6** 1, 2, -3
6.8 0, 2.68, 131, -38.6 **6.10** 0.825 rad   **6.12** 0.862 rad
6.14 1.76   **6.20** $-1, 2.1i, -2.1i$ **6.23** 0.5, 1.18, $4.2 + 2.1i$, $4.2 - 2.1i$
6.40 $K\theta = (Wl \cos \theta)/2 + Pl \sin \theta$ **6.42** $E_2 = 1450$ volts, $I = 27.3$ amp
7.2 $\frac{41}{13}, \frac{8}{13}$ **7.4** $\frac{29}{24}, \frac{19}{12}, -\frac{43}{24}$ **7.6** 1, 2, 1, 2

7.8 1, 1, 1   **7.10** 1, 2, 3, 4   **7.12** $-\frac{118}{11}$, 2, $\frac{40}{11}$, $\frac{4}{11}$

7.14 $8\frac{5}{6}$, -7, $9\frac{1}{4}$, $-14\frac{1}{3}$, $-2\frac{1}{12}$, $15\frac{2}{3}$, $-6\frac{1}{3}$

7.16 $\begin{bmatrix} -1 & 4 & -15 \\ 6 & 5 & -2 \\ 9 & 10 & 5 \end{bmatrix}$ **7.23** $\begin{bmatrix} -7 & 10 & 10 & -9 \\ 67 & 2 & 92 & -9 \\ 20 & 4 & 34 & -6 \\ 23 & 16 & 61 & -18 \end{bmatrix}$

7.29 $\begin{bmatrix} \sqrt{2}/2 & -\sqrt{2}/2 \\ \sqrt{2}/2 & \sqrt{2}/2 \end{bmatrix}$ **7.35** $\begin{bmatrix} -\frac{3}{14} & -\frac{1}{2} & -\frac{19}{14} \\ \frac{5}{28} & \frac{1}{4} & \frac{13}{28} \\ \frac{2}{7} & 0 & \frac{1}{7} \end{bmatrix}$

7.43 $\begin{bmatrix} -0.0223 & 0.0483 & 0.0892 & 0.212 & -0.361 \\ -0.922 & 0.331 & 0.688 & -0.242 & -0.238 \\ -0.561 & 0.383 & 0.245 & -0.167 & -0.242 \\ 1.14 & -0.465 & -0.550 & 0.193 & 0.390 \\ -0.257 & 0.0558 & 0.0260 & -0.0632 & 0.353 \end{bmatrix}$

7.57 -8 **7.59** -10 **7.61** 22

7.64 -13 **7.69** 0   **7.74** 249

7.84 7.4, -3.4 **7.87** 2, 3.73, 0.27   **7.90** 3.88, 1.65, 3, 0.48

7.93 0.227, 0.0742, 0.126, 0.175, 0.0515, 0.301

7.97 1.04 rad/sec, $A_2/A_1 = 0.339$; 4.61 rad/sec, $A_2/A_1 = -0.738$

8.2 11,550; 16,800; 5800; -4500; -3400 **8.4** 0.38267

8.6 0.1785; 1.0176; 1.6914; 2.3040   **8.9** 0.4175   **8.11** 56,938

8.15 $p = 0.42262 + 0.01548(x - 25) - 0.0000644(x - 25)(x - 30)$; 0.37463

8.23 31,472; 32,054; 5920   **8.28** -50; 0; 75; 25

8.33 $2x^2 - 4x + 4$; $8x$; 16

8.36 $f(u) = -10 + 2u - 0.5u(u - 1)(u - 2)$; $u = (x + 1)/2$

8.41 $x = 80,000u + 20,000u(u - 1) + 3333u(u - 1)(u - 2) - 5833u(u - 1)$
$(u - 2)(u - 3) + 2083u(u - 1)(u - 2)(u - 3)(u - 4)$; $u = t/100$

8.44 290,000; 993,000; 994,000   **8.51** 0.92717   **8.57** 11.9, 7.7

8.63 141   **8.68** 25.8   **8.69** 25.86

8.75 $1,900,000; $2,400,000; $3,000,000$

8.77 0.000015   **8.84** $20.3 < x < 29.7$ **8.86** $47°$

9.3 -0.008056; -0.006676; -0.007366

9.10 -0.006026; -0.006562; -0.008706; -0.010046 **9.16** -0.005986

9.29 0.0255; 0.0606; 0.3327; 0.8419   **9.30** 0.0740; 0.0721; 0.0672; 0.0592

9.31 5.26 mps; 5.14 mps; 4.84 mps; 4.33 mps

9.37 -0.0002640; -0.0002856; -0.0002760 **9.41** 0.00000216; 0.000000096

9.49 -0.000278; 0.000512; 0.001302; 0.002092

9.50 0.00002146; 0.00001134; 0.00000122; -0.00019098

9.51 0.387 mile/sec²; 0.370 mile/sec²; 0.326 mile/sec²; 0.257 mile/sec²

9.54 $(12f_6 - 75f_5 + 200f_4 - 240f_3 + 300f_2 - 137f_1)/60h$ **9.59** 0.00133

9.66 By Simpson's rule: 2 increments, 57.427; 4 increments, 57.303; 8 increments, 57.296; true area, 57.296

9.69 22.485   **9.80** 26,700 in./sec   **9.88** 0.02020; 0.000173

9.102 25.4 in.   **9.108** 0.705; 0.552; 0.267; 0.148; 0.161   **9.111** 15.4

10.3 5196 ft; 1732 ft/sec; 1950 ft/sec   **10.7** 0.372 sec   **10.9** 0.316 sec

10.11 $1 - x^2/2! + x^4/4! - x^6/6! + \cdots$ **10.16** 0.92720; 0.70357

10.22 -0.0059958 **10.25** -0.00027608 **10.26** 54.94

10.36 0.00005   **10.52** 0.940 (3 terms)   **10.55** 4 terms (error $= 0.0000023$)

10.57 $20.2 < x < 29.8$ **10.60** 46,950 psi   **10.62** $48°$

10.64 3884 ft   **10.66** $\theta = 0.8 + 5.75t^2 + 5.41t^4 + 1.51t^6$

11.1 (a) 1.24280267; (b) 1.27866133   **11.2** (a) 1.24280533; (b) 1.27942933

11.4 $\dfrac{dB}{dt} = 2k_1A_0e^{-k_1t} - k_2B^2$

Time, min	A	B
5	0.861	0.276
10	0.741	0.500
15	0.638	0.671
20	0.549	0.794
25	0.472	0.877
30	0.407	0.927
35	0.350	0.951
40	0.301	0.958
45	0.259	0.950
50	0.223	0.934
55	0.192	0.911
60	0.166	0.884

11.7 $e_p = O(h^5)$; $e_c = O(h^6)$

11.17 $h = 14.8$ ft

11.20   At $t = 0.212$ sec, $\phi/\phi_0 = 2$; the total time $= 1.06$ sec

12.1 $a = 8.056$, $b = 0.015$, $c = -4.05 \times 10^{-6}$; maximum deviation $= -0.19$; standard deviation $= 0.11$

12.4 $a = 2.2846 \times 10^7$, $b = 2.1949 \times 10^3$; maximum deviation $= 0.0615$; standard deviation $= 0.0511$

12.6   Four terms can be dropped

INDEX